Period Spaces for p-divisible Groups

by

M. Rapoport and Th. Zink

PRINCETON UNIVERSITY PRESS

PRINCETON, NEW JERSEY

1996

The Annals of Mathematics Studies are edited by
Luis A. Caffarelli, John N. Mather, and Elias M. Stein

Princeton University Press books are printed on acid-free paper and meet the guidelines for permanence and durability of the Committee on Production Guidelines for Book Longevity of the Council on Library Resources

Printed in the United States of America by Princeton Academic Press

10 9 8 7 6 5 4 3 2 1

Library of Congress Cataloging-in-Publication Data

A CIP catalog record for this book is available from the Library of Congress
ISBN 0-601-02782-X
ISBN 0-691-02781-1 (pbk.)

The publisher would like to acknowledge the authors of this volume for providing the camera-ready copy from which this book was printed

Contents

Introduction

Let E be a p-adic field and let Ω_E^d be the complement of all E-rational hyperplanes in the projective space $\mathbf{P}^{d-1}$. This is a rigid-analytic space over E equipped with an action of $GL_d(E)$. Drinfeld [Dr 2] has constructed a system of unramified coverings $\tilde{\Omega}_E^d$ of Ω_E^d to which the action of $GL_d(E)$ is lifted. These covering spaces are interesting for at least two reasons. Firstly, these spaces can be used to p-adically uniformize the rigid-analytic spaces corresponding to Shimura varieties associated to certain unitary groups. This uniformization looks formally very similar to the complex uniformization by the open unit ball which gives rise to these Shimura varieties. Secondly, Drinfeld conjectured that the ℓ-adic cohomology groups with compact supports $H_c^i(\tilde{\Omega}_E^d \otimes_E \bar{E}, \bar{\mathbf{Q}}_\ell)$, $\ell \neq p$, give a realization of all supercuspidal representations of $GL_d(E)$ which would give a construction of these representations analogous to the construction of the discrete series representations of semi-simple Lie groups through L^2-cohomology (Griffiths, Schmid, Langlands, ...).

In the present work we generalize Drinfeld's construction to other p-adic groups. This construction is based on the moduli theory of p-divisible groups of a fixed isogeny type. The moduli spaces obtained in this way are formal schemes over the ring of integers O_E whose generic fibres yield rigid-analytic spaces generalizing Ω_E^d. The covering spaces are then obtained by trivializing the Tate modules of the universal p-divisible groups over these formal schemes. Furthermore, we show how these spaces may be used to uniformize (an open part of !, see below) the rigid-analytic spaces associated to general Shimura varieties. We will also exhibit a rigid-analytic period map from the covering spaces to one of the p-adic symmetric spaces associated to the p-adic group.

Before describing in some more detail our main results we sketch the background of the problems considered here and our motivation. The subject of p-adic uniformization starts with the paper of Mumford [M2] which was inspired by Tate's work on the uniformization of elliptic curves with multiplicative reduction over a discretely valued field. In this paper Mumford introduced the one-dimensional formal scheme $\hat{\Omega}^2_E$ and showed that an algebraic curve with completely split reduction over $Spec\, O_E$ is uniformized by a suitable subset of $\hat{\Omega}^2_E$. Cherednik [Ch] discovered that Shimura curves associated to quaternion algebras which ramify at the prime p admit a p-adic uniformization in the sense of Mumford by the whole of $\hat{\Omega}^2_E$. Drinfeld [Dr2] subsequently gave an algebro-geometric proof using the moduli theory of p-divisible groups. In his paper Drinfeld formulates for any $d \geq 2$ a moduli problem of p-divisible groups and shows that it is representable by the formal scheme $\hat{\Omega}^d_E$, a higher-dimensional analogue of Mumford's formal scheme which had been introduced by Deligne and Mustafin [Mu]. For higher-dimensional Shimura varieties a p-adic uniformization by Ω^d_E is possible only in rare cases, comp. theorem IV below (comp. (6.50), cf. also [R1]). For instance, the naive hope that if the group giving rise to the Shimura variety is anisotropic at the place p one should have uniformization at the places of the Shimura field lying above p turns out to be quite false, as was first observed by Langlands [La]. Indeed, the special fibre of the Shimura variety usually is not totally degenerate, comp. [Z1], [R1]. A closely related observation is that there may be infinitely many isogeny classes in the special fibre. This explains why our theorem III below which applies to a general Shimura variety exhibits a uniformization only of the tubular neighbourhood of a fixed isogeny class.

Another motivation for us was Drinfeld's conjecture on the ℓ-adic cohomology groups of $\tilde{\Omega}^d_E$. This conjecture was made more precise by Carayol [Ca]. His version also involves an action of the multiplicative group of the division algebra D with center E and invariant $1/d$ on the covering space $\tilde{\Omega}^d_E$. It roughly states that the resulting action of the triple product $W_E \times D^\times \times GL_d(E)$, where W_E denotes the Weil group of E, is a Langlands correspondence. There recently has been a flurry of activity concerning this conjecture (we mention the work of H. Carayol, of G. Faltings, of A. Genestier and of M. Harris). Carayol [Ca] also pointed out that a similar conjecture can be made in the case where the Drinfeld moduli problem is replaced by the formal deformation problem of Lubin and Tate. Shortly after Kottwitz formulated a very elegant recipe for such correspondences for arbitrary reductive p-adic groups, cf. [R2], §5. From this point of view our construction of the covering spaces yields the rigid-analytic spaces for which his recipe should describe their ℓ-adic cohomology. The conjecture of Kottwitz is the analogue in this purely local context of the global problem of determining the reciprocity laws describing the correspondence between automorphic representations and ℓ-adic representations of Galois groups of number fields defined by Shimura varieties [Ko2].

The third motivation for us was to elucidate in this context the role of p-adic period morphisms. This subject starts with Dwork's investigation (comp. [Ka1]) of the formal deformation space of an ordinary elliptic curve (comp. also [Ka 3], [DI]). His period morphism π maps the open unit disc D to the affine line $\mathbf{A}^1$ and is given by the famous formula $\pi^*(\tau) = \log q$ where $q = T + 1$ in terms of the coordinates τ on $\mathbf{A}^1$ resp. T on D. It describes the variation of the Hodge filtration of the deformed elliptic curve. Grothendieck [Gr2] introduced a new point of view through his rigidity theorem for p-divisible groups up to isogeny. The rigid-analytic point of view (which was present in Dwork's original work) was re-introduced by Gross and Hopkins [HG2] when they defined a period mapping in the case of the formal deformation space of a supersingular elliptic curve. Their period morphism maps the open unit disc to the projective line. Although it cannot be expressed in terms of elementary functions, a great deal is known about it, comp. [HG2], [Yu]. In the general case the period morphism maps one of our covering spaces to a Grassmann variety and describes the variation of the Hodge filtration induced by the universal p-divisible group. Our construction of it is closest in spirit to Grothendieck's approach. In addition, the question of determining the image of a period morphism touches on one of the fundamental open problems in the domain of p-adic cohomology, namely the conjectures of Fontaine [Fo2]. They constitute the p-adic analogue of Riemann's theorem characterizing the classical periods coming from abelian varieties. Assuming his conjectures to hold it turns out that the image is a p-adic symmetric space, either in the more elementary sense as defined by Fontaine's condition, or as defined by van der Put and Voskuil [PV] through geometric invariant theory (they are identical, as proved by Totaro).

We will now give an overview of our main results. We will first describe the moduli problems of p-divisible groups and the *representability theorem* which yields the formal schemes generalizing $\hat{\Omega}^d_E$ above. Next we will describe the *covering spaces* and the rigid-analytic *period morphism*. Finally we shall explain our non-archimedean *uniformization theorems* for Shimura varieties.

To formulate our representability theorem we introduce some notations. We fix a prime number p. If O is a complete discrete valuation ring of unequal characteristic $(0,p)$ we denote by $Nilp_O$ the category of locally noetherian schemes S over $Spec\ O$ such that the ideal sheaf $p\cdot\mathcal{O}_S$ is locally nilpotent. We denote by $\bar{S}$ the closed subscheme defined by $p\cdot\mathcal{O}_S$. A locally noetherian formal scheme over $Spf\ O$ will be identified with the set–valued functor on $Nilp_O$ it defines. A morphism $\mathcal{X}\to\mathcal{Y}$ of formal schemes is called *locally formally of finite type* if the induced morphism $\mathcal{X}_{\text{red}}\to\mathcal{Y}_{\text{red}}$ between their underlying reduced schemes of definition is locally of finite type.
In what follows we call a *quasi–isogeny between p–divisible groups X and Y* over a scheme $S\in Nilp_{\mathbf{Z}_p}$ an isogeny multiplied by a power of $1/p$.
The moduli problems of p–divisible groups which we want to consider are of two types. The type (EL) will parametrize p–divisible groups with endomorphisms and with level structures within a fixed isogeny class. The type (PEL) will parametrize p–divisible groups with polarizations, endomorphisms and level structures within a fixed isogeny class. The moduli problems depend on certain *rational* and *integral* data which we now formulate in both cases in a simplified form where the level structures are absent. Let L be an algebraically closed field of characteristic p and let $W(L)$ be its ring of Witt vectors. Let $K_0=K_0(L)=W(L)\otimes_{\mathbf{Z}}\mathbf{Q}$ and let σ be the Frobenius automorphism of K_0.

Case (EL): The *rational data* consists of a 4-tuple (B,V,b,μ), where B is a finite–dimensional semi–simple algebra over $\mathbf{Q}_p$ and V a finite left B–module. Let $G=GL_B(V)$ (algebraic group over $\mathbf{Q}_p$). Then b is an element of $G(K_0)$. The final datum μ is a homomorphism $\mathbf{G}_m\to G_K$ defined over a finite extension K of K_0. Let $V\otimes_{\mathbf{Q}_p}K=\bigoplus V_i$ be the corresponding eigenspace decomposition and $V^j_K=\bigoplus_{i\geq j}V_i$ the associated decreasing filtration. We require that the *filtered isocrystal over* K, $(V\otimes_{\mathbf{Q}_p}K_0, b(\mathrm{id}\otimes\sigma), V^\bullet_K)$, is the filtered isocrystal associated to a p–divisible group over $Spec\,O_K$ ([Gr1], [Fo1], [Me]). The *integral data* consists of a maximal order O_B in B and an O_B–lattice Λ in V.

Case (PEL): In this case we assume $p\neq 2$. The *rational data* are given by a

6–tuple $(B, *, V, (\ ,\ ), b, \mu)$. Here B and V are as before. Furthermore, B is endowed with an anti–involution $*$ and V is endowed with a non–degenerate alternating bilinear form $(\ ,\ ) : V \otimes_{\mathbf{Q}_p} V \to \mathbf{Q}_p$ such that

$$(dv, v') = (v, d^* v'), \quad d \in B.$$

The remaining data are as before relative to the algebraic group G over $\mathbf{Q}_p$ whose values in a $\mathbf{Q}_p$–algebra R are

$$G(R) = \{g \in GL_B(V \otimes R);\ \ (gv, gv') = c(g)(v, v'),\ \ c(g) \in R^\times\}.$$

We require that the rational data define the filtered isocrystal associated to a p–divisible group over $Spec\, O_K$ endowed with a polarization (= symmetric isogeny to its dual). The *integral data* are as before. We assume that O_B is stable under $*$ and that Λ is self–dual with respect to the alternating form $(\ ,\ )$.

In either case let E be the field of definition of the conjugacy class of μ, a finite extension of $\mathbf{Q}_p$ contained in K. Let $\breve{E} = E.K_0$, with ring of integers $O_{\breve{E}}$. The representability theorem in rough outline may then be formulated as follows (3.25).

Theorem I *We fix data of type (EL) or (PEL). Let* $\mathbf{X}$ *be a p–divisible group with action of* O_B *over* $Spec\, L$ *with associated isocrystal isomorphic to* $(V \otimes_{\mathbf{Q}_p} K_0, b(\mathrm{id} \otimes \sigma))$. *In the case (PEL) we endow* $\mathbf{X}$ *with a* O_B*–polarization defined by the alternating form on* $V \otimes_{\mathbf{Q}_p} K_0$. *We consider the functor* $\breve{\mathcal{M}}$ *on* $Nilp_{O_{\breve{E}}}$ *which associates to* S *the set of isomorphism classes of the following data.*

1.) *A p–divisible group* X *over* S *with* O_B*–action.*

2.) *An* O_B*–quasi–isogeny* $\varrho : \mathbf{X} \times_{Spec\, L} \bar{S} \to X \times_S \bar{S}$.

These data are required to satisfy the following conditions.

(i) *We have* $\det_{\mathcal{O}_S}(d; Lie X) = \det_K(d; V_K^0 / V_K^1)$ *as polynominal functions in* $d \in B$ *(Kottwitz condition* [Ko3]*).*

(ii) *Let* $M(X)$ *be the Lie algebra of the universal extension of* X. *Then locally on* S *there is an* O_B*–isomorphism* $M(X) \simeq \Lambda \otimes_{\mathbf{Z}} \mathcal{O}_S$.

(iii) *In the case (PEL) there exists an isomorphism* $p : X \to X^\vee$ *into the dual p–divisible group such that* $\hat{\varrho} \circ p \circ \varrho : \mathbf{X} \times_{Spec\, L} \bar{S} \to \mathbf{X} \times_{Spec\, L} \bar{S}$

differs from the quasi–isogeny induced by the fixed O_B–polarization on $\mathbf{X}$ *by a constant in* $\mathbf{Q}_p^{\times}$.

The functor $\breve{\mathcal{M}}$ *is representable by a formal scheme locally formally of finite type over* $Spf\, O_{\breve{E}}$.

This representability theorem is more general than Drinfeld's but it is also less precise in that it does not identify the formal scheme $\breve{\mathcal{M}}$. In fact we know very little about $\breve{\mathcal{M}}$ in the most general case, not even about its local structure. For instance, we do not know if $\breve{\mathcal{M}}$ is flat over $Spf\ O_{\breve{E}}$, although this has been proved in numerous special cases ([CN1], [CN2], [DP], [dJ1], [R1], [St], [Z1]). We reduce here this conjecture to the corresponding statement on the *local model* of our moduli problem ((3.26)). Namely, we define an explicit closed subscheme $\mathbf{M}^{loc}$ of a finite product of Grassmannian varieties over $Spec\ O_E$ such that $\breve{\mathcal{M}}$ is locally for the étale topology isomorphic to the completion of $\mathbf{M}^{loc}$ along a closed subscheme (this generalizes and makes more precise a concept introduced in [R1], cf. also [dJ1], [DP]). We hope that the conjecture is of interest in commutative algebra.
The above representability theorem is reduced by standard techniques to the *universal case*. In the proof of the universal case we have to allow the field L appearing above to be an arbitrary perfect field of characteristic p. A p–divisible group over L is called *decent* ((2.13.)) if the K_0–vector space underlying its isocrystal $(N, \mathbf{F})$ is generated by elements n satisfying an equation $\mathbf{F}^s n = p^r n$ for some integers r and $s > 0$. If L is algebraically closed any p–divisible group is decent. The definition of decency is implicit in Kottwitz [Ko1], comp. (1.8). Using this concept the representability theorem in the universal case may be formulated as follows (2.16).

Theorem II *Let* $\mathbf{X}$ *be a decent p–divisible group over Spec L. We consider the functor on* $Nilp_{W(L)}$ *which associates to* $S \in Nilp_{W(L)}$ *the set of isomorphism classes of pairs* (X, ϱ) *consisting of a p–divisible group X over S and a quasi–isogeny* $\varrho : \mathbf{X} \times_{Spec\, L} \bar{S} \to X \times_S \bar{S}$ *of p–divisible groups over* $\bar{S}$. *This functor is representable by a formal scheme locally formally of finite type over* $Spf\, W(L)$.

Just as in Drinfeld's case, the moduli problem appearing here is not of the type usually considered in algebraic geometry, hence we cannot directly apply standard methods. We ultimately use a finiteness theorem in the Bruhat-Tits building of GL_n (2.18) which seems interesting in its own right. Before turning to the next circle of ideas we mention that the formal scheme

$\breve{\mathcal{M}}$ associated to the moduli problem of type (EL) or (PEL) comes with additional structure. To the pair (G, b) there is associated (1.12) the algebraic group J over $\mathbf{Q}_p$ with points in a $\mathbf{Q}_p$-algebra R

$$J(R) = \{g \in G(R \otimes_{\mathbf{Q}_p} K_0);\ \sigma(g) = b^{-1} g b\}.$$

The group $J(\mathbf{Q}_p)$ of quasi-isogenies of $\mathbf{X}$ acts on the left of $\breve{\mathcal{M}}$, via

$$g \cdot (X, \varrho) = (X, \varrho \circ g^{-1}).$$

The formal scheme $\breve{\mathcal{M}}$ can be broken up into a disjoint sum of open and closed formal subschemes as follows. Let Δ be the abelian group dual to the group of $\mathbf{Q}_p$-rational characters of G. The group $J(\mathbf{Q}_p)$ acts on Δ by translations. There is a canonical $J(\mathbf{Q}_p)$-equivariant map $\varkappa : \breve{\mathcal{M}} \to \Delta$ (3.52). In essence, the morphism $\varkappa$ associates to an S-valued point (X, ϱ) of $\breve{\mathcal{M}}$ the height of ϱ. It is not clear to us under which conditions the fibres of $\varkappa$ are connected. (A similar question may be asked in the rigid-analytic context, cf. below). We also remark that the formal scheme $\breve{\mathcal{M}}$ depends only on the equivalence class (3.18) of the data of type (EL) or (PEL). We finally mention that $\breve{\mathcal{M}}$ is equipped with a natural Weil descent datum from $Spf\, O_{\breve{E}}$ to $Spf\, O_E$ (3.48), i.e. an isomorphism

$$\alpha : \breve{\mathcal{M}} \longrightarrow \breve{\mathcal{M}}^{\tau}.$$

Here $\tau \in Gal(\breve{E}/E)$ is the relative Frobenius automorphism. Although this descent datum is not effective, it becomes effective after suitably completing $\breve{\mathcal{M}}$: a suitable completion of $\breve{\mathcal{M}}$ can be written in a canonical way as $\mathcal{M} \times_{Spf\, O_E} Spf\, O_{\breve{E}}$ for a pro-formal scheme $\mathcal{M}$ over $Spf\, O_E$ (3.51). This will be used in the uniformization theorems below.

We next turn to the rigid-analytic aspects of the situation. We continue with the formal scheme $\breve{\mathcal{M}}$ associated to our moduli problem of type (EL) or (PEL). Let $\breve{\mathcal{M}}^{rig}$ be its associated rigid-analytic space over $\breve{E}$ (its *generic fibre*). It should be pointed out that, contrary to what happens in Drinfeld's case, the formal scheme $\breve{\mathcal{M}}$ may not be p-adic, i.e. $p\mathcal{O}_{\breve{\mathcal{M}}}$ may not be an ideal of definition. (Indeed, we believe that Drinfeld's moduli problem and trivial variants of it are the only ones yielding a p-adic formal scheme but we were unable to prove something along these lines.) Therefore Raynaud's construction of the associated rigid space no longer applies. That $\breve{\mathcal{M}}$ should still have a generic fibre was suggested to us by the paper of Gross and Hopkins [HG2]; it turned out that this construction had occurred simultaneously for somewhat different reasons in work of Berthelot [Ber]. With hindsight, the generic fibre of a formal scheme may be interpreted in a very natural framework in terms of Huber's *adic spaces* [Hu].

The flatness conjecture alluded to above would imply that $\breve{\mathcal{M}}^{rig}$ is non-empty. Using now level structures on the universal p–divisible group over $\breve{\mathcal{M}}$ we may imitate the procedure of Drinfeld to construct a tower of non-trivial étale coverings of $\breve{\mathcal{M}}^{rig}$ (5.34). If the algebraic group G is connected (this excludes the orthogonal groups), the covering group of this tower is $G(\mathbf{Q}_p)$ which acts through Hecke correspondences, and commutes with the action of $J(\mathbf{Q}_p)$ which is lifted from $\breve{\mathcal{M}}^{rig}$ to the tower of coverings. For the layers in the tower $\{\breve{\mathbf{M}}_K;\ K \subset G(\mathbf{Q}_p)\}$ which correspond to parahoric subgroups $K \subset G(\mathbf{Q}_p)$ we have models as formal schemes $\breve{\mathcal{M}}_K$ defined by moduli problems of p–divisible groups. All we have said about $\breve{\mathcal{M}}$ remains valid for $\breve{\mathcal{M}}_K$. That the covering group is $G(\mathbf{Q}_p)$ is due to the following observation (5.33). Let (X, ϱ) be a point of $\breve{\mathcal{M}}$ over the formal spectrum of a complete discrete valuation ring of unequal characteristic. Then the rational p–adic Tate module of the generic fibre of X_Λ is isomorphic to V, where the isomorphism respects the structures of B–modules and the alternating bilinear forms of both spaces. When G is no longer connected we obtain coverings whose covering group is an inner form of G. The general problem of determining the rational p–adic Tate module of the generic fibre of a p–divisible group over the ring of integers of a p–adic field in terms of its filtered isocrystal has been solved to a large degree by Fontaine [Fo2], but important questions remain unsettled, cf. (1.20).

The next ingredient of the rigid–analytic picture is the period morphism (5.16)

$$\breve{\pi} = \breve{\pi}^1 \times \breve{\pi}^2 : \breve{\mathcal{M}}^{rig} \longrightarrow \breve{\mathcal{F}}^{rig} \times \Delta.$$

Here $\breve{\pi}^2 = \varkappa^{rig}$ is derived from the morphism of formal schemes introduced above, and $\breve{\mathcal{F}} = \mathcal{F} \times_{Spec\ E} Spec\ \breve{E}$ is the homogeneous projective algebraic variety under $G_{\breve{E}}$ defined by the conjugacy class of the one–parameter subgroup μ. The definition of the first component $\breve{\pi}^1$ of the period morphism is based on the observation that, if (X, ϱ) denotes the universal object over $\breve{\mathcal{M}}$, there is a *canonical* isomorphism

$$V \otimes_{\mathbf{Q}_p} \mathcal{O}_{\breve{\mathcal{M}}^{rig}} = M(X_\Lambda) \otimes_{\mathcal{O}_{\breve{\mathcal{M}}}} \mathcal{O}_{\breve{\mathcal{M}}^{rig}}.$$

This is essentially merely a re-interpretation of Grothendieck's rigidity theorem for crystalline Dieudonné theory up to isogeny but the new feature here, the passage to the rigid category, originates with the paper of Gross and Hopkins mentioned above. The first example of such a period morphism is due to Dwork (comp.[Ka1]), in the context of abelian varieties with ordinary reduction ((3.80) and (5.51)).

The period morphism $\breve{\pi}$ is étale (5.17) and $J(\mathbf{Q}_p)$–equivariant. Furthermore, at least if G is connected, the fibre of $\breve{\pi}$ through a point may be identified with $G(\mathbf{Q}_p)^1/K_\Lambda$ where $G(\mathbf{Q}_p)^1$ is the group of points of $G(\mathbf{Q}_p)$ where the values of all $\mathbf{Q}_p$–rational characters of G are units and where K_Λ is the subgroup which fixes the lattice Λ. Something similar holds for the period morphism induced on the tower of coverings mentioned above ((5.37)). In Drinfeld's example, the first component of the period morphism coincides with the inclusion of Ω^d_E in $\mathbf{P}^{d-1}$. The proof of this comparison result (5.47) was communicated to us by Faltings.

The question of the *image* of the period morphism, or rather of its first component $\breve{\pi}^1$, leads directly to the concept of a *p–adic symmetric space* (or a *p–adic period domain*). Namely, by a conjecture of Fontaine, this image is described by the *weakly admissible* points in $\breve{\mathcal{F}}$ (5.28). Here a point x of $\breve{\mathcal{F}}(K)$ is called weakly admissible if for every $\mathbf{Q}_p$–rational representation (V,ϱ) of G the filtered isocrystal $(V \otimes_{\mathbf{Q}_p} K_0, \varrho(b)\cdot(\mathrm{id}_V \otimes \sigma), \mathcal{F}^\bullet_x)$ over K, where $\mathcal{F}^\bullet_x$ is the filtration of $V \otimes_{\mathbf{Q}_p} K$ defined by x, is weakly admissible [Fo2]. It is easy to see (1.36) that the set $\mathcal{F}^{wa}$ of weakly admissible points in $\breve{\mathcal{F}}$ is a rigid–analytic admissible open subset of $\breve{\mathcal{F}}^{rig}$ stable under the action of $J(\mathbf{Q}_p)$. The admissible open subsets of generalized flag varieties obtained in this way from a triple $(G, b, \{\mu\})$ where G is a connected reductive group over $\mathbf{Q}_p$, where $b \in G(K_0)$ and where $\{\mu\}$ is a conjugacy class of one–parameter subgroups of G, are called *p–adic symmetric spaces.* It turns out that, as proved by Totaro in response to a conjecture in a first version of our paper, this definition gives the same spaces as the definition in terms of geometric invariant theory due to van der Put and Voskuil [PV] (comp.(1.51)). (However, it should be pointed out that these spaces do not in general satisfy the axioms imposed on p–adic symmetric spaces in [PV], as these authors prove themselves in loc. cit.). Summarizing this part of our paper, it may be said that the system of étale coverings of $\breve{\mathcal{M}}^{rig}$ bears many resemblances to the tower of varieties over a number field defined in the theory of Shimura varieties, and this analogy seems to go quite far (comp. [R2]). In fact, we conjecture ((1.37), cf. also (5.53)) that there is a rigid–analytic space $(\mathcal{F}^{wa})'$ mapping in an étale and bijective way to $\mathcal{F}^{wa}$, and a local system in $\mathbf{Q}_p$–vector spaces over $(\mathcal{F}^{wa})'$ with typical fibre V such that $\breve{\mathbf{M}}_K$ is the space of level structures of level K of this local system. It was pointed out by de Jong that $(\mathcal{F}^{wa})'$ may not be isomorphic to $\mathcal{F}^{wa}$.

We finally turn to the description of the non–archimedean uniformization theorems for Shimura varieties. We slightly change our notations. Let B be a finite–dimensional algebra over $\mathbf{Q}$ equipped with a *positive* anti–involution $*$. Let V be a finite B–module with a non–degenerate alternating bilinear form $(\ ,\ )$ with values in $\mathbf{Q}$ satisfying the identity appearing in the description of the case PEL above. We define the algebraic group G over $\mathbf{Q}$ in complete analogy with the case (PEL). Let $h : Res_{\mathbf{C}/\mathbf{R}}\mathbf{G}_m \rightarrow G_{\mathbf{R}}$ be such that (G, h) satisfies the axioms of Deligne defining a Shimura variety over the Shimura field $E \subset \mathbf{C}$. We fix an order O_B of B such that $O_B \otimes \mathbf{Z}_p$ is a maximal order of $B \otimes_{\mathbf{Q}} \mathbf{Q}_p$ stable under $*$, and a self–dual $O_B \otimes_{\mathbf{Z}} \mathbf{Z}_p$–lattice Λ in $V \otimes_{\mathbf{Q}} \mathbf{Q}_p$. We fix an open compact subgroup $C^p \subset G(\mathbf{A}_f^p)$. Let $\bar{\mathbf{Q}}$ be the field of algebraic numbers in $\mathbf{C}$ and fix an embedding $\nu : \bar{\mathbf{Q}} \rightarrow \bar{\mathbf{Q}}_p$. We denote by the same symbol the corresponding place of E above p and let E_ν the completion of E in ν.

These data define a moduli problem of PEL–type parametrizing triples $(A, \bar{\lambda}, \bar{\eta}^p)$ consisting of a O_B–abelian variety, a $\mathbf{Q}$–homogeneous principal O_B–polarization and a C^p–level structure and which is representable by a quasi–projective scheme $\mathcal{A}_{C^p}$ over $Spec\ O_{E_\nu}$ (cf. §6 for details). Let C_p be the fix group of Λ in $G(\mathbf{Q}_p)$ and $C = C^p.C_p$. The Shimura variety $Sh(G, h)_C$ is contained as an open and closed subscheme in the generic fibre of $\mathcal{A}_{C^p}$. We take for L the algebraic closure of the residue field of O_{E_ν}. We fix a point $(A_0, \bar{\lambda}_0, \bar{\eta}_0^p)$ of $\mathcal{A}_{C^p}(L)$. Let N_0 be the isocrystal associated to A_0. We fix an isomorphism $N_0 \simeq V \otimes_{\mathbf{Q}_p} K_0$ which respects the actions of $B \otimes K_0$ and the alternating bilinear forms on both sides. This allows us to write the Frobenius operator on N_0 as $b(\mathrm{id} \otimes \sigma)$, with $b \in G(K_0)$. Let $\mathcal{M}$ be the (pro-) formal scheme over $Spf\ O_{E_\nu}$ associated to the data of type (PEL), $(B \otimes \mathbf{Q}_p, *, V \otimes \mathbf{Q}_p, (\ ,\ ), b, \mu, O_B \otimes \mathbf{Z}_p, \Lambda)$. It is acted on by the group $J(\mathbf{Q}_p)$. Here μ denotes a 1–parameter subgroup of G defined over a finite extension K of K_0 in the conjugacy class defined by h. The methods employed to relate $\breve{\mathcal{M}}$ and the local model $\mathbf{M}^{loc}$ may be used to show that locally for the étale topology $\mathcal{A}_{C^p}$ is isomorphic to $\mathbf{M}^{loc}$. The formulation of one uniformization theorem is as follows (6.30).

Theorem III *Assume that $(A_0, \bar{\lambda}_0, \bar{\eta}_0^p)$ is basic, i.e. the corresponding element $b \in G(K_0)$ is basic* [Kol]. *Then*
(i) *The set of points $(A, \bar{\lambda}, \bar{\eta}^p)$ of $\mathcal{A}_{C^p}(L)$ such that $(A, \bar{\lambda})$ is isogenous to $(A_0, \bar{\lambda}_0)$ is a closed subset Z of $\mathcal{A}_{C^p}$.*
(ii) *Let $\mathcal{A}_{C^p/Z}$ denote the formal completion of $\mathcal{A}_{C^p}$ along Z. There is an isomorphism of formal schemes over $Spf\, O_{E_\nu}$,*

$$I(\mathbf{Q}) \setminus [\mathcal{M} \times G(\mathbf{A}_f^p)/C^p] \xrightarrow{\sim} \mathcal{A}_{C^p/Z}.$$

Here I is an inner form of G such that $I(\mathbf{Q})$ is the group of quasi-isogenies of $(A_0, \bar{\lambda}_0)$, which acts diagonally through suitable natural embeddings of groups,

$$I(\mathbf{Q}) \longrightarrow J(\mathbf{Q}_p), \;\; I(\mathbf{Q}) \longrightarrow G(\mathbf{A}_f^p).$$

The source of this isomorphism is a finite disjoint sum of formal schemes of the form $\Gamma \setminus \mathcal{M}$, where $\Gamma \subset J(\mathbf{Q}_p)$ is a discrete subgroup which is cocompact modulo center.

Heuristically, Z should be thought of as an isogeny class in $\mathcal{A}_{C^p}(L)$ which is the most supersingular. In the Siegel case (principally polarized abelian varieties with level structure prime to p) the subscheme Z is the *supersingular locus.* It may be conjectured that such isogeny classes always exist in the special fibre. The above uniformization theorem for formal schemes implies a corresponding rigid–analytic uniformization theorem, cf. (6.36). This represents an admissible open subset of the rigid–analytic variety over E_ν associated to the Shimura variety $Sh(G, h)_C$ (the tubular neighbourhood of Z) as the finite disjoint sum of quotients of $\mathcal{M}^{rig}$, resp. of one of its coverings, by a discrete subgroup of a p–adic group. In this rigid–analytic version the open compact subgroup $C \subset G(\mathbf{A}_f)$ is completely arbitrary.
We prove in fact a uniformization theorem even for non–basic isogeny classes but since these do not form in general a closed subset the formulation is more technical. Indeed, we prove this more general but somewhat formal version first (6.24) and then deduce the above theorem from it. In the deduction we use the fact that Tate's theorem on endomorphism of abelian varieties over

a finite field becomes extremely simple in the basic case, as well as results of Katz [Ka2] on the constancy of isocrystals over a complete discrete valuation ring (these are also behind the results of [RR] which are used here as well). In general the set of points in a basic isogeny class makes up only a small part of the special fibre of a Shimura variety. (This supports the earlier statement that most often the formal schemes $\breve{\mathcal{M}}$ described above are not p-adic). There are, however, examples of Shimura varieties where the points in the special fibre form one basic isogeny class in which case we obtain a *p-adic* uniformization theorem (6.50). As explained above we believe that it is not simply coincidental that the uniformizing formal scheme is in these cases one of Deligne's formal schemes $\hat{\Omega}$ (or products of them). We content ourselves with stating the following special case of such a p-adic uniformization theorem which generalizes a result in [R1].

Theorem IV *In the above notation we assume that B is a division algebra over $\mathbf{Q}$ and that the involution $*$ is of the second kind, i.e. induces a non-trivial automorphism of its center K. We further assume that the B-module V is of rank 1. Let F denote the field of invariants under $*$ in K. We assume that there is precisely one prime ideal $\mathbf{p}$ above p in F and that $\mathbf{p} = \mathbf{q}\cdot\bar{\mathbf{q}}$ splits in K. We assume that*

$$\begin{aligned} \mathrm{inv}_{\mathbf{q}}B &= 1/d \\ \mathrm{inv}_{\bar{\mathbf{q}}}B &= -1/d. \end{aligned}$$

Let $\Phi \subset \mathrm{Hom}(K, \bar{\mathbf{Q}})$ be the unique CM-type of K such that $\nu\circ\varphi$ defines the place $\mathbf{q}$ of K for all $\varphi \in \Phi$. For any $\varphi\in\Phi$ there is an isomorphism

$$B\otimes_{K,\varphi}\mathbf{C} \simeq M_d(\mathbf{C})$$

such that the tensor product of $*$ with complex conjugation becomes the involution $X \mapsto {}^t\bar{X}$. We may write $V\otimes_{K,\varphi}\mathbf{C} = \mathbf{C}^d\otimes\mathbf{C}^d$ in such a way that the action of $M_d(\mathbf{C})$ is via the first factor and such that

$$(Z_1\otimes W_1,\ Z_2\otimes W_2) = \mathrm{Tr}_{\mathbf{C}/\mathbf{R}}({}^t\bar{Z}_1 Z_2\cdot \bar{W}_1\cdot H_\varphi W_2)$$

and where

$$H_\varphi = \mathrm{diag}\,(-\sqrt{-1},\ldots,-\sqrt{-1};\ \sqrt{-1},\ldots,\sqrt{-1}).$$

Let r_φ be the number of places where $-\sqrt{-1}$ appears in H_φ. Let $J_\varphi : V \otimes_{K,\varphi} \mathbf{C} \to V \otimes_{K,\varphi} \mathbf{C}$ be the endomorphism given by $\mathrm{id}_{\mathbf{C}^d} \otimes H_\varphi$, and let $J = \bigoplus J_\varphi : V \otimes \mathbf{R} \to V \otimes \mathbf{R}$. The homomorphism $h : Res_{\mathbf{C}/\mathbf{R}} \mathbf{G}_m \to \mathbf{G}_\mathbf{R}$ defining the Shimura variety $Sh(G,h)$ is defined by the condition that $h(r)$ for $r \in \mathbf{R}^\times$ acts on $V_\mathbf{R}$ by multiplication by r and $h(\sqrt{-1})$ acts as J.
We fix an element $\alpha \in \Phi$ and assume that

$$\begin{aligned} r_\alpha &= 1 \\ r_\varphi &= 0, \quad \varphi \in \Phi \setminus \{\alpha\}. \end{aligned}$$

Let $C_p \subset G(\mathbf{Q}_p)$ be the unique maximal compact subgroup and let $C^p \subset G(\mathbf{A}_f^p)$ be a sufficiently small open compact subgroup and put $C = C^p.C_p$. Then there is a model Sh_C of the Shimura variety of level C over O_{E_ν} which is open and closed in $\mathcal{A}_{C^p}$ and there is an equivariant isomorphism of p-adic formal schemes

$$I(\mathbf{Q}) \setminus \left(\hat{\Omega}^d_{F_\mathbf{p}} \times_{Spf\ O_{F_\mathbf{p}}} Spf\ O_{\breve{E}_\nu}\right) \times G(\mathbf{A}_f)/C \simeq Sh_C^\wedge \times_{Spf\ O_{E_\nu}} Spf\ O_{\breve{E}_\nu}.$$

Here $Sh_C^\wedge$ denotes the completion of Sh_C along its special fibre. Furthermore $I(\mathbf{Q})$ is the group of $\mathbf{Q}$-rational points of an inner form of G such that $I(\mathbf{Q}_p) = \{(a,b) \in GL_d(F_\mathbf{p}) \times GL_d(F_\mathbf{p})^{opp};\ ab \in \mathbf{Q}_p\}$ and with $I(\mathbf{A}_f^p) \simeq G(\mathbf{A}_f^p)$. We used α to identify $F_\mathbf{p}$ with E_ν. The natural descent datum on the right hand side induces on the left hand side the natural descent datum on the first factor multiplied with the action of

$$g = (\Pi^{-1}, p^f \Pi) \in B_\mathbf{q}^\times \times B_\mathbf{q}^{\times opp}$$

on $G(\mathbf{A}_f)/C$. Here Π is a uniformizing element of $D_\mathbf{q}$ and f is the index of inertia of $F_\mathbf{p}$.

The rigid -analytic version of this theorem represents the Shimura variety Sh_C as a finite disjoint sum of quotients of one of Drinfeld's covering spaces of $\Omega^d_{E_\nu}$ by a discrete cocompact subgroup of $I(\mathbf{Q}_p)$. The rigid-analytic version is also considered in [V].

This concludes our brief description of the subject matter of this paper. We refer to the report [R2] for further remarks on general p–adic period domains and their cohomology.

We now describe briefly the contents of the various chapters. Section 1 besides defining the p–adic symmetric spaces assembles various facts about (filtered) isocrystals. We point out in particular our conjecture (1.20) on the Fontaine functor which we prove in a special case by an extension of an argument of Kottwitz. We also mention that the present formulation of the Harder–Narasimhan filtration in this context is due to Faltings and is an improvement on our first version. This chapter also contains various examples which are considered again in later parts of the manuscript from other points of view. These various examples are (in our opinion) fun but they also form the backbones of the theory. Section 2 is devoted to the proof of the representability theorem in the absolute case. In chapter 3 we formulate the moduli problems and prove their representability. We also construct the local models $\mathbf{M}^{loc}$ mentioned above. In the appendix to chapter 3 we prove the existence of normal forms for polarized chains of lattices over a general base scheme. This existence theorem is more or less standard when the base is a complete discrete valuation ring (Bruhat-Tits theory) but we were unable to find it in this form in the literature. In chapter 4 which may be omitted at a first reading we define the Hecke correspondences on the formal schemes of chapter 3. In chapter 5 we treat the rigid–analytic aspects of the situation and in chapter 6 we prove the uniformization theorems for Shimura varieties.

It remains for us to thank all those who helped us during the long period of gestation of this work (our main results were presented at the Oberwolfach meeting on arithmetic algebraic geometry in July 1992). It is a pleasure to acknowledge the influence of many conversations with R. Kottwitz on the subject (comp. also [R2]). We thank B. Gross and M. Hopkins for communicating to us their ideas long before their publication. We also thank P. Berthelot for providing us with his construction of the generic fibre mentioned above. We are grateful to G. Faltings for giving us access to some of his unpublished manuscripts and for his various contributions. We also thank J.-L. Waldspurger for solving some subtle questions on classical groups over p–adic fields. We are grateful to J. de Jong and R. Huber for giving us generous advice on rigid–analytic geometry. We furthermore thank J. de Jong for his careful reading of parts of the manuscript and his many suggestions. We are thankful for helpful discussions to M. Aschbacher, J.-F. Boutot, P. Deligne, J.-M. Fontaine, M. Harris, G. Laumon, W. Messing, A. Ogus, D. Ramakrishnan, U. Stuhler and B. Totaro. Finally it is a pleasure to acknowledge the support and hospitality of the following institutions: MIT,

IAS, MSRI, the University of Minnesota, the Newton Institute, Harvard University and the University of Chicago. We are especially grateful to the Deutsche Forschungsgemeinschaft for its continual financial support. We also thank P. Herdieckerhoff and B. Rahner for the typesetting of the manuscript.

Period Spaces for p-divisible Groups

1. p–adic symmetric domains

The aim of this chapter is to introduce the *p–adic symmetric domains* and to discuss the conjectural *local systems* on them. In chapter 5 we will show that in many cases the p–adic symmetric domains are the *conjectural* image of the *period morphism*. In addition, we introduce the concept of a *decent isocrystal*, cf. (1.8).

1.1 We first recall some concepts of σ–linear algebra. Let L be a perfect field of characteristic p. Let $W(L)$ be its ring of Witt vector and $K_0 = K_0(L) = W(L)_{\mathbf{Q}}$ its fraction field. We denote by σ the Frobenius automorphism. An *isocrystal over* L is a finite–dimensional K_0–vector space V equipped with a bijective σ–linear endomorphism Φ. The dimension of V is called the *height* of the isocrystal. The isocrystals over L form in an obvious way a $\mathbf{Q}_p$–linear category. Let L be algebraically closed. Then the category of isocrystals over L is a noetherian, artinian semi–simple abelian category. Its simple objects are parametrized by the elements of $\mathbf{Q}$. To $\lambda \in \mathbf{Q}, \lambda = r/s, (r,s) = 1, s > 0$ $(r, s \in \mathbf{Z})$ there corresponds the simple object

$$E_\lambda = (K_0^s, \begin{bmatrix} 0 & 1 & & \\ & & \ddots & 1 \\ p^r & & & 0 \end{bmatrix} \cdot \sigma)$$

and $D_\lambda = \mathrm{End}(E_\lambda)$ is a division algebra with center $\mathbf{Q}_p$ and invariant $-\lambda$. If (V, Φ) is an isocrystal we will write

$$V = \bigoplus V_\lambda$$

for its isotypical or slope decomposition.

Over an arbitrary perfect field L, the category of isocrystals is no longer semi–simple, but the isotypical decomposition continues to hold, compatible with base change $L \to L'$, [Z2]. An isocrystal (V, Φ) over L is isotypical iff there are integers r, s with $s > 0$ and a $W(L)$–lattice M in V such that

$$\Phi^s(M) = p^r M.$$

There is an obvious variant of these concepts where instead of a σ–linear endomorphism one considers a σ^r–linear endomorphism, for $r \neq 0$ fixed.

1.2 Let L be a perfect field of characteristic p. Let $K_0 = K_0(L)$ and let K be a finite extension of K_0. A *filtered isocrystal over* K (Fontaine uses the name "filtered module" [Fo2] - neither terminology is very good, ours not since these are not filtered objects in the category of isocrystals) is given by an isocrystal (V, Φ) over L and a decreasing filtration $\mathcal{F}^\bullet$ on the K–vector space $V \otimes_{K_0} K$ such that $\mathcal{F}^r = (0)$ and $\mathcal{F}^s = V \otimes_{K_0} K$ for suitable $r, s \in \mathbf{Z}$. The filtered isocrystals over K form a $\mathbf{Q}$–linear category with $\otimes$ and internal Hom. It is an exact category, but not an abelian category. An admissible monomorphism $(V, \Phi, \mathcal{F}^\bullet)$, also called a subobject, is given by a subvector space V' which is Φ–stable, for which $V' \otimes_{K_0} K$ is equipped with the induced filtration.

1.3 A filtered isocrystal $(V, \Phi, \mathcal{F}^\bullet)$ over K is called *weakly admissible* ([Fo2], §4) if for every subobject $(V', \Phi', \mathcal{F}'^\bullet)$ we have

$$\sum i \cdot \dim gr^i_{\mathcal{F}'}(V' \otimes_{K_0} K) \leq \mathrm{ord}_p \det(\Phi')$$

and if for $(V', \Phi', \mathcal{F}'^\bullet) = (V, \Phi, \mathcal{F}^\bullet)$ we have equality in this relation. It is known that the full subcategory of weakly admissible filtered isocrystals over K is an abelian category which is closed under extensions and under passage to the dual object. A theorem of Faltings ([Fal], comp. also [T]), proving a conjecture of Fontaine [Fo2], 5.2.6, states that it is closed under $\otimes$. If $K \subset K'$, the obvious base change functor from the category of filtered isocrystals over K into the category of filtered isocrystals over K' preserves the corresponding subcategories of weakly admissible objects. We shall need to analyze how a filtered isocrystal can fail to be weakly admissible, by introducing the analogue of the Harder–Narasimhan filtration.

Let $(V, \Phi, \mathcal{F}^\bullet) \neq 0$ be a filtered isocrystal over K. We define its *HN-slope*

$$\mu(V) = \mu(V, \Phi, \mathcal{F}^\bullet) = \frac{\sum i \cdot \dim gr^i_{\mathcal{F}}(V \otimes_{K_0} K) - \mathrm{ord}_p \det(\Phi)}{\dim V}$$

In analogy with Mumford's definition for vector bundles over a curve we call $(V, \Phi, \mathcal{F}^\bullet)$ *semi-stable* if for every subobject $(V', \Phi', \mathcal{F}'^\bullet) \neq (0)$ we have

$$\mu(V') \leq \mu(V).$$

Therefore, $(V, \Phi, \mathcal{F}^\bullet)$ is weakly admissible if and only if it is semi-stable and $\mu(V) = 0$. The following proposition is the analogue of the canonical filtration of Harder–Narasimhan–Quillen–Tjurin in the context of vector bundles. The proof of this proposition is almost word – for – word the same ([HN]), the main point being that for a morphism of filtered isocrystals $V' \to V$ which induces an isomorphism of the underlying vector spaces we have $\mu(V') \leq \mu(V)$, and will therefore be omitted.

Proposition 1.4 (Faltings) *Let $V = (V, \Phi, \mathcal{F}^\bullet)$ be a filtered isocrystal over K. Then V possesses a unique decreasing filtration by subobjects $V^\bullet$ parametrized by $\mathbf{Q}$, called its canonical filtration, with the following property..*
Let $V^{\alpha+} = \sum_{\beta > \alpha} V^\beta$. If $V^{\alpha+} \subsetneqq V^\alpha$, then $V^\alpha / V^{\alpha+}$ is semi-stable of HN-slope α.
Furthermore, if $V^\alpha \subsetneqq V$, then

$$\mu(V^\alpha) > \mu(V).$$

In particular, V is semi-stable if and only if its associated canonical filtration is trivial.

Remarks 1.5 (i) It is obvious that any morphism of filtered isocrystals over K is strictly compatible with the canonical filtrations. The canonical filtration is also compatible with passage to the dual and with the formation of the tensor product of two filtered isocrystals. This last fact follows from the theorem of Faltings mentioned above.

(ii) We recall the definition of the Tate object $\mathbf{1}(n)$, $n \in \mathbf{Z}$. In the context of isocrystals,

$$\mathbf{1}(n) = (K_0, \ p^n \sigma).$$

In the context of filtered isocrystals over K, we filter $\mathbf{1}(n)$ such that

$$\mathrm{Fil}^i(\mathbf{1}(n)) = \begin{cases} K & i \leq n \\ (0) & i > n \end{cases}.$$

A Tate twist of a (filtered) isocrystal V is defined as

$$V(n) = V \otimes \mathbf{1}(n).$$

It is obvious that $\mu(V(n)) = \mu(V)$.

1.6 We also recall Fontaine's functors. Let L be a perfect field of characteristic p. Let $K_0 = K_0(L)$ and let K be a finite extension of K_0. We denote by B_{crys} Fontaine's crystalline period field ([Fo3]). It is a K_0-algebra, equipped with a continuous action of $Gal(\bar{K}/K)$, a σ-linear endomorphism and with a filtration of the K-algebra $B_{crys} \otimes_{K_0} K$. A p-adic Galois representation U of $Gal(\bar{K}/K)$ is called crystalline if the dimension of the K_0-vector space

$$\mathcal{G}(U) = (U \otimes_{\mathbf{Q}_p} B_{crys})^{Gal(\bar{K}/K)}$$

is equal to the dimension of U. From B_{crys} the K_0-vector space $\mathcal{G}(U)$ inherits the structure of a filtered isocrystal relative to the extension K/K_0. Fontaine has shown [Fo2] that the functor $\mathcal{G}$ induces a fully faithful exact $\otimes$-functor from the category of crystalline Galois representations of $Gal(\bar{K}/K)$ to a full subcategory of the category of weakly admissible filtered isocrystals over K. An object of the essential image over $\mathcal{G}$ is called *admissible*. We denote by $\mathcal{F}$ Fontaine's quasi-inverse $\otimes$-functor to $\mathcal{G}$ from the category of admissible filtered isocrystals over K to the category of crystalline Galois representations,

$$\mathcal{F}(V) = (\mathrm{Fil}^0(V \otimes_{K_0} B_{crys}))^{\Phi}.$$

1.7 Let G be a linear algebraic group over $\mathbf{Q}_p$. Let L be a perfect field of characteristic p and $K_0 = K_0(L)$ and let

$$b \in G(K_0).$$

Then to any $\mathbf{Q}_p$-rational representation V of G we associate an isocrystal over L,

$$(V \otimes K_0,\ b\,(\mathrm{id} \otimes \sigma)).$$

In this way we obtain an exact $\otimes$–functor from the category $\mathcal{REP}(G)$ of finite–dimensional rational representations of G over $\mathbf{Q}_p$ to the category of isocrystals over L. Let $g \in G(K_0)$ and put

$$b' = g\, b\, \sigma(g)^{-1}.$$

Then multiplication by g defines an isomorphism between the $\otimes$–functor associated to b and the $\otimes$–functor associated to b'. If L is algebraically closed and G is connected we use the notation $B(G)$ to denote the set of σ–conjugacy classes of $G(K_0)$ (cf. [Ko1]). The fact that L does not appear in this notation is justified by proposition (1.16) below.
We denote by $\mathbf{D}$ the algebraic torus over $\mathbf{Q}_p$, whose character group is $\mathbf{Q}$. Kottwitz [Ko1] associates to b a morphism of algebraic groups defined over K_0,

$$\nu : \mathbf{D} \longrightarrow G_{K_0}.$$

We will call this the *slope morphism.* If V is a $\mathbf{Q}_p$–rational representation of G, the morphism ν defines a $\mathbf{Q}$–grading on the vector space $V \otimes K_0$. The morphism ν is characterized by the property that this grading is the slope decomposition of the isocrystal associated to b and V. We say that $\lambda \in \mathbf{Q}$ is a slope of V, if the isotypic component of slope λ is not equal to zero. The property that λ is a slope of V does not depend on the choice of b in the σ–conjugacy class $\bar{b}$.
The group $\mathbf{Q}^{\times}$ acts on $\mathbf{D}$, since it acts on the character group $\mathbf{Q}$. For $s \in \mathbf{Q}^{\times}$ we use the notation $s\nu$ for the composite $\mathbf{D} \xrightarrow{s} \mathbf{D} \xrightarrow{\nu} G$. Let $\mathbf{D} \to \mathbf{G}_m$ be the projection to the multiplicative group induced by the inclusion of the character groups $\mathbf{Z} \subset \mathbf{Q}$. Then for a suitable s the morphism $s\nu$ factors through this projection,

$$s\nu : \mathbf{G}_m \longrightarrow G.$$

Since both sides of the slope morphism are defined over $\mathbf{Q}_p$ the conjugate ν^{σ} by the Frobenius morphism is defined. We have the formula:

$$b\nu^{\sigma}b^{-1} = \nu.$$

To check this we may replace ν by $s\nu$. Then it is enough to check that for $a \in \mathbf{G}_m(K_0)$ we have:

$$b\sigma(s\nu(\sigma^{-1}a))b^{-1} = s\nu(a)$$

We interpret both sides as endomorphisms of $V \otimes K_0$. Let $\Phi = b\sigma$ be the Frobenius on $V \otimes K_0$. Then the assertion is:

$$\Phi s\nu(\sigma^{-1}a) = s\nu(a)\Phi.$$

This is obvious if we restrict to an isotypic component.

Definition 1.8 *We call a σ–conjugacy class $\bar{b}$ in $G(K_0)$ decent if there exists an element $b \in \bar{b}$ such that for some natural number s:*

$$(b\sigma)^s = s\nu(p)\sigma^s.$$

We suppose here that $s\nu$ factors through a morphism $\mathbf{G}_m \to G$, which is also denoted by $s\nu$. The identity is between elements of the semi–direct product $G(K_0) \rtimes < \sigma >$. We will call b a *decent element* in $\bar{b}$ and the above equation a *decency equation* for b.

Corollary 1.9 *Assume that $\bar{b}$ is decent, and that b and s are from the definition (1.8). Then $b \in G(\mathbf{Q}_{p^s})$, and ν is defined over $\mathbf{Q}_{p^s}$.*

Here $\mathbf{Q}_{p^s}$ denotes the unramified extension of degree s of $\mathbf{Q}_p$.

Proof: Let us first prove the second assertion. We set $b_s = b\sigma(b)\ldots\sigma^{s-1}(b)$. Then we get from the formula above

$$b_s\nu^{\sigma^s}b_s^{-1} = \nu.$$

By definition of decent we have $b_s = s\nu(p)$. Inserting this in the equation above we get the desired $\nu^{\sigma^s} = \nu$.
To prove the first assertion we consider the equation:

$$(b\sigma)^s(b\sigma) = (b\sigma)(b\sigma)^s.$$

We obtain $s\nu(p)\sigma^s b\sigma = b\sigma s\nu(p)\sigma^s$. Taking into account that $s\nu(p)$ commutes with $b\sigma$, the assertion follows.

Corollary 1.10 *Let $b_1, b_2 \in \bar{b}$ which satisfy a decency equation for the same integer s. Then the elements b_1 and b_2 are σ–conjugate in $G(K_0 \cap \mathbf{Q}_{p^s})$.*

Proof: There is an element $g \in G(K_0)$, such that $b_2 = gb_1\sigma(g^{-1})$ and $\nu_2 = g\nu_1 g^{-1}$. The decency equations for b_1 and b_2 are:

$$(b_1\sigma)^s = s\nu_1(p)\sigma^s, \qquad g(b_1\sigma)^s g^{-1} = gs\nu_1(p)g^{-1}\sigma^s.$$

Comparing these equations we see that g commutes with σ^s, so that $g \in G(K_0 \cap \mathbf{Q}_{p^s})$.

1.11 Let $\bar{b}$ be a decent σ–conjugacy class and let $b \in \bar{b}$ be decent. Then $b \in G(\mathbf{Q}_{p^s})$ defines for every $\mathbf{Q}_p$–rational representation V of G an isocrystal over the field $L_s = \mathbf{F}_{p^s} \cap L$. The corollary (1.10) says that this isocrystal only depends on $\bar{b}$, up to isomorphism. Its base change under $L_s \to L$ is $(V \otimes K_0,\ b\sigma)$.
Assume that G is connected and that L is algebraically closed. Then by Kottwitz [Ko1] any σ–conjugacy class is decent.

Proposition 1.12 *Let L a perfect field of characteristic p, and $b \in G(W(L)_{\mathbf{Q}})$. Then the following functor on the category of $\mathbf{Q}_p$–algebras is representable by a smooth affine group scheme over $\mathbf{Q}_p$,*

$$J(R) = \{g \in G(R \otimes_{\mathbf{Q}_p} W(L)_{\mathbf{Q}});\quad g(b\sigma) = (b\sigma)g\}.$$

Assume that $b \in G(W(L')_{\mathbf{Q}})$, where L' is an algebraically closed subfield of L. We denote by J' the corresponding functor defined with L'. Then the canonical morphism $J' \to J$ is an isomorphism.

For the proof we need a lemma.

Lemma 1.13 *Let V be a finite dimensional vector space over $W(L)_{\mathbf{Q}}$. Assume we are given a σ^s–linear isomorphism $\phi : N \to N$, where s is some nonzero integer. Then the functor on the category of $\mathbf{Q}_p$–algebras*

$$F(R) = \{n \in N \otimes_{\mathbf{Q}_p} R;\ \ \phi(n) = n\}$$

is representable by an affine space over $\mathbf{Q}_p$.

Proof: Choosing a basis of the $\mathbf{Q}_p$–vector space R, we see that

$$F(R) = N^{\phi} \otimes_{\mathbf{Q}_p} R.$$

Here N^{ϕ} denotes the invariants of ϕ. Hence the assertion is that N^{ϕ} is a finite dimensional $\mathbf{Q}_p$–vector space. To see this we may assume that L is algebraically closed and then apply a theorem of Dieudonné (e.g. [Z2] 6.29), which tells us that the dimension of N^{ϕ} over $\mathbf{Q}_p$ is the dimension of the part of slope zero of V over $W(L)_{\mathbf{Q}}$. This proves the lemma.

Let us assume for a moment that L is algebraically closed. Let L'' be a field extension of L. Then the argument of the proof shows that the functor F'' defined by $N \otimes_{W(L)_{\mathbf{Q}}} W(L'')_{\mathbf{Q}}$ and the σ^s–linear operator $\phi \otimes \sigma^s$ coincides with F.

Proof of proposition (1.12): We choose an embedding $G \subset GL(V)$. We denote by B the endomorphism of V induced by b. We consider the following functor:

$$F(R) = \{g \in End\, V \otimes W(L)_{\mathbf{Q}} \otimes R;\ \ B^{-1}gB = \sigma(g)\}$$

By the lemma above, applied to the σ–linear map $B\sigma(g)B^{-1}$, it is representable.

More precisely there is a finite–dimensional $\mathbf{Q}_p$–subspace $W \subset End\, V \otimes W(L)_{\mathbf{Q}}$, such that $F(R) = W \otimes R$. We choose a basis $A_1, \ldots, A_m$ of W, where the A_i are endomorphisms of V with coefficients in $W(L)_{\mathbf{Q}}$. Hence we have an identification of F with an affine space:

$$(r_1, \ldots, r_m) \in R^m \mapsto r_1A_1 + \ldots + r_mA_m.$$

Let $\{f_k\}$ be the equations of the subset G in $GL(V)$. Then the subfunctor $J \subset F$ is given by the following conditions:

$$f_k(\sum_{i=1}^{i=m} r_iA_i) = 0, \quad \det(\sum_{i=1}^{i=m} r_iA_i) \neq 0$$

Using a basis of $W(L)_{\mathbf{Q}}$ over $\mathbf{Q}_p$, we may rewrite these conditions in terms of polynomials in $r_1, \ldots, r_m$ with coefficients in $\mathbf{Q}_p$. Hence J is a locally closed subfunctor of F.

Finally, if the coefficients of B lie in $W(L')$, then we have remarked that the functor F does not change if we replace L by L'. Then J does not change because it is defined inside F by the same conditions for L and L'. □

Corollary 1.14 *In the notation of proposition (1.12), assume that b satisfies a decency equation of the form*

$$(b\sigma)^s = s\nu(p)\sigma^s$$

(cf. (1.8)), where $s\nu$ *factors through a homomorphism* $\mathbf{G}_m \to G$ *which is also denoted by* $s\nu$ *and which is defined over* $\mathbf{Q}_{p^s}$. *Then* J *is an inner form of the centralizer* $G_{s\nu(p)}$, *a Levi subgroup of* $G_{\mathbf{Q}_{p^s}}$.

Proof: By the decency equation $b\sigma$ defines a 1–cocycle of the adjoint group $(G_{s\nu(p)})_{ad}(\mathbf{Q}_{p^s})$. Hence

$$J'(R) = \{g \in G_{s\nu(p)}(R \otimes_{\mathbf{Q}_p} \mathbf{Q}_{p^s});\;\; g(b\sigma) = (b\sigma)g\}$$

defines an inner form of $G_{s\nu(p)}$. It remains to be checked that any element in $J(R)$ lies in $J'(R)$. For this it is enough to remark that an element in $J(R)$ commutes with $s\nu(p)$ (cf. (1.9)). By the decency equation it therefore commutes with σ^s.

Remark 1.15 Let G be a connected reductive group and let L be algebraically closed. By Kottwitz [Ko1] the following conditions on $b \in G(W(L)_{\mathbf{Q}})$ are equivalent:

(i) The slope homomorphism ν factors through the center of G.

(ii) The element b is σ–conjugate to an element in $T(W(L)_{\mathbf{Q}})$ where T is an elliptic maximal torus of G.

(iii) The algebraic group J of (1.12) is an inner form of G.

In this case the element b respectively its class $\bar{b} \in B(G)$ is called *basic*. Proposition (1.12) admits the following variant.

Proposition 1.16 *Let* b_1 *and* b_2 *be two elements of* $G(W(L)_{\mathbf{Q}})$. *Consider the functor*

$$J(R) = \{g \in G(R \otimes_{\mathbf{Q}_p} W(L)_{\mathbf{Q}});\;\; g(b_1\sigma) = (b_2\sigma)g\}.$$

Then this functor is representable by a smooth affine scheme over $\mathbf{Q}_p$. *Assume that* $b_1, b_2 \in G(W(L')_{\mathbf{Q}})$, *where* L' *is an algebraically closed subfield*

of L and let J' be the corresponding functor. Then the canonical morphism $J' \to J$ is an isomorphism.
In particular the map from the set of σ-conjugacy classes in $G(W(L')_{\mathbf{Q}})$ to the set of σ-conjugacy classes in $G(W(L)_{\mathbf{Q}})$ is injective. The map is bijective if both L and L' are algebraically closed and G is connected.

Proof: Indeed, the surjectivity part of the last assertion follows from (1.9) since by Kottwitz every σ-conjugacy class is decent, if G is connected and L algebraically closed.

1.17 Let K_0 be the fraction field of the Witt vectors of an algebraically closed field L. Let K be a finite extension of K_0. Let G be an algebraic group over $\mathbf{Q}_p$. Let us consider a cocharacter

$$\mu : \mathbf{G}_m \longrightarrow G$$

defined over K, and an element

$$b \in G(K_0).$$

Then to any $\mathbf{Q}_p$-rational representation V of G we have associated a filtered isocrystal

$$\mathcal{I}(V) = (V \otimes K_0, b\sigma, V_K^i)\,, \tag{1.1}$$

where the filtration V_K^i is given by the weight spaces $V_{K,i}$ with respect to μ:

$$V_K^i = \bigoplus_{j \geq i} V_{K,j}\,. \tag{1.2}$$

Definition 1.18 *Let G be a reductive group. We call the pair (μ, b) admissible, if one of the following equivalent conditions is fullfilled:*

(i) For any $\mathbf{Q}_p$-rational representation V of G the filtered isocrystal $\mathcal{I}(V)$ is admissible.

(ii) There is a faithful $\mathbf{Q}_p$-rational representation of G, such that $\mathcal{I}(V)$ is admissible.

We make the same definition for weakly admissible.

Proof: If V is a faithful representation, then any $\mathbf{Q}_p$–rational representation appears as a direct summand of $V^{\otimes n} \otimes \hat{V}^{\otimes m}$. Hence the equivalence of the conditions follows from the following facts (Fontaine [Fo2]). A direct sum of filtered isocrystals is admissible, iff each summand is admissible. A tensor product of admissible filtered isocrystals is admissible. The same is true for weakly admissible filtered isocrystals, but the last fact is then a theorem of Faltings (cf. (1.3)). □

1.19 Let (μ, b) be an admissible pair in a reductive algebraic group G. Consider Fontaine's functor $\mathcal{F}$ from the category of admissible filtered isocrystals over K to the category of crystalline representations of the Galois group $Gal(\bar{K}/K)$. We denote by $\mathcal{F}_v$ the composite of $\mathcal{F}$ with the natural forgetful functor to the category of finite–dimensional $\mathbf{Q}_p$–vector spaces. Let $\mathcal{REP}(G)$ be the category of finite dimensional rational representations of G over $\mathbf{Q}_p$. Then the composite of $\mathcal{F}_v$ with the functor (1.1) defines a fibre functor

$$\mathcal{F}_v \circ \mathcal{I} : \mathcal{REP}(G) \longrightarrow (\mathbf{Q}_p - \text{vector spaces}).$$

Let Ver be the standard fibre functor. Then $\mathrm{Hom}(Ver, \mathcal{F}_v\mathcal{I})$ is a right torsor under the group G and hence defines a cohomology class:

$$\mathrm{cls}(\mu, b) \in H^1(\mathbf{Q}_p, G).$$

We have a conjecture to compute this cohomology class in the case that G is a connected reductive group, as follows. Denote by $\hat{G}$ the connected component of the L–group of G. We denote the center by $Z(\hat{G})$. Let Γ be the group $Gal(\bar{\mathbf{Q}}_p/\mathbf{Q}_p)$. Then Kottwitz [Ko2] defines a commutative diagram:

$$\begin{array}{ccc} H^1(\mathbf{Q}_p, G) & \longrightarrow & B(G) \\ \big\downarrow & & \big\downarrow \kappa \\ X^*(Z(\hat{G})^\Gamma)_{tor} & \longrightarrow & X^*(Z(\hat{G})^\Gamma). \end{array}$$

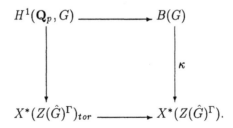

The left vertical arrow is an isomorphism.

Proposition 1.20 *Assume that the derived group of the connected reductive group G is simply connected. Let (μ, b) be an admissible pair. The cocharacter μ defines in a canonical way a character of $Z(\hat{G})$. We denote its restriction to $Z(\hat{G})^\Gamma$ by $\mu^\sharp$. Then we have:*

$$\mathrm{cls}(\mu, b) = \kappa(b) - \mu^\sharp.$$

Here we consider the left hand side as an element of $X^*(Z(\hat{G})^\Gamma_{tors}$.

We conjecture that this proposition holds without the assumption that the derived group is simply connected. It shows that for a fixed b the invariant $\mathrm{cls}(\mu, b)$ depends only on the conjugacy class of μ, if it is defined.
Before proving proposition (1.20) we note that for a torus T over $\mathbf{Q}_p$ all admissible pairs may be described in an elementary way:

Proposition 1.21 *The following conditions for a pair (μ, b) with respect to the torus T are equivalent:*

(i) *(μ, b) is weakly admissible*

(ii) *$\mu - \nu$ is orthogonal to all $\mathbf{Q}_p$-rational characters of T. Here we view ν as an element of $X_*(T) \otimes \mathbf{Q}$.*

(iii) *For any $\mathbf{Q}_p$-rational character χ of T, we have*

$$\mathrm{ord}_p\ \chi(b) \ = < \chi, \mu > .$$

(iv) *(μ, b) is admissible.*

Proof: Clearly the conditions (ii) and (iii) are equivalent. The first condition implies the third. Indeed, let V be the one–dimensional representation given by χ. Then the isocrystal $N = (V \otimes K_0, b\sigma)$ is isotypic of slope $\mathrm{ord}_p(\chi(b))$. The only non zero weight space of V_K is $V_{K,j}$ for $j =< \chi, \mu >$.
Next we have to show that (iii) implies (i). Let V be an irreducible representation of T. Let

$$N = \bigoplus_{\alpha \in \mathbf{Q}} N_\alpha$$

be the decomposition of the associated isocrystal into isotypic components. We need to verify:

$$\sum_{\alpha} \alpha \dim N_\alpha = \sum i \dim V_{K,i}. \tag{1.3}$$

The characters of T appearing in V_K form an orbit under the Galois group $\mathrm{Gal}(\bar{\mathbf{Q}}_p/\mathbf{Q}_p)$. Let χ be a particular character of this orbit and $\mathrm{Nm}\,\chi$ be the product of elements of this orbit. Clearly the right hand side of (1.3) is

$$\sum_{\chi' \in \mathrm{orbit}} < \chi', \mu > = < \mathrm{Nm}\,\chi, \mu > .$$

Assume that $s\nu$ factors through $\mathbf{G}_m$. Then the left hand side of (1.3) is

$$\frac{1}{s} \sum i \dim W_i \ ,$$

where $W_i \subset V_{\bar{\mathbf{Q}}_p}$ is the weight space with respect to $s\nu$. But since $V_{\bar{\mathbf{Q}}_p} = \oplus \chi'$ the last expression is equal to

$$\frac{1}{s} \sum_{\chi'} < \chi', s\nu > \; = \; < \mathrm{Nm}\,\chi, \nu > .$$

Since $\mathrm{Nm}\,\chi$ is rational over $\mathbf{Q}_p$ this proves the claim.
It remains to be shown that $\mathcal{I}(V)$ is semistable. We may take b to be decent and $K_0 = W(\mathbf{F}_q)_{\mathbf{Q}}$ for some finite field, cf (1.9). We consider the Harder–Narasimhan filtration of $\mathcal{I}(V)$. If $\mathcal{I}(V)$ is not semistable we have a subisocrystal $M \subset N$ belonging to this filtration with HN–slope strictly bigger than zero. But by the uniqueness of the HN–filtration M is $T(\mathbf{Q}_p)$–invariant. Since $T(\mathbf{Q}_p) \subset T(K_0)$ is Zariski dense, we get that M is $T(K_0)$–invariant and in particular $bM = M$. On the other hand $b\sigma M = M$ since M is a subisocrystal. Therefore $\sigma M = M$, i.e. M is defined over $\mathbf{Q}_p$. From the irreducibility of V we get the contradiction $M = N$.
Finally we show that the first three conditions of the proposition are equivalent to (μ, b) being admissible. Our argument follows that of Kottwitz [Ko3] §12. We fix a finite Galois extension F of $\mathbf{Q}_p$, such that the torus T splits over F. Then μ is defined over F and we get a map:

$$F^\times \longrightarrow T(F) \xrightarrow{\mathrm{Nm}} T(\mathbf{Q}_p).$$

By local class field theory its restriction to the units in F defines a p-adic representation $V(\mu)$ of $Gal(\bar{K}_0/K_0F)$ on V. This representation is crystalline and hence the image of some admissible filtered isocrystal $N(V(\mu)) = (N, \Phi, Fil)$ under the functor $\mathcal{F}$. The filtration of V is given over the composite $K = FK_0$.
One can give an explicit description of the isocrystal $N(V(\mu))$. Let $\bar{b}_\mu$ be the image of μ under the map (Kottwitz [Ko1]):

$$X_*(T) \longrightarrow B(T).$$

Lemma 1.22 *Let $b_\mu \in \bar{b}_\mu$. Denote by $\mathcal{I}_{\mu,b_\mu}(V)$ the isocrystal associated by (1.17) to the pair (μ, b_μ). Then the image of this isocrystal by the Fontaine functor is $V(\mu)$,*

$$\mathcal{F}(\mathcal{I}_{\mu,b_\mu}(V)) = V(\mu).$$

Proof of lemma (1.22): The functor $V \mapsto N$ is a fibre functor on the tannakian category $\mathcal{REP}(T)$ over K_0. We may choose an isomorphism of fibre functors

$$N \cong V \otimes K_0.$$

By Kottwitz this isomorphism may be taken in such a way that it maps the Frobenius Φ to $b_\mu\sigma$. We claim that the filtration on $V \otimes K$ induced by this isomorphism is the filtration given by μ.
Indeed, the isomorphism above defines a filtration of the category $\mathcal{REP}_K(T)$, which is independent of the choice of that isomorphism. The filtration is given by a unique cocharacter $\mu' \in X_*(T)$.
Let us consider the category of tori over $\mathbf{Q}_p$ that split over F. Then the assignment $\mu \mapsto \mu'$ is an automorphism of the functor $T \mapsto X_*(T)$. We have to show that this automorphism is the identity. But the functor $X_*(T)$ is represented by the torus $Res_{F/\mathbf{Q}_p}\mathbf{G}_{m,F}$ and its universal cocharacter μ_{univ}. The elements $\tau \in Gal(F/\mathbf{Q}_p)$ define in a natural way characters χ_τ of $Res_{F/\mathbf{Q}_p}\mathbf{G}_{m,F}$. This is a basis. We denote the dual basis for the cocharacters by λ_τ. Then $\lambda_1 = \mu_{univ}$ is the universal cocharacter.
The p-adic representation associated to the representation $V \doteq F$ of the torus $Res_{F/\mathbf{Q}_p}\ \mathbf{G}_{m,F}$ and the cocharacter λ_1 is isomorphic to the rational Tate module of the Lubin–Tate formal group associated to F, by the main

theorem of formal complex multiplication ([Se]). As a representation of $Res_{F/\mathbf{Q}_p}\mathbf{G}_{m,F}$ the isocrystal of the Lubin–Tate formal group is a direct sum of the characters χ_τ. By definition of the Lubin–Tate formal group the Lie algebra corresponds to χ_1. Hence the filtration defined by the Lie algebra is given by λ_1. This proves the equality $\mu_{univ} = \mu'_{univ}$. By the universality we get our lemma. □

Hence we have shown that the pair (μ, b_μ) is admissible. We are now able to prove the proposition (1.20) for tori. We have an exact sequence:

$$1 \to H^1(Gal(\mathbf{Q}_p^{nr}/\mathbf{Q}_p), T(\mathbf{Q}_p^{nr})) \longrightarrow B(T) \longrightarrow X_*(T)_\Gamma \otimes \mathbf{Q} \to 1.$$

Since b and b_μ have the same image μ in $X_*(T)_\Gamma \otimes \mathbf{Q}$, their σ–conjugacy classes differ by a cocycle t_σ, which is defined over a finite unramified extension L of $\mathbf{Q}_p$.
Hence we have for a suitable choice of b_μ in its σ–conjugacy class that $b = t_\sigma b_\mu$.
We set $(V \otimes L)^{t_\sigma \sigma} = W$. Then we have an equality of filtered isocrystals

$$(V \otimes K_0, t_\sigma b_\mu \otimes \sigma, \mathrm{Fil}) = (W \otimes K_0, b_\mu \otimes \sigma, \mathrm{Fil})\,.$$

Since we know that the right hand side is the filtered isocrystal associated to the crystalline representation $W(\mu)$, it follows that (μ, b) is admissible. Moreover we have an isomorphism of functors:

$$\mathcal{FI}(V) = W(\mu).$$

Especially the proposition (1.20) holds for T. □

From this proposition (1.20) follows in the general case. Indeed, let D be the quotient of G by the derived group G^{der}. Since the cohomology of G^{der} vanishes we get an injection

$$H^1(\mathbf{Q}_p, G) \longrightarrow H^1(\mathbf{Q}_p, D).$$

Hence $\mathrm{cls}(\mu, b)$ is uniquely determined by the image of (μ, b) in D. Since we know the proposition for D, the proof is finished. □

Definition 1.23 *Let G be a connected reductive group. We fix a finite extension K of $W(L)_{\mathbf{Q}}$. We call two pairs (μ, b) and (μ', b') equivalent,* *iff there is an element $g \in G(K_0)$ such that $b' = gb\sigma(g)^{-1}$ and such that the cocharacters μ' and $g\mu g^{-1}$ define the same filtration on the category $\mathcal{REP}(G)$ (compare Milne [Mi]).*

One checks easily that two pairs are equivalent, iff the corresponding functors $\mathcal{I}_{\mu,b}$ and $\mathcal{I}_{\mu',b'}$ are isomorphic.

Definition 1.24 *We call a pair (μ, b) special, iff there is a subtorus T of G, which is defined over $\mathbf{Q}_p$, such that there is a pair (μ', b') equivalent to (μ, b) with the property that μ' factors through T and $b' \in T(K_0)$ and that (μ', b') is an admissible pair with respect to T.*

A special pair is admissible. The proposition (1.20) holds for a special pair, since it holds for the torus T. We have a weak assertion of existence of special pairs:

Proposition 1.25 *Assume that G is a connected reductive group. Let (μ, b) be a weakly admissible pair in G, such that b is basic. Then there is a cocharacter μ' in the $G(K)$– conjugacy class of μ such that (μ', b) is special.*

Proof: By Kottwitz [Ko1] there is an elliptic maximal torus T over $\mathbf{Q}_p$ in G, such that the σ –conjugacy class of b is in the image of the map

$$B(T) \longrightarrow B(G).$$

Changing (μ, b) in its equivalence class, we may assume that $b \in T(K_0)$. Choose a μ' in the conjugacy class of μ, that factors through T.
Let again denote by D the factor group of G by the derived group. The image of (μ', b) in D is weakly admissible since it coincides with the image of (μ, b). Since the groups of cocharacters that are defined over $\mathbf{Q}_p$ of T and D coincide up to torsion, it follows from the proposition (1.20) that (μ', b) is admissible for T.

1.26 We also mention the following compatibility of our conjecture for the connected reductive group G and an inner form which was pointed out to us by J. de Jong. We assume that G is connected and fix a cohomology class $\eta \in H^1(\mathbf{Q}_p, G)$. By Steinberg's theorem we may represent η by an unramified cocycle $\sigma \mapsto g_\sigma$, $g_\sigma \in G(K_0)$. Let G' be the inner form of G defined by the image of this cocycle in the adjoint group. Hence we have an identification

$$G_{K_0} = G'_{K_0}.$$

Also, the cocycle defines a tensor equivalence

$$\mathcal{REP}(G) \longrightarrow \mathcal{REP}(G'),$$

such that the fiber functors of both tensor categories over K_0 are the same under the identification of G_{K_0} with G'_{K_0}.
An admissible pair (μ, b) for G defines an admissible pair (μ', b') for G' as follows. The functor $\mathcal{I}$ corresponding to (μ, b) defines by composing with the tensor equivalence above a functor $\mathcal{I}'$ from $\mathcal{REP}(G')$ into the category of admissible filtered isocrystals over K. The composition with the obvious tensor functor into the category of K_0–vector spaces is the standard fibre functor over K_0 and hence $\mathcal{I}'$ is given by a unique pair (μ', b'). It is easy to see that under the identification of G_{K_0} with G'_{K_0} we have

$$b' = bg_\sigma^{-1}, \quad \mu' = \mu.$$

Since G' is an inner form of G there is a canonical identification

$$X^*(Z(\hat{G})^\Gamma) = X^*(Z(\hat{G}')^\Gamma).$$

It induces an identification (comp. diagram before (1.20))

$$H^1(\mathbf{Q}_p, G) = H^1(\mathbf{Q}_p, G').$$

Lemma 1.27 *Under the above identification we have*

$$\begin{aligned} \operatorname{cls}(\mu', b') &= \operatorname{cls}(\mu, b) - \eta \\ \kappa(b') - \mu'^{\natural} &= \kappa(b) - \mu^{\natural} - \eta. \end{aligned}$$

Proof: By definition $\operatorname{cls}(\mu', b')$ measures the difference between the standard fibre functor Ver' on $\mathcal{REP}(G')$ and $\mathcal{F}_v \circ \mathcal{I}'$, and similarly for $\operatorname{cls}(\mu, b)$. The first assertion follows since η measures the difference between Ver and Ver'. The definition of κ ([Ko2]) implies

$$\kappa(b') = \kappa(b) - \eta.$$

Since obviously $\mu'^{\natural} = \mu^{\natural}$, the second assertion follows. □

We therefore see that our conjectures for G and for G' are equivalent; by choosing η suitably we may assume in proving it that $\mathrm{cls}(\mu', b') = 0$. Using this remark J. de Jong has checked our conjecture in some cases where the derived group of G is not simply–connected.

1.28 We can make our conjecture explicit in the example of the special orthogonal group. Since this group (which is semi–simple) is not simply connected this case is not covered by proposition (1.20) and in fact we cannot prove our conjecture in this case. Let F be a finite extension of $\mathbf{Q}_p$ and let V be a quadratic space over F, i.e. a finite F–vector space with a non–degenerate symmetric F–bilinear form. Let G be the special orthogonal group of V, considered as an algebraic group over $\mathbf{Q}_p$ by restriction of scalars from F to $\mathbf{Q}_p$. Then $H^1(\mathbf{Q}_p, G) \simeq \mathbf{Z}/2$ classifies the isomorphism classes of quadratic spaces over F of the same dimension and with the same discriminant as V. Let (μ, b) be an admissible pair and let $V' = \mathcal{F}_v\mathcal{I}(V)$ be the associated quadratic space over F. An explicit way of describing the cohomology class $\mathrm{cls}(\mu, b) \in H^1(\mathbf{Q}_p, G)$ is as follows. Let $w(V)$, resp. $w(V')$ be the Witt invariant of V resp. V' (take any of the various definitions to be found in the literature). Then

$$\mathrm{cls}(\mu, b) = w(V') - w(V) \in \mathbf{Z}/2.$$

Therefore, our conjecture in this case may be considered as a formula for the Witt invariant of V'. This formula is as follows. To μ corresponds its class $\mu^\sharp \in \mathbf{Z}/2$ which is trivial or non–trivial according as to whether μ lifts to the spin group or not. Similarly, we associate to b the trivial resp. the nontrivial element of $\mathbf{Z}/2$ according as to whether b may be lifted to the spin group $\tilde{G}(K_0)$ or not. Our conjecture then states that

$$w(V') = w(V) + (\kappa(b) - \mu^\sharp).$$

1.29 It is also interesting to consider certain non–connected groups. As a representative example we consider the orthogonal groups. We retain the previous notations, except that we denote now by G the orthogonal group of the quadratic space V. Now $H^1(\mathbf{Q}_p, G)$ classifies the isomorphism classes of quadratic spaces over F of the same dimension as V. There is a bijection

$$H^1(\mathbf{Q}_p, G) = \mathbf{Z}/2 \oplus F^\times/F^{\times 2}$$

which associates to a quadratic space V' of the same dimension as V the differences of the Witt invariants and of the discriminants of V and V'. This decomposition is a splitting of the short exact sequence induced on cohomology

$$0 \longrightarrow H^1(\mathbf{Q}_p, G^0) \longrightarrow H^1(\mathbf{Q}_p, G) \longrightarrow H^1(F, \mathbf{Z}/2) \longrightarrow 0.$$

Here G^0 denotes the connected component of the identity, the special orthogonal group. Let (μ, b) be an admissible pair and let V' be the corresponding quadratic space over F. We wish to define an analogue of the right hand side of proposition (1.20). Since μ factors through G^0, the invariant $\mu^\sharp \in \mathbf{Z}/2$ is defined. We consider $\mu^\sharp$ as an element of $H^1(\mathbf{Q}_p, G)$ with trivial component in $F^\times/F^{\times 2}$.
We now define the component of $\kappa(b)$ in $F^\times/F^{\times 2}$. On the maximal exterior power $\wedge^{\max} V$ we have an induced symmetric F–bilinear form,

$$(\ ,\ ) : \wedge^{\max} V \otimes_F \wedge^{\max} V \longrightarrow F.$$

Since $(\det b)\sigma$ commutes with the action of F it is easy to see that there is a generator $w \in \wedge^{\max} V \otimes_{\mathbf{Q}_p} K_0$ with

$$(\det b)\sigma(w) = w.$$

Then $(w, w) \in F \otimes_{\mathbf{Q}_p} K_0$ is invariant under $\mathrm{id} \otimes \sigma$, i.e. defines an element of $F^\times$. Its image in $F^\times/F^{\times 2}$ is independent of the choice of w. The difference between this element and the discriminant of V is the component of $\kappa(b)$ in $F^\times/F^{\times 2}$. To define the component of $\kappa(b)$ in $\mathbf{Z}/2$ we distinguish two cases.

First case: $\dim V$ even. In this case there is a surjective homomorphism

$$\tilde{G}(K_0) \longrightarrow G(K_0)$$

where $\tilde{G}$ denotes the Clifford group of V (Bourbaki, Algèbre IX, §9). If $\tilde{b} \in \tilde{G}(K_0)$ is any lifting of b, the component of $\kappa(b)$ in $\mathbf{Z}/2$ is defined as

$$\mathrm{ord}_p \mathrm{Nm}_{F/\mathbf{Q}_p}(\mathrm{SNm}(\tilde{b})) \in \mathbf{Z}/2.$$

Here $\mathrm{SNm}(\tilde{b}) \in F \otimes K_0$ denotes the spinor norm of $\tilde{b}$, of which we take its norm $\mathrm{Nm}_{F/\mathbf{Q}_p}$ down to K_0. (In loc.cit. the spinor norm is only defined on a subgroup of $\tilde{G}(K_0)$; however, when $\dim V$ is even, the definition given there extends to all of $\tilde{G}(K_0)$.)

Second case: dim V odd. In this case G is the direct product of G^0 and its center Z,

$$G = G^0 \times Z.$$

This decomposition induces in this case the above decomposition of $H^1(\mathbf{Q}_p, G)$. If $b = b^0 \cdot z$ with $b^0 \in G^0(K_0)$ and $z \in Z(K_0)$, then $\kappa(b^0) \in \mathbf{Z}/2$ is defined and may be considered as an element of $H^1(\mathbf{Q}_p, G)$, and z defines an unramified cocycle with class $\kappa(z)$ in $H^1(\mathbf{Q}_p, Z) = F^\times/F^{\times 2}$ which we also consider as an element of $H^1(\mathbf{Q}_p, G)$. We put $\kappa(b) = \kappa(b^0) + \kappa(z)$. It is easy to see that this element has the component in $F^\times/F^{\times 2}$ described earlier in general.

We now conjecture that again, with these definitions,

$$\mathrm{cls}(\mu, b) = \kappa(b) - \mu^\sharp \in H^1(\mathbf{Q}_p, G).$$

The following fact is well–known and easy to prove.

Proposition 1.30 *(Fontaine, Messing, Ogus): The images of* $\mathrm{cls}(\mu, b)$ *and of* $\kappa(b) - \mu^\sharp$ *in* $F^\times/F^{\times 2}$ *coicide.*

In terms of the generator w of $(\wedge^{\max} V) \otimes K_0$ above this proposition states that (w, w) is the discriminant of V'.
Another case of interest is the case of the group of orthogonal similitudes (with similitude factor in $F^\times$ or in $\mathbf{Q}_p^\times$). In this case we do not even have a conjectural description of $\mathrm{cls}(\mu, b)$ (except under special hypotheses), much less a proof.

1.31 Let G be an algebraic group over $\mathbf{Q}_p$. We fix a conjugacy class of cocharacters

$$\mu : \mathbf{G}_m \longrightarrow G.$$

To fix ideas we consider the cocharacters defined over subfields of the completion $\mathbf{C}_p$ of a fixed algebraic closure $\bar{\mathbf{Q}}_p$ of $\mathbf{Q}_p$. Let E be the field of definition of the conjugacy class. Then E is a finite extension of $\mathbf{Q}_p$ contained in $\bar{\mathbf{Q}}_p$. It is a local analogue of the Shimura field.

Two cocharacters will be considered *equivalent* if they define the same filtration on the category $\mathcal{REP}(G)$. The equivalence classes of cocharacters form (the $\mathbf{C}_p$–valued points) of a projective algebraic variety $\mathcal{F}$ defined over E, a partial flag variety. Let V be a $\mathbf{Q}_p$–rational faithful representation of G. To the cocharacter μ defined over K we associate the filtration $\mathcal{F}^i_\mu(V) = V^i_K$ given by the weight spaces $V_{K,j}$ with respect to μ, cf. (1.2). This filtration only depends on the equivalence class of μ and this defines a closed immersion of $\mathcal{F}$ into a flag variety of V,

$$\mathcal{F} \hookrightarrow \mathcal{F}lag(V) \otimes_{\mathbf{Q}_p} E.$$

Here the points with values in a $\mathbf{Q}_p$–algebra R of $\mathcal{F}lag(V)$ are the filtrations of $\mathcal{F}^i$ of $V \otimes R$ by R–submodules, which are direct summands, and such that rk $\mathcal{F}^i$ = dim V^i_K. The variety $\mathcal{F}$ is a homogeneous space under the algebraic group G_E. Our next aim will be to construct rigid–analytic subsets of $\mathcal{F}_{\breve{E}} = \mathcal{F} \otimes_E \breve{E}$. Here we denote by $\breve{E} = EK_0 = EK_0(\bar{\mathbf{F}}_p)$ the completion of the maximal unramified extension of E.

1.32 We will first give a general method of construction of a rigid–analytic structure on certain subsets of projective algebraic varieties. Let F be a locally compact discretely valued field and let $\mathbf{C}_p$ be the completion of an algebraic closure of F. For $z \in \mathbf{P}^n(\mathbf{C}_p)$ we will denote by $\tilde{z}$ a unimodular representative, i.e.

$$|\tilde{z}_i| \leq 1, \text{ all } i; \ \ |\tilde{z}_i| = 1, \text{ at least one } i; \ \ i = 0, \dots, n.$$

Such a representative is unique up to a unit in $\mathbf{C}_p$. Let $X \subset \mathbf{P}^n$, $T \subset \mathbf{P}^m$ be projective schemes defined over F. Let K be a subfield of $\mathbf{C}_p$ which is complete in its induced topology and which contains F. Let

$$\mathcal{H} \subset (X \times T) \otimes_F K$$

be a closed subscheme. Let $\mathcal{H}_t \subset X$ be the fibre through $t \in T(F)$. We wish to put the structure of an admissible open subset (in the sense of the Grothendieck topology) of the rigid–analytic space over K underlying $X(\mathbf{C}_p)$ on the set

$$X(\mathbf{C}_p) \setminus \bigcup_{t \in T(F)} \mathcal{H}_t.$$

We imitate the procedure of [SS], §1. Represent $\mathcal{H}$ as the common zero set of a finite set of bi–homogeneous polynomials *with integral coefficients*,

$$f_\alpha(X_0,\ldots,X_n;T_0,\ldots,T_m),\ \alpha\in A.$$

The real–valued function on $X(\mathbf{C}_p)\times T(\mathbf{C}_p)$

$$(x,t)\mapsto |f_\alpha(\tilde{x},\tilde{t})|$$

is independent of the choice of unimodular representatives of x and t respectively and will be denoted by $|f_\alpha(x,t)|$. For $\epsilon>0$ and $t\in T(F)$ consider the tubular neighbourhood of $\mathcal{H}_t$,

$$\mathcal{H}_t(\epsilon)=\{x\in X(\mathbf{C});|f_\alpha(x,t)|\le\epsilon,\quad \alpha\in A\}.$$

Let $\tilde{t}$ and $\tilde{t'}$ be unimodular representatives of points t and t' in $T(F)$. We use the triangular inequality

$$|f_\alpha(\tilde{x},\tilde{t'})|\le\max\{|f_\alpha(\tilde{x},\tilde{t'})-f_\alpha(\tilde{x},\tilde{t})|,|f_\alpha(\tilde{x},\tilde{t})|\}.$$

Observing that the coefficients of f_α and the coordinates of $\tilde{x}$ are integral it follows that for $\epsilon>0$ there exists $\delta>0$ such that for $||\tilde{t'}-\tilde{t}||\le\delta$ (maximum norm) we have

$$|f_\alpha(\tilde{x},\tilde{t'})-f_\alpha(\tilde{x},\tilde{t})|\le\epsilon,\text{i.e.}$$

$$\mathcal{H}_{t'}(\epsilon)\subset\mathcal{H}_t(\epsilon).$$

Using the fact that the set of unimodular representatives in $F^{m+1}\setminus\{0\}$ of $T(F)$ is compact we therefore deduce the following lemma.

Lemma 1.33 *For every $\epsilon>0$ there exists a finite subset $S\subset T(F)$ with*

$$\bigcup_{t\in T(F)}\mathcal{H}_t(\epsilon)=\bigcup_{t\in S}\mathcal{H}_t(\epsilon).$$

□

Proposition 1.34 *For every $\epsilon > 0$,*

$$X(\mathbf{C}_p) \setminus \mathcal{H}_t(\epsilon), \quad t \in T(F)$$

is an admissible open subset of X_K hence so is also

$$X_\epsilon = X(\mathbf{C}_p) \setminus \bigcup_{t \in T(F)} \mathcal{H}_t(\epsilon),$$

a finite intersection of subsets of the previous kind. On $X(\mathbf{C}_p) \setminus \bigcup_{t\in T(F)} \mathcal{H}_t$ there is a structure of admissible open subset of X_K (considered as a rigid space over K) characterized by the fact that for any sequence $\epsilon_1 > \epsilon_2 \ldots \to 0$ the covering

$$X_{\epsilon_1} \subset X_{\epsilon_2} \subset \ldots$$

is an admissible covering.

Proof: For the first assertion it suffices to establish the following fact. Let $f_\alpha(X_0, \ldots, X_n), \alpha \in A$, be a finite set of homogeneous polynomials. Then the subset of $\mathbf{P}^n$,

$$\{x \in \mathbf{P}^n ; |f_\alpha(x)| > \epsilon, \text{ some } \alpha\}$$

is an admissible open. This follows easily from the fact that for $i = 0, \ldots, n$ the intersection of this set with the set of points $x \in \mathbf{P}^n(\mathbf{C}_p)$ where the unimodular representative $\tilde{x}$ satisfies $|\tilde{x}_i| = 1$ is obviously isomorphic to the following admissible open subset of the closed polydisc in $\mathbf{A}^n$:

$$\left\{ (\frac{\tilde{x}_0}{\tilde{x}_i}, \ldots, \frac{\tilde{x}_n}{\tilde{x}_i}); |\frac{\tilde{x}_j}{\tilde{x}_i}| \leq 1, j = 0, \ldots, n; |f_\alpha(\frac{\tilde{x}_0}{\tilde{x}_i}, \ldots, \frac{\tilde{x}_n}{\tilde{x}_i})| > \epsilon, \text{ some } \alpha \in A \right\}.$$

The argument also shows that for $\epsilon_1 > \epsilon_2 \ldots \to 0$

$$X(\mathbf{C}_p) \setminus \mathcal{H}_t(\epsilon_i), i = 1, 2, \ldots$$

is an admissible open covering of $X \setminus \mathcal{H}_t$. To show the second assertion, let $f : Y \to X_K$ be a morphism of an affinoid variety Y into X_K such that $f(Y) \subset X(\mathbf{C}_p) \setminus \bigcup_{t \in T(F)} \mathcal{H}_t$. We have to see that f factors through X_{ϵ_i} for suitable i. But by the above remark, for every $t \in T(F)$ there exists a minimal $i(t)$ such that

$$f(Y) \subset X(\mathbf{C}_p) \setminus \mathcal{H}_t(\epsilon_{i(t)}).$$

By the remarks preceding lemma (1.33) the function $t \mapsto i(t)$ is continuous. Since $T(F)$ is compact it follows that it assumes its maximum, which proves the assertion. □

1.35 Let G be an algebraic group over $\mathbf{Q}_p$. We fix a conjugacy class of cocharacters $\mu : \mathbf{G}_m \to G$ and the corresponding flag variety $\mathcal{F}$ over E, cf. (1.31). Let $K_0 = W(\bar{\mathbf{F}}_p)_{\mathbf{Q}}$ and fix an element $b \in G(K_0)$. Let K be a finite extension of $\breve{E}$ and let $\mathcal{F}^\bullet_\mu \in \mathcal{F}(K)$ be a point corresponding to the cocharacter μ. Then $\mathcal{F}^\bullet_\mu$ will be called *weakly admissible* (with respect to b) if the pair (μ, b) is weakly admissible, cf. (1.18) (this condition is independent of the representative μ of $\mathcal{F}_\mu$). We denote by $\mathcal{F}^{wa}(K) = \mathcal{F}^{wa}_b(K)$ the set of weakly admissible filtrations. Let $J = J_b$ be the algebraic group associated to b, cf. (1.12). Then $J(\mathbf{Q}_p) \subset G(K_0)$ and hence operates on $\mathcal{F}(K)$. This operation preserves $\mathcal{F}^{wa}_b(K)$. In what follows we again sometimes take the naive point of view identifying $\mathcal{F}$ with the set of its $\mathbf{C}_p$-valued points. We denote by $E.\mathbf{Q}_{p^s}$ the join.

Proposition 1.36 *(i) Let $b \in G(K_0)$. The set of weakly admissible filtrations with respect to b in $\mathcal{F}$ has a natural structure of an admissible open rigid-analytic subset of $\mathcal{F} \otimes_E \breve{E}$. If $b' = gb\sigma(g)^{-1}$, $g \in G(K_0)$, then $\mathcal{F}^\bullet_\mu \mapsto g^{-1}(\mathcal{F}^\bullet_\mu) = \mathcal{F}^\bullet_{g^{-1}\mu g}$ induces an isomorphism between the weakly admissible subset corresponding to b and the weakly admissible subset corresponding to b'.*
(ii) Let $\bar{b}$ be a decent σ-conjugacy class in $G(K_0)$ and let $b \in \bar{b}$ be a decent element, satisfying a decency equation with the integer $s > 0$ (cf.(1.8)). Then the subset of weakly admissible filtrations in $\mathcal{F}(\mathbf{C}_p)$ with respect to b has a natural structure of an admissible open rigid-analytic subset defined over $E.\mathbf{Q}_{p^s}$. If $b' \in \bar{b}$ satisfies a decency equation with the same integer s, then any $g \in G(\mathbf{Q}_{p^s})$ with $b' = gb\sigma(g)^{-1}$ (cf.(1.9)) induces via $\mathcal{F}^\bullet \mapsto g^{-1}(\mathcal{F}^\bullet)$ an isomorphism defined over $E.\mathbf{Q}_{p^s}$ between the corresponding admissible open subsets of $\mathcal{F} \otimes_E E_s$.

Proof: We first prove (ii). Let V be a faithful $\mathbf{Q}_p$-rational representation of G. The set of conditions on $\mathcal{F}^\bullet \in \mathcal{F}(K)$ to be weakly admissible is parametrized by the set of Φ-stable subspaces $V_0' \subset V_0 = V \otimes K_0$. Let

$V_s = V \otimes \mathbf{Q}_{p^s}$ and equip V_s with the σ-linear operator $\Phi_s = b(\mathrm{id}_V \otimes \sigma)$. Then $(V_0, \Phi) = (V \otimes K_0, b\sigma) = (V_s, \Phi_s) \otimes_{\mathbf{Q}_{p^s}} K_0$ and there is a bijection between the Φ-invariant subspaces of V_0 and the Φ_s-invariant subspaces of V_s. Let T' be the projective algebraic variety over $\mathbf{Q}_{p^s}$ parametrizing the subspaces of V_s which are compatible with the isotypical decomposition of V_s. More precisely, let $V_0 = \bigoplus V_\lambda$ be the isotypical decomposition and $V_s = \bigoplus (V_s \cap V_\lambda)$ be the induced decomposition of V_s. For a $\mathbf{Q}_{p^s}$-algebra R,

$$T'(R) = \{V' \subset V_s \otimes_{\mathbf{Q}_{p^s}} R;\ V' \quad \text{a direct summand with} \quad V' = \bigoplus (V' \cap ((V_s \cap V_\lambda) \otimes_{\mathbf{Q}_{p^s}} R))\}.$$

Then T' is a disjoint sum of closed subschemes of Grassmannians of V_s. We are going to put a $\mathbf{Q}_p$-rational structure on T' such that for the descended variety T over $\mathbf{Q}_p$ we have

$$T(\mathbf{Q}_p) = \{\Phi_s\text{-stable subspaces of } V_s\}.$$

The $\mathbf{Q}_p$-structure is defined via a descent datum,

$$\alpha : T' \longrightarrow T'^{\sigma}.$$

We may interpret T'^{σ} as the functor which to a $\mathbf{Q}_{p^s}$-algebra R associates the set of direct summands of $V_s \otimes_{\mathbf{Q}_{p^s}, \sigma} R$ compatible with the isotypical decomposition. We have an isomorphism

$$\Phi_s : V_s \otimes_{\mathbf{Q}_{p^s}, \sigma} R \longrightarrow V_s \otimes_{\mathbf{Q}_{p^s}} R.$$

We define α to be the map which associates to the direct summand V' of $V_s \otimes_{\mathbf{Q}_{p^s}} R$ the direct summand $\Phi_s^{-1}(V')$ of $V_s \otimes_{\mathbf{Q}_{p^s}, \sigma} R$. To see that α is a descent datum we have to check that $\alpha^{\sigma^{s-1}} \circ \ldots \circ \alpha : T' \to T'$ is the identity. Let $V = \oplus V_\lambda$ be the isotypical decomposition. Then $V_s = \oplus (V_s \cap V_\lambda)$ is a decomposition into Φ-stable subspaces. From the definition of the slope morphism ν it follows that

$$(\Phi_s)^s | (V_s \cap V_\lambda) = p^r \, \mathrm{id}_{V_s \cap V_\lambda}, \quad \lambda = r/s.$$

Any direct summand V' of $V_s \otimes R$ in $T'(R)$ decomposes as $V' = \oplus V'_\lambda$ where V'_λ is a direct summand of $(V_s \cap V_\lambda) \otimes_{\mathbf{Q}_{p^s}} R$ and

$$(\Phi_s)^s(V'_\lambda) = p^r\ V'_\lambda = V'_\lambda,$$

as required.
Consider the variety $\mathcal{F}_1 = \mathcal{F}lag(V)$ of flags $\mathcal{F}^\bullet$ of V with dim $\mathcal{F}^i =$ dim $\mathcal{F}^i_\mu$. This is a projective variety over $\mathbf{Q}_p$ and there is a closed embedding defined over E, $\mathcal{F} \subset \mathcal{F}_1 \otimes_{\mathbf{Q}_p} E$. Consider the closed subscheme defined over $\mathbf{Q}_{p^s}$,

$$\mathcal{H} \subset (\mathcal{F}_1 \times T) \otimes_{\mathbf{Q}_p} \mathbf{Q}_{p^s}$$

given by pairs $(\mathcal{F}^\bullet, V')$ where $\mathcal{F}^\bullet$ is a filtration of $V_s \otimes_{\mathbf{Q}_{p^s}} R$ in $\mathcal{F}_1$ and $V' = \oplus V'_\lambda$ is a direct summand of $V_s \otimes_{\mathbf{Q}_{p^s}} R = \oplus(V_s \cap V_\lambda) \otimes_{\mathbf{Q}_{p^s}} R$ such that

$$\sum_i i\ \mathrm{rk}(\mathrm{gr}^i_{\mathcal{F}\cap V'}(V')) > \sum_\lambda \lambda\ \mathrm{rk}(V'_\lambda).$$

Noting that the left hand side is equal to

$$\sum_{i=-N}^{\infty} \mathrm{rk}(\mathcal{F}^i \cap V') - N \cdot \mathrm{rk}(V')$$

for some $N \gg 0$, we see that $\mathcal{H}$ is indeed a closed subscheme defined by conditions of Schubert type. Furthermore, the above considerations show that the set of weakly admissible filtrations in $\mathcal{F}_1(\mathbf{C}_p)$ is given by

$$\mathcal{F}_1(\mathbf{C}_p)^{wa} = \mathcal{F}_1(\mathbf{C}_p) \setminus \bigcup_{t\in T(\mathbf{Q}_p)} \mathcal{H}_t.$$

Applying proposition (1.34) we see that this is an admissible open subset of $\mathcal{F}_1 \otimes_{\mathbf{Q}_p} \mathbf{Q}_{p^s}$ defined over $\mathbf{Q}_{p^s}$. The result follows from intersecting $\mathcal{F}_1(\mathbf{C}_p)^{wa}$ with $\mathcal{F} \subset \mathcal{F}_1 \otimes_{\mathbf{Q}_p} E$. The last assertion of (ii) is obvious.
We now turn to the proof of (i). The case when b is decent follows at once from (ii). Let G be connected. Then by Kottwitz [Ko1] there exists $g \in G(K_0)$ such that $b' = gb\sigma(g)^{-1}$ is decent. Identifiying $\mathcal{F}_b(\mathbf{C}_p)^{wa}$ with $g\ \mathcal{F}_{b'}(\mathbf{C}_p)^{wa}$, the result for b' implies the assertion for b. Let $G_1 = GL(V)$. Then G is a closed subgroup of G_1 and b induces a σ-conjugacy class $\bar{b}_1$ which is decent since G_1 is connected. Let $b_1 = g_1 b\sigma(g_1)^{-1}$ be a decent element in $\bar{b}_1$ and let $\mathcal{F}_1^{wa} \subset \mathcal{F}_1 \otimes_E \breve{E}$ be the corresponding admissible open subset of weakly admissible filtrations. Then $\mathcal{F}^{wa} = (\mathcal{F} \otimes_E \breve{E}) \cap g_1 \mathcal{F}_1^{wa}$ is an admissible open subset of $\mathcal{F} \otimes_E \breve{E}$, as required. The last assertion of (i) is trivial. □

1.37 In the rest of this chapter we will consider $\mathcal{F}^{wa} = \mathcal{F}_b^{wa}$ as an admissible open subset of $\mathcal{F}_{\breve{E}}$ with the rigid analytic structure given by proposition (1.36). We call this subset the *p-adic symmetric space* or *p-adic period domain* associated to the triple formed by G, the conjugacy class $\{\mu\}$ of μ and the element $b \in G(K_0)$. (We refer to (5.45) for the problem of lowering the field of definition to E). We are now in a position to state our basic conjecture on the existence of *local systems* on $\mathcal{F}^{wa}$. Here we adopt the following definition of a *local system in* $\mathbf{Q}_p$*-vector spaces* on a rigid-analytic space X (it was suggested to us by J. de Jong). Recall (cf. e.g. [SS]) that the big étale site in the category of rigid spaces has as *coverings* morphisms $Y \to X$ which are étale and such that there exists an admissible covering of Y by affinoids such that its image is an admissible covering of X. Locally constant sheaves in $\mathbf{Z}/p^n$-modules are defined exactly like in algebraic geometry, and so are smooth $\mathbf{Z}_p$-sheaves. However, the category of local systems of $\mathbf{Q}_p$-vector spaces is defined starting with smooth $\mathbf{Z}_p$-sheaves, tensoring the Hom groups with $\mathbf{Q}_p$ and *then imposing descent with respect to étale coverings.* (We mention that this last condition is automatic in the algebraic case, if the base is normal.) In particular, a local system in $\mathbf{Q}_p$-vector spaces on X is not necessarily defined by a smooth $\mathbf{Z}_p$-sheaf on X, but merely by a smooth $\mathbf{Z}_p$-sheaf over an étale covering Y of X. The local systems in $\mathbf{Q}_p$-vector spaces on X form a $\otimes$-category and every point $x \in X$ defines a fibre functor in the category of finite $\mathbf{Q}_p$-vector spaces. Before stating our conjecture we mention that an étale surjective morphism $Y \to X$ factors in a unique way as $Y \to X' \to X$ such that $Y \to X'$ is an étale covering and $X' \to X$ is étale and a bijection on points (this fact was communicated to us by J. de Jong).

We conjecture the existence of an étale bijective morphism $(\mathcal{F}_b^{wa})' \to \mathcal{F}_b^{wa}$ and of a $\otimes$-functor from the category $\mathcal{REP}(G)$ to the category of local systems in $\mathbf{Q}_p$-vector spaces on $(\mathcal{F}_b^{wa})'$ with the following property: Let (μ, b) be a weakly admissible pair of G and $\mathcal{F}_\mu^\bullet \in \mathcal{F}_b^{wa}(K)$ the corresponding point. Then the pair (μ, b) is admissible and the fibre functor on $\mathcal{REP}(G)$ which associates to a representation the fibre in $\mathcal{F}_\mu^\bullet$ of the corresponding local system is isomorphic to the fibre functor considered in (1.19) with respect to (μ, b),

$$\mathcal{F}_v \circ \mathcal{I} : \mathcal{REP}(G) \longrightarrow (\mathbf{Q}_p\text{-vector spaces}).$$

We note that the conjecture that every weakly admissible pair (μ, b) is admissible is due to Fontaine. The validity of this conjecture would imply the existence of interesting étale coverings of $(\mathcal{F}^{wa})'$. We shall exhibit in cases related to p–divisible groups plausible candidates for these coverings, cf. (5.34). For further remarks on the tower of étale coverings comp. (5.53)

1.38 We now introduce the algebraic groups G over $\mathbf{Q}_p$ which will occupy us in this paper. We will distinguish two cases. The first case will be related to classifiying p–divisible groups with given endomorphisms and level structures, the second one will be related to classifying p–divisible groups with given endomorphisms and polarization and level structures. Accordingly we will name these cases (EL) resp. (PEL). We will fix data of the following type.

Case (EL):

Let F be a finite direct product of finite field extensions of $\mathbf{Q}_p$.
Let B be a finite central algebra over F.
Let V be a finite dimensional B–module.
Let $G = GL_B(V)$ considered as an algebraic group over $\mathbf{Q}_p$.

Case(PEL):

Let F, B, V be as in case (EL).
Let $(\ ,\ )$ be a nondegenerate alternating $\mathbf{Q}_p$–bilinear form on V.
Let $b \mapsto b^*$ be an involution on B which satisfies:

$$(bv, w) = (v, b^* w), \qquad v, w \in V.$$

Let G be the algebraic group over $\mathbf{Q}_p$, whose points with values in a $\mathbf{Q}_p$–algebra R are given by:

$$G(R) = \{g \in GL_B(V \otimes R);\ (gv, gw) = c(g)(v, w), \quad c(g) \in R\}.$$

We denote by F_0 the elements of F, which are fixed by the involution $*$.

Let $b \in G(K_0)$ where again $K_0 = K_0(\bar{\mathbf{F}}_p)$. Then we obtain an isocrystal associated to the natural representation of G on V,

$$N(V) = V \otimes K_0, \quad \Phi = b\,(\mathrm{id}_V \otimes \sigma).$$

It is equipped with an action of B and in the case (PEL) with an alternating bilinear form of isocrystals,

$$\psi : N(V) \otimes N(V) \longrightarrow \mathbf{1}(n).$$

Here $n = \text{ord } c(b)$. Indeed, since $\bar{\mathbf{F}}_p$ is algebraically closed we may write $c(b) = p^n \cdot u\sigma(u)^{-1}$ with a unit u. Then the pairing is defined as

$$\psi(v, v') = u^{-1} \cdot (v, v'), \quad v, v' \in N(V).$$

Any other choice of u multiplies ψ by an element in $\mathbf{Q}_p^\times$. Furthermore, the isocrystal with B-action (resp. and its $\mathbf{Q}_p$-*homogeneous polarization* ψ in case (PEL)) depends up to isomorphism only on the σ-conjugacy class of b.

We now fix in addition to the data above a conjugacy class of cocharacters $\mu : \mathbf{G}_m \to G$ and the corresponding homogeneous algebraic variety $\mathcal{F}$ defined over E. Let us make these objects more explicit. We fix a cocharacter μ_o in our conjugacy class defined e.g. over $\bar{\mathbf{Q}}_p$ and let $\mathcal{F}_o^\bullet$ be the corresponding B-invariant filtration of $V \otimes \bar{\mathbf{Q}}_p$. In the case (PEL) we have

$$\mathcal{F}_o^i = (\mathcal{F}_o^{m-i+1})^\perp.$$

Here m denotes the composite $c \circ \mu \in Hom(\mathbf{G}_m, \mathbf{G}_m) = \mathbf{Z}$.

Lemma 1.39 *(i) The field E may be described as the field of definition of the isomorphism class of $\mathcal{F}_o^\bullet$ as D-invariant filtration, or equivalently as the finite extension of $\mathbf{Q}_p$ generated by the traces*

$$\mathrm{Tr}(d;\ gr^i_{\mathcal{F}_o}(V \otimes \bar{\mathbf{Q}}_p)), \quad d \in B, \quad i \in \mathbf{Z}.$$

(ii) The variety $\mathcal{F}$ may be described as follows. Let R be a E-algebra. Then $\mathcal{F}(R)$ is the set of filtrations $\mathcal{F}^\bullet$ of $V \otimes_{\mathbf{Q}_p} R$ by R-submodules, which are direct summands, and such that

$$\mathrm{Tr}(d;\ gr^i_{\mathcal{F}}(V \otimes R)) = \mathrm{Tr}(d;\ gr^i_{\mathcal{F}_o}(V \otimes \bar{\mathbf{Q}}_p)), \quad d \in B, \quad i \in \mathbf{Z}$$

and in case (PEL) such that in addition

$$\mathcal{F}^i = (\mathcal{F}^{m-i+1})^{\perp}.$$

Proof: The equivalence of the two descriptions of E in (i) holds because two representations of D are isomorphic iff they have identical traces. Let $\mathcal{F}_1$ be the scheme whose R–valued points are described in (ii). Then it is easy to see that the remaining assertions follow if we can prove that $\mathcal{F}_1$ is a homogeneous variety under G_E. We restrict ourselves to the case (PEL), the case (EL) being similar but simpler. After extending scalars from $\mathbf{Q}_p$ to $\bar{\mathbf{Q}}_p$, the data $(F, B, *, V, (\ ,\ ))$ decompose as a direct product of data of one of the following kind.

(A) $B = End(W) \times End(\check{W})$, where W is a finite–dimensional vector space over $\bar{\mathbf{Q}}_p$ and $\check{W}$ its dual, and $(u, v)^* = ({}^t v, {}^t u)$.

$V = W \otimes V' + \check{W} \otimes \check{V}'$, where V' is a finite–dimensional vector space over $\bar{\mathbf{Q}}_p$.

$$\begin{aligned}(w_1 \otimes v_1' + \check{w}_1 \otimes \check{v}_1', w_2 \otimes v_2' + \check{w}_2 \otimes \check{v}_2') \quad = \quad & < w_1, \check{w}_2 >< v_1', \check{v}_2' > - \\ & < \check{w}_1, w_2 >< \check{v}_1', v_2' >\end{aligned}$$

$$G = \{(1 \otimes g', c(1 \otimes {}^t g'^{-1})),\ \ g' \in GL(V'),\ \ c \in \mathbf{G}_m\}$$

(C) $B = End(W)$, where W is a finite–dimensional vector space over $\bar{\mathbf{Q}}_p$ equipped with a symmetric bilinear form $(\ ,\ )_W$, and $*$ is the transposition with respect to $(\ ,\ )_W$.

$V = W \otimes V'$, where the vector space V' is equipped with an alternating bilinear form $(\ ,\ )_{V'}$ and $(\ ,\ ) = (\ ,\ )_W \otimes (\ ,\ )_{V'}$.

$$G = \{cg';\ \ g' \in \mathrm{Sp}(V', (\ ,\ )_{V'}),\ \ c \in \mathbf{G}_m\}.$$

(BD) As in (C), except that $(\ ,\ )_W$ is anti–symmetric and $(\ ,\ )_{V'}$ is symmetric. Furthermore, in this case

$$G = \{cg';\ \ g' \in \mathrm{SO}(V', (\ ,\ )_{V'}),\ \ c \in \mathbf{G}_m\}.$$

Corresponding to this decomposition, the variety $\mathcal{F}_1 \otimes_E \bar{\mathbf{Q}}_p$ also decomposes into a product of varieties of partial flags of V of either of the following type (as does the adjoint group of G).

(A) $\mathcal{F}^i = W \otimes \mathcal{F}'^i \oplus \check{W} \otimes (\mathcal{F}'^{m+1-i})^\perp$.

The correspondence $\mathcal{F}^\bullet \mapsto \mathcal{F}'^\bullet$ identifies $\mathcal{F}$ with the variety of partial flags of V' with fixed dimensions $\dim(\mathcal{F}'^i)$.

(C, BD) $\mathcal{F}^i = W \otimes \mathcal{F}'^i$, and the correspondence $\mathcal{F}^\bullet \mapsto \mathcal{F}'^\bullet$ identifies $\mathcal{F}$ with the variety of partial flags of V' with fixed dimensions $\dim(\mathcal{F}'^i)$ and with $\mathcal{F}'^i = (\mathcal{F}'^{m+1-i})^\perp$ (with respect to $(\ ,\ )_{V'}$).

In the case (A) it is obvious that the flags $\mathcal{F}'^\bullet$ form a homogeneous space under G. In the cases (C) and (BD) note that $\mathcal{F}'^i$ is totally isotropic for $i \geq 1/2(m+1)$ and that these determine all the other members of the flag uniquely. Therefore Witt's theorem shows that again these flags form a homogeneous space under the adjoint group of G.

1.40 Before discussing examples of weakly admissible subsets we point out that the whole theory explained so far has a variant where we replace $\mathbf{Q}_p$ by a finite extension F of $\mathbf{Q}_p$, cf. [Dr2] or (3.56) below. More precisely, let L be a perfect field extension of the residue field κ_F of F. Let

$$K_F(L) = F \otimes_{F^t} K_0(L),$$

where $F^t = K_0(\kappa_F)$ is the maximal subfield of F, which is unramified over $\mathbf{Q}_p$. Let τ be the automorphism over F induced by the Frobenius automorphism of L relative to κ_F. A $\tau - K_F(L)$-space is a finite-dimensional $K_F(L)$-vector space N equipped with a bijective τ-linear endomorphism Φ.

The theory of these objects (N, Φ) (called $\tau - K_F(L)$-spaces) is completely analogous to the theory of isocrystals. We leave it to the reader to formulate the corresponding results in this context. In particular, if L is algebraically closed, then (N, Φ) is isotypical of slope $\lambda = r/s$ if and only if there is a $O_F \otimes_{O_{F^t}} W(L)$-lattice M in N such that

$$\Phi^s(M) = \pi^r M,$$

where $\pi \in F$ is a uniformizer. Then $\operatorname{End}(N, \Phi)$ is a central division algebra over F with invariant $-\lambda$.

We have also an obvious notion of a filtered $\tau - K_F(L)$-space relative to a finite extension $K/K_F(L)$.

There is a natural equivalence between the category of isocystals over L and the category of $\tau - K_F(L)$-spaces. We describe this in a more general setting.
We consider an algebraic group G' over F and the group $G = \mathrm{Res}_{F/\mathbf{Q}_p}G'$ obtained by Weil restriction. Then there is a natural bijection between the σ-conjugacy classes in $G(K_0(L))$ and the τ-conjugacy classes in $G'(K_F(L))$. Indeed one has an equality

$$G(K_0(L)) = \prod_{\varrho} G'(F \otimes_{F^t,\varrho} K_0(L)),$$

where ϱ runs through the elements of $Gal(F^t/F)$. The bijection associates to the τ-conjugacy class of an element $b' \in G'(K_F(L))$ the σ-conjugacy class of the element $b \in G(K_0(L))$, which has component b' for $\varrho = 1$ and component 1 for all other ϱ in the decomposition above. Let V' be a rational representation of G' on an F-vector space. The natural equivalence of categories above associates to the $\tau - K_F(L)$-space $(V' \otimes_F K_F(L), b'\tau)$ the isocrystal $(V' \otimes_{\mathbf{Q}_p} K_0(L), b\sigma)$.
We may extend this equivalence to a functor from the category of filtered $\tau - K_F(L)$-spaces relative to an extension $K/K_F(L)$ to the category of filtered isocrystals relative to $K/K_0(L)$. Indeed, let us assume for simplicity that K contains the Galois closure of F. We extend a filtration on $V' \otimes_F K$ to a filtration on

$$V' \otimes_{\mathbf{Q}_p} K = \oplus_{\iota: F \to K} V' \otimes_{F,\iota} K,$$

by taking the given filtration for $\iota = 1$ and the trivial filtration on the summands corresponding to $\iota \neq 1$. Here the trivial filtration is the filtration whose graded module is $V' \otimes_{F,\iota} K$ in degree zero.
In group theoretic terms let μ' be a cocharacter of G'_K corresponding to the filtration on $V' \otimes_F K$. Consider the decomposition:

$$G_K = \prod_{\iota: F \longrightarrow K} G'_K.$$

Then the cocharacter μ of G_K which defines the filtration on $V' \otimes_{\mathbf{Q}_p} K$ above has component μ' at the entry corresponding to the $\iota = 1$ and trivial component at all other entries. If we fix an algebraic closure $\bar{F}$ of F and consider $K_F(\bar{\kappa}_F)$ and K as subfields of the completion $\mathbf{C}_p$ of $\bar{F}$, the algebraic variety $\mathcal{F}(G', \mu')$ is defined over the local Shimura field E', where E' is a finite extension of F contained in $\bar{F}$. It is obvious that the local Shimura field E of $\mathcal{F}(G,\mu)$ coincides with E' and that we have a natural identification of algebraic varieties over $Spec\, E$,

$$\mathcal{F}(G,\mu) = \mathcal{F}(G',\mu').$$

Since E contains F we have

$$\breve{E} = EK_0(\bar{\kappa}_F) = E'K_F(\bar{\kappa}_F) = \breve{E}'.$$

The above identification induces an isomorphism on the weakly admissible subsets,

$$(\mathcal{F}(G,\mu)\otimes_E \breve{E})^{wa} = (\mathcal{F}(G',\mu')\otimes_{E'} \breve{E}')^{wa}.$$

We also have

$$J_b = Res_{F/\mathbf{Q}_p} J'_{b'}.$$

We may twist the correspondence defined above by any central cocharacter $w' : \mathbf{G}_m \to G'$ defined over F in the following sense. Let $\pi \in F$ be a fixed prime element. We redefine $b \in G(K_0)$ as the element whose entry b_1 corresponding to $\varrho = 1$ in the decomposition for $G(K_0)$ above is given by

$$w'(\pi)b_1 = w'(p)b',$$

while the entries corresponding to $\varrho \neq 1$ are all equal to $w'(p)$.
We redefine μ to be the cocharacter of G_K, whose entry at $\iota = 1$ is μ', and whose other entries are all equal to w'.
For this twisted correspondence between filtered $\tau - K_F(L)$-spaces and filtered isocrystals with an F-action all remains valid what was said in the untwisted case $w' = 1$ above.

1.41 We shall now discuss a few examples to give an idea of the various possibilities of the admissible open subset $\mathcal{F}^{wa} \subset \mathcal{F}\otimes_E \breve{E}$. In each case we shall consider a situation of type (EL) or (PEL) and fix an element $b \in G(K_0)$. The conjugacy class of cocharacters will be given by fixing a model filtration $\mathcal{F}_o^\bullet$ of $V\otimes\bar{\mathbf{Q}}_p$, as in (1.39). In these examples we will denote the $\mathbf{Q}_p$-vector space by V and the corresponding isocrystal $(V\otimes K_0, b\sigma)$ by (N,Φ).

1.42 We shall first discuss the following example of type (PEL). Let $B = D$ be the quaternion division algebra over $\mathbf{Q}_p$ and let $*$ be a neben involution. Explicitly, present D in the following form

$$D = \mathbf{Q}_{p^2}[\Pi];\ \Pi^2 = p,\ \Pi a = \sigma(a)\Pi.$$

Here $\mathbf{Q}_{p^2}$ is the unramified extension of degree 2 in K_0. Then the involution may be given as follows,

$$\begin{aligned} a^* &= \sigma(a),\ \ a \in \mathbf{Q}_{p^2} \\ \Pi^* &= \Pi \end{aligned}$$

Let $V = (V, \iota)$ be a free D–module of rank n, with a non–degenerate alternating bilinear form satisfying the conditions in (1.38). In this case G is a non–trivial inner form of the group Gp_{2n} of symplectic similitudes. For $b \in G(K_0)$ we take any element with $c(b) = p$ and such that the corresponding isocrystal (N, Φ) is isotypical with all slopes equal to $\lambda = 1/2$. Using the action of D on $N = V \otimes K_0$ and the decomposition $\mathbf{Q}_{p^2} \otimes K_0 = K_0 \oplus K_0$ we obtain a direct sum decomposition (a $\mathbf{Z}/2$–grading)

$$N = N_0 \bigoplus N_1$$

with

$$\deg \Pi = \deg \Phi = 1.$$

Furthermore, N_i is totally isotropic with respect to (,), as follows from the identity

$$\begin{aligned} a \cdot (v, v') = (av, v') = (\iota(\sigma^i(a))v, v') = \\ = (v, \iota(\sigma^{i+1}(a))v') = (v, \sigma(a) \cdot v') = \sigma(a) \cdot (v, v'),\ \ v, v' \in N_i,\ a \in \mathbf{Q}_{p^2}. \end{aligned}$$

We define a new non–degenerate alternating bilinear form

$$\begin{aligned} < , > &: N_0 \otimes N_0 \longrightarrow K_0 \\ < v, v' > &= (v, \Pi v'). \end{aligned}$$

We also introduce the σ–linear endomorphism of N_0,

$$\Phi_0 = \Pi^{-1} \cdot \Phi | N_0.$$

Then, since

$$\text{orddet}\,\Phi = \text{orddet}\,\Pi = 2n,$$

it follows that orddet $\Phi_0 = 0$. Furthermore, since Φ has all its slopes equal to $1/2$ it follows that Φ_0 has all its slopes equal to 0. Also

$$< \Phi_0\, v, \Phi_0\, v' >= \sigma(< v, v' >).$$

Indeed,

$$\begin{aligned} < \Phi_0\, v, \Phi_0\, v' > &= (\Pi^{-1}\Phi\, v, \Phi\, v') = \sigma(v, \Pi v') \\ &= \sigma(< v, v' >). \end{aligned}$$

It follows that there exists a unique $\mathbf{Q}_p$-rational structure for the pair $(N_0, <, >)$ such that the Frobenius acts through Φ_0. Writing $(V_0, <,>)$ for this symplectic space over $\mathbf{Q}_p$ we obtain easily an identification

$$J = \text{Gp}(V_0, <,>).$$

In particular, $J(\mathbf{Q}_p)$ is the set of $\mathbf{Q}_p$-rational points of the split inner form of G.

We consider now the case of the Siegel Grassmannian, i.e. we take as the model filtration $\mathcal{F}_o^\bullet$ the one with

$$(0) = \mathcal{F}_o^2 \subset \mathcal{F}_o^1 \subset \mathcal{F}_o^0 = V \otimes \bar{\mathbf{Q}}_p$$

with $\mathcal{F}_o = \mathcal{F}_o^1$ a D-invariant totally isotropic subspace of dimension $2n$. Hence $\mathcal{F}$ is the $\mathbf{Q}_p$-variety of D-invariant Lagrangian subspaces of V. If $\mathcal{F} \in \mathcal{F}(K)$ is such a subspace there is a direct sum decomposition, compatible with the one of N,

$$\mathcal{F} = \mathcal{F}_0 \bigoplus \mathcal{F}_1.$$

Furthermore, $\mathcal{F}_0 \subset N_0 \otimes_{K_0} K$ is a maximal totally isotropic subspace (with respect to $<,>$) and in fact the map sending $\mathcal{F}$ to $\mathcal{F}_0$ identifies $\mathcal{F}(K)$ with the set of K-valued points of the Grassmannian of Lagrangian subspaces of $(V_0, <,>)$.

Proposition 1.43 *Under the above identification, the subset $\mathcal{F}^{wa}$ of the Grassmannian of Lagrangian subspaces of $(V_0, <\ ,>\ )$ is characterized by the following condition:*

For all totally isotropic subspaces $W_0 \subset V_0$ we have

$$\dim \mathcal{F} \cap (W_0 \otimes K) \leq 1/2 \cdot \dim W_0.$$

Proof: Clearly the the HN–slope $\mu(N, \Phi, \mathcal{F}) = 0$, hence weak admissibility is equivalent to semi–stability.
The uniqueness of the canonical filtration of $(N, \Phi, \mathcal{F})$ implies its D–invariance. It follows that $(N, \Phi, \mathcal{F})$ is weakly admissible if and only if for all subspaces $P \subset N$ stable under Φ *and D–invariant* we have

$$\dim(\mathcal{F}_0 \cap (P \otimes_{K_0} K)) \leq \operatorname{orddet}(\Phi; P).$$

However, since all slopes of Φ are equal to $1/2$ the right hand side of this inequality is equal to

$$\operatorname{orddet}(\Phi; P) = 1/2 \cdot \dim P.$$

By the D–invariance of P we obtain a direct sum decomposition $P = P_0 \oplus P_1$ and the Φ–invariance of P is equivalent to the Φ_0–invariance of P_0, i.e. to the fact that P_0 is a $\mathbf{Q}_p$–rational subspace of N_0. The above condition is therefore equivalent to the condition appearing in the statement of the proposition, but for *all* $\mathbf{Q}_p$–rational subspaces $W_0 \subset V_0$. However, this is equivalent to the apparently weaker requirement that the inequality hold only for totally isotropic subspaces, as the following argument shows. Put

$$W_0^{'} = W_0 \cap W_0^{\perp}.$$

Then this is a totally isotropic subspace and on

$$W_0^{''} = W_0 / W_0^{'}$$

we have a natural non–degenerate alternating bilinear form. Therefore we have for the image $\mathcal{F}_0^{''}$ in $W_0^{''} \otimes K$ of the totally isotropic subspace $\mathcal{F}_0 \cap (W_0 \otimes K)$ that

$$\dim \mathcal{F}_0^{''} \leq 1/2 \cdot \dim W_0^{''}.$$

Therefore, if $\dim \mathcal{F}_0 \cap (W_0' \otimes K) \leq 1/2 \cdot \dim W_0'$, we obtain

$$\dim \mathcal{F}_0 \cap (W_0 \otimes K) \leq 1/2 \cdot \dim W_0.$$

□

1.44 The next class of examples is of type (EL) and arises from the following set-up. Let F be a finite extension of degree n of $\mathbf{Q}_p$ and let $B = D$ be a central simple algebra of dimension d^2 over F and with invariant

$$\mathrm{inv}(D) = s \,(\mathrm{mod}\, d).$$

We fix s with $0 \leq s \leq d-1$ in its congruence class mod d. Let $V = (V, \iota)$ be a free D-module of rank 1. Let us fix an isomorphism $D \cong V$. In this case G is the multiplicative group of D^{opp} (the opposite algebra of D). Obviously $G = Res_{F/\mathbf{Q}_p} G'$, where G' is the multiplicative group of D^{opp} considered as an algebraic group over F. We will describe this example by making use of the procedure of (1.40), i.e. b (resp. μ or $\mathcal{F}_0$) will be induced by that procedure from b' (resp. from μ' or $\mathcal{F}_0^{\bullet\,\prime}$) and the obvious central cocharacter $w' : \mathbf{G}_m \to G'$, $t \mapsto t \cdot 1 \in D$.
Let $\tilde{F}$ be an unramified extension of degree d of F contained in D and let $\tau \in \mathrm{Gal}(\tilde{F}/F)$ be the relative Frobenius automorphism. We may present D in the form

$$D = \tilde{F}[\Pi]; \quad \Pi^d = \pi^s, \Pi x = \tau(x)\Pi,$$

where $\pi \in F$ is a prime element. We fix once and for all an embedding $\varepsilon : F \to \bar{\mathbf{Q}}_p$, i.e., an algebraic closure of F. We use the notation $K_0' = K_F(\bar{\mathbf{F}}_p)$. We set $b' \in G'(K_0')$ equal to

$$b' = \pi\Pi^{-1}.$$

It is easy to see that the corresponding $\tau - K_0'$-space $(N', \Phi') = (V \otimes_F K_0', b'\tau)$ is isotypical with all slopes $\lambda = (d-s)/d$. The corresponding isocrystal $(N, \Phi) = (V \otimes_{\mathbf{Q}_p} K_0, b\sigma)$ under the procedure of (1.40) is then isotypical of slope $(d-s)/nd$. Explicitly the element b of

$$D^{opp} \otimes K_0 = \prod_{\varrho : F^t \to K_O} D^{opp} \otimes^{F^t, \varrho} K_0$$

has the component $p\Pi^{-1}$ for $\varrho = 1$ and the component p for all other ϱ. As model filtration $\mathcal{F}_o'^{\bullet}$ we take one with $\mathcal{F}_o'^2 = (0), \mathcal{F}_o'^0 = V \otimes_F \bar{\mathbf{Q}}_p$ and with $\mathcal{F}_o' = \mathcal{F}_o'^1$ equal to

$$\dim(\mathcal{F}_o') = (d-s)d.$$

This determines $\mathcal{F}_o'$ up to isomorphism. The field of definition of the corresponding algebraic variety $\boldsymbol{\mathcal{F}'}$ is $E = \varepsilon(F)$. Consider the decomposition (cf. 1.40))

$$V \otimes_{\mathbf{Q}_p} K = \prod_{\iota: F \to K} V_\iota.$$

Under the correspondence of (1.40) we are therefore considering filtrations $\mathcal{F}^\bullet$ with $\prod_{\iota \neq \varepsilon} V_\iota \subset \mathcal{F}^1$.

We fix an extension $\varepsilon' : \tilde{F} \to K_0'$ of ε. We have a decomposition

$$\tilde{F} \otimes_F K_0' = \bigoplus K_0'.$$

The sum ranges over the powers $\tau^i (i \in \mathbf{Z}/d)$ of $\tau \in Gal(\tilde{F}/F)$. The action of $\tilde{F}$ on V induces a decomposition of N',

$$N' = \bigoplus_{i \in \mathbf{Z}/d} N_i'.$$

Here

$$N_i' = \{v \in N';\ \iota(f)v = \varepsilon' \circ \tau^i(f)v,\ f \in \tilde{F}\}.$$

The operators Φ' and Π of N' are homogeneous of respective degrees

$$\begin{aligned} deg\, \Phi' &= 1 \\ deg\, \Pi &= -1. \end{aligned}$$

The restriction of $\Phi'(\pi\Pi^{-1})^{-1}$ to N_0' is a τ-linear operator with all slopes equal to 0,

$$\Phi_0' = \Phi'(\pi\Pi^{-1})^{-1}|N_0' : N_0' \longrightarrow N_0'.$$

We therefore obtain a rational structure of N_0' over E : Putting $V_0' = N_0'^{\Phi_0'}$ (fixed module), we have an isomorphism

$$V_0' \otimes_E K_0' = N_0'$$

such that under this isomorphism

$$\mathrm{id}_{V_0'} \otimes \tau = \Phi_0'.$$

Let $\mathcal{F}' \in \mathcal{F}'(\mathbf{C}_p)$ and let $\mathcal{F}_0' \subset N_0' \otimes_{E.K_0} \mathbf{C}_p$ be its image under the obvious projection map. Then $\mathcal{F}_0'$ is a $(d-s)$–dimensional subspace in the d–dimensional $\mathbf{C}_p$–vector space $V_0' \otimes_E \mathbf{C}_p$. It is easy to see that we obtain in this way an identification with the Grassmannian

$$\mathcal{F}' = \mathrm{Grass}_s(V_0').$$

Proposition 1.45 *Under this identification the subset of weakly admissible points of $\mathcal{F}'(K)$ corresponds to the set of $(d-s)$-dimensional subspaces $\mathcal{F}_0'$ of $V_0' \otimes_E K$ satisfying the following condition:*
For every rational subspace $W_0' \subset V_0'$ we have

$$\dim(\mathcal{F}_0' \cap (W_0' \otimes_E K)) \leq (d-s)/d \cdot \dim V_0'.$$

Proof: Just as in the proof of proposition (1.43) the conditions on weak admissibility are parametrized by the Φ'–stable D–invariant subspaces $P' \subset N'$. However, these correspond precisely to the Φ_0'–stable subspaces $P_0' \subset V_0'$, i.e. to the E–rational subspaces $W'_0 \subset V_0'$. Furthermore, the condition corresponding to P' may be reexpressed in terms of W'_0. Indeed,

$$\dim(\mathcal{F}' \cap (P' \otimes_{K_0'} K)) = d \cdot \dim(\mathcal{F}_0' \cap (W'_0 \otimes_E K))$$

and

$$\begin{aligned} \mathrm{orddet}(\Phi'; P') &= (d-s)/d \cdot \dim(P') \\ &= (d-s) \cdot \dim(W_0'). \end{aligned}$$

Therefore the condition becomes

$$\dim(\mathcal{F}_0' \cap (W'_0 \otimes_E K)) \leq (d-s)/d \cdot \dim(W_0').$$

which proves the proposition. In exactly the same way one sees that

$$J' = GL(V_0').$$

□

Corollary 1.46 *Assume $s = 1$ in the preceding proposition. Then the weakly admissible subset of $\mathcal{F}$ corresponds bijectively to the set of points of the projective space $\mathbf{P}(V)$ not containing a F-rational line.* □

1.47 The next example will again be of type (EL), with $F = B$ an extension of degree n of $\mathbf{Q}_p$. Let $V = F^d$ be the standard d–dimensional F–vector space. Then $G = Res_{F/\mathbf{Q}_p} GL_d$. We again describe this example through the procedure of (1.40) and the obvious central cocharacter $w' : \mathbf{G}_m \to GL_d$, $t \mapsto t \cdot 1$.
Let $G' = GL_d$ over $Spec\, F$. We fix once and for all an embedding $\varepsilon : F \to \bar{\mathbf{Q}}_p$ and again use the notation $K_0' = K_F(\bar{\mathbf{F}}_p)$. We fix $b' \in G'(K_0')$ equal to

$$b' = \pi \begin{pmatrix} 0 & 1 & & \\ & \ddots & \ddots & \\ & & & 1 \\ \pi & & & 0 \end{pmatrix}^{-1}$$

Here π denotes a uniformizer in F. The corresponding $\tau - K_0'$–space (N', Φ') is isotypical with all slopes $\lambda = (d-1)/d$.
As model filtration subspace $\mathcal{F}_o' = \mathcal{F}_o'^1$ we take one with

$$\dim \mathcal{F}_o' = d - 1.$$

Again $\mathcal{F}_o'^2 = (0)$ and $\mathcal{F}_o'^0 = V \otimes_F \bar{\mathbf{Q}}_p$. The Shimura field is equal to $E = \varepsilon(F)$. We obviously have an identification with the projective space of V over $Spec\, F$,

$$\mathcal{F}' = \mathbf{P}(V).$$

Proposition 1.48 *All points of $\mathcal{F}'(K)$ are weakly admissible.*

Proof: This is obvious since the $\tau - K_0'$–space (N', Φ') is simple. □

In this case J is the multiplicative group of the central division algebra with invariant $= 1 \pmod d$ over F.

Remark 1.49 We mention briefly some variants of example (1.44). For simplicity we take $F = \mathbf{Q}_p$.

(i) For the first variant we consider a central simple algebra of dimension d^2 over $\mathbf{Q}_p$, with invariant $= s(\mathrm{mod}\ d)$. We again take V to be a free D–module of rank 1. We fix an integer r, $0 \leq r \leq d-1$ which is a multiple of s modulo d,

$$i \cdot s = r + j \cdot d \ .$$

As element $b \in G(K_0) = (D^{\text{opp}} \otimes K_0)^\times$ we take

$$b = \pi^{-j} \cdot \Pi^i,$$

where $\Pi \in D$ is as before. Then the corresponding isocrystal (N, Φ) has all its slopes $\lambda = r/d$. We consider D–invariant filtrations $\mathcal{F}$ with $\mathcal{F}^2 = (0)$, $\mathcal{F}^0 = V \otimes \bar{\mathbf{Q}}_p$ and with $\mathcal{F} = \mathcal{F}^1$ of dimension $r \cdot d$. We analyze this case following the same principles. Let $\mathbf{Q}_{p^d}$ be an unramified extension of degree d of $\mathbf{Q}_p$ contained in D which we assume embedded in K_0. The decomposition $\mathbf{Q}_{p^d} \otimes K_0 = \bigoplus_{\varsigma : \mathbf{Q}_{p^d} \to K_0} K_0$ yields decompositions

$$\begin{aligned} N &= \bigoplus N_\varsigma \\ \mathcal{F} &= \bigoplus \mathcal{F}_\varsigma \end{aligned}$$

and $\Phi(N_\varsigma) = N_{\varsigma \circ \sigma}$, $\Pi(N_\varsigma) = N_{\varsigma \circ \sigma^{-1}}$. In this case $\Phi \cdot \Pi$ preserves the summand N_0 corresponding to the chosen embedding and induces there a σ–linear endomorphism Φ_0 which is isotypical with all its slopes $\lambda = (r+s)/d$. Now let us assume that $r+s$ is prime to d, and let us determine the weakly admissible filtrations. Let $P \subset N$ be a Φ–stable D–invariant subspace. Then $P = \bigoplus P_\varsigma$ and P_0 is stable under Φ_0. However,

$$\operatorname{orddet}(\Phi_0; P_0) = (r+s)/d \cdot \dim(P_0)$$

and since $r+s$ was assumed prime to d we conclude that P is trivial, i.e. (0) or N. We therefore see that in this case all points of $\mathcal{F}$ are weakly admissible. In the same way one sees that $J(\mathbf{Q}_p)$ is the multiplicative group of the central division algebra of invariant $= r + s (\operatorname{mod} d)$ over $\mathbf{Q}_p$.

(ii) We want to do one more example, one with filtrations with a higher number of steps. Let again D be a central simple algebra of dimension d^2 over $\mathbf{Q}_p$, with invariant $= s (\operatorname{mod} d)$, with $0 \le s \le d-1$. Let V be a free D–module of rank 1. As element $b \in G(K_0) \simeq GL_n(D^{opp})$ we take $b = \Pi^j$, where $j = 1 (\operatorname{mod} d)$.
Then the corresponding isocrystal (N, Φ) has all its slopes $\lambda = js/d$. We fix integers

$$\delta^1 \geq \delta^2 \geq \ldots \geq \delta^a \geq 0$$

with

$$\sum_{i=1}^{a} \delta^i = js.$$

We consider D–invariant filtrations $\mathcal{F}^\bullet$ with $\mathcal{F}^{a+1} = (0), \mathcal{F}^0 = V \otimes \bar{\mathbf{Q}}_p$ and $\dim \mathcal{F}^i = \delta^i \cdot d, i = 1, \ldots, a$. We may analyze this example in the same way as the preceding ones and obtain the following results. The $\mathbf{Q}_p$–variety $\mathcal{F}$ may be identified with the variety of incomplete flags on the standard vector space $\mathbf{Q}_p^d$,

$$(0) \subset \mathcal{F}^a \subset \mathcal{F}^{a-1} \subset \ldots \subset \mathcal{F}^1 \subset \mathbf{C}_p{}^d$$

with $\dim \mathcal{F}^i = \delta^i, i = 1, \ldots, a$. The subset of weakly admissible elements over K is characterized by the following condition: For any rational subspace $W \subset \mathbf{Q}_p^d$ we have

$$\sum_{i=1}^{a} \dim (\mathcal{F}^i \cap (W \otimes K)) \leq (js/d) \cdot \dim W.$$

In this case $J \simeq GL_d$.

1.50 In the preceding examples the element b was basic. We now give one example (of type (EL)) where b is not basic. Let $B = \mathbf{Q}_p$ and let $V = \mathbf{Q}_p^{2n}$. We denote the canonical base by $e_1, \ldots, e_{2n}$ and let

$$V_- = \text{span } < e_1, \ldots, e_n >, V_+ = \text{span } < e_{n+1}, \ldots, e_{2n} > .$$

As the element $b \in G(K_0) = GL_{2n}(K_0)$ we take $b = p \cdot \mathrm{id}_{V_-} \oplus \mathrm{id}_{V_+}$. Then the slope decomposition of the isocrystal (N, Φ) has the form $N = N_0 \oplus N_1$ with

$$N_0 = V_+ \otimes K_0 \, , N_1 = V_- \otimes K_0 .$$

We consider the space $\mathcal{F} = \mathrm{Grass}_n(V)$ of subspaces $\mathcal{F}$ of dimension n. It is then easy to see that $\mathcal{F}$ is weakly admissible if and only if

$$\mathcal{F} \cap (V_+ \otimes K) = (0).$$

In other words, $\mathcal{F}^{wa}$ may be identified with the big cell defined by V_+, an open algebraic subvariety of $\mathcal{F}$. It may be identified with the variety of splittings of the exact sequence of vector spaces,

$$0 \to V_+ \to V \to V_- \to 0,$$

i.e. with an affine space of dimension $n^2 = \dim \operatorname{Hom}(V_-, V_+)$. In this case J is the Levi subgroup of G,

$$J(\mathbf{Q}_p) = GL(V_+) \times GL(V_-).$$

1.51 We now wish to relate the weakly admissible subset to Geometric Invariant Theory. We assume that G is connected. We fix a conjugacy class of cocharacters μ and the corresponding projective algebraic variety $\mathcal{F}$ over E. We fix an invariant inner product on the Lie algebra of a maximal torus of G and use it to interpret μ as a conjugacy class of a character. To this character there is associated an ample line bundle $\mathcal{L}$ on $\mathcal{F}$ which is homogeneous under the derived group G_E^{der}, after perhaps replacing $\mathcal{L}$ by a positive tensor power. Let $b \in G(K_0)$ be such that *the set of weakly admissible points in* $\mathcal{F}_{\breve{E}}$ *is non-empty*. Let J be the corresponding algebraic group over $\mathbf{Q}_p$. Then $J \cap G^{der}$ is a subgroup of J defined over $\mathbf{Q}_p$ which will be denoted by J^{der}. Then $J_{\breve{E}}^{der}$ is a subgroup of $G_{\breve{E}}^{der}$ and hence acts on $(\mathcal{F}, \mathcal{L})_{\breve{E}}$. For any maximal $\mathbf{Q}_p$-split torus $T \subset J^{der}$ let

$$\mathcal{F}_b(T)^{ss} \subset \mathcal{F}_{\breve{E}}$$

be the set of points which are semi-stable for the restriction to $T_{\breve{E}}$ of the action of $J_{\breve{E}}^{der}$ on $(\mathcal{F}, \mathcal{L})_{\breve{E}}$. For any finite extension K of $\breve{E}$, let

$$\mathcal{F}_b^{ss}(K) = \bigcap_T \mathcal{F}_b(T)^{ss}(K),$$

where the intersection is over all maximal $\mathbf{Q}_p$-split tori of J^{der}. It is obvious that this set is stable under the action of $J(\mathbf{Q}_p)$. This set was considered by van der Put, Voskuil [PV]. It is easy to see that this set is unchanged if

$\mathcal{L}$ is replaced by a positive tensor power and does not depend on the choice of the invariant inner product above.
The following theorem was proved by B. Totaro in response to a conjecture in a first version of this book.

Theorem 1.52 (B. Totaro): *We have*

$$\mathcal{F}_b^{ss}(K) = \mathcal{F}_b^{wa}(K),$$

for any finite extension K of $\breve{E}$.

The assumption that $\mathcal{F}$ contain weakly admissible points is only made to ensure that semi-stable points in the sense of (1.3) are weakly admissible. We now illustrate this statement in the previous examples.

1.53 We first consider example (1.42), with the notation introduced there. In order to analyze the subset $\mathcal{F}^{ss}(K)$ we use the Hilbert–Mumford criterion. Note that the representation of J on $\Gamma(\mathcal{F}, \mathcal{L})$ is, under the identification of J with $\mathrm{Gp}(V_0, < \ , > \ )$ a positive tensor power of the n–th exterior power of V_0. Choose a basis of V_0,

$$e_{\pm 1}, e_{\pm 2}, \ldots, e_{\pm n}$$

with

$$\begin{aligned} < e_i, e_{-i} > \ &= \ 1, \quad i = 1, \ldots, n \\ < e_i, e_j > \ &= \ 0, \quad i \neq -j. \end{aligned}$$

Let T be the diagonal torus and let $\lambda \in X_*(T)$ be a 1–parameter subgroup,

$$\lambda(t)e_i = t^{r_i} e_i, \quad i \in \{\pm 1, \ldots, \pm n\}; \quad r_{-i} = -r_i.$$

We investigate the Mumford criterion in case λ lies in the positive Weyl chamber, i.e.

$$r_1 > r_2 > \ldots > r_n > 0.$$

Let $I \subset \{\pm 1, \ldots, \pm n\}$ and let

$$L_I = \mathrm{span}\,\{e_i; \ \ i \in I\}$$

be the subspace spanned by the corresponding basis vectors. Then L_I is totally isotropic if $I \cap (-I) = \emptyset$. Consider the self–dual standard flag fixed by T,

$$(0) \subset L_1 \subset L_2 \subset \ldots \subset L_n \subset L_{-n} \subset \ldots \subset L_{-2} \subset L_{-1} = V_0$$

with

$$\begin{aligned} L_i &= L_{\{1,\ldots,i\}} \\ L_{-i} &= L_{\{1,\ldots,i-1,\pm i,\ldots,\pm n\}}, \quad 1 \le i \le n. \end{aligned}$$

Let $\mathcal{F}_0 \subset V \otimes K$ be a Lagrangian subspace and let $I(\mathcal{F}_0)$ be the set of jumps in the intersection of $\mathcal{F}_0$ with the standard flag, i.e.

$$\mu \in I(\mathcal{F}_0) \Leftrightarrow \exists\, v \in \mathcal{F}_0 \cap (L_\mu \otimes K)$$

with $v \notin$ preceding member of the chain. Then it follows easily for the action of T on the corresponding points in the Grassmannian,

$$\lim_{t \to 0} \lambda(t) \cdot \mathcal{F}_0 = L_{I(\mathcal{F}_0)}.$$

Furthermore, $\lambda(t)$ operates on the fibre in $L_{I(\mathcal{F}_0)}$ of the homogeneous line bundle defined by the n–th exterior power representation through the character Σr_μ, $\mu \in I(\mathcal{F}_0)$. Therefore the point corresponding to $\mathcal{F}_0$ satisfies Mumford's criterion with respect to T, λ if and only if

$$\sum_{\mu \in I(\mathcal{F}_0)} r_\mu \le 0.$$

By convexity this condition then holds for all λ in the closure of the positive Weyl chamber. It suffices to check this condition on the extremal 1–parameter subgroups i.e. the fundamental co–weights,

$$\lambda_i := (r_1 = \ldots = r_i = 1, r_{i+1} = \ldots = r_n = 0), \quad 1 \le i \le n.$$

Therefore the above condition for all λ in the closure of the positive Weyl chamber is equivalent to the condition

$$|\{1, \ldots, i\} \cap (-I(\mathcal{F}_0))| \ge |\{1, \ldots, i\} \cap I(\mathcal{F}_0)|, \quad i = 1, \ldots, n.$$

Looking back at the definition of the flag and recalling that $(-I(\mathcal{F}_0)) \cap I(\mathcal{F}_0) = \emptyset$ this condition is in turn equivalent to

$$\dim(\mathcal{F}_0 \cap L_i) \leq i/2, \quad i = 1, \ldots, n.$$

Since any totally isotropic subspace of V_0 is conjugate under $J(\mathbf{Q}_p)$ to L_i for suitable i $(1 \leq i \leq n)$ and since $\mathcal{F}^{ss}(K)$ is stable under the action of $J(\mathbf{Q}_p)$ it follows that the points in $\mathcal{F}^{ss}(K)$ satisfy the condition appearing in the statement of proposition (1.52). Conversely, since all maximal split tori and all 1–parameter subgroups may be conjugated under $J(\mathbf{Q}_p)$ into (T, λ) as above with λ in the closure of the positive Weyl chamber it follows that all points satisfying the condition in proposition (1.52) lie in $\mathcal{F}^{ss}(K)$.

1.54 We now discuss the other examples. In example (1.44) the group J is $GL(V_0')$ acting on the appropriate Grassmannian. The corresponding subset $\mathcal{F}^{ss}(K)$ was determined in [PV], 2.8.2. and was found to be described by exactly the same condition as the one appearing in proposition (1.45). Something analogous applies to the last example of (1.49) (use [PV], 2.8.1.). The first example of (1.49) and example (1.47) are also compatible with theorem (1.52) Indeed, in these cases the group J is an inner form of G anisotropic modulo center so that the condition describing $\mathcal{F}^{ss}(K)$ is empty. It follows that $\mathcal{F}^{ss}(K) = \mathcal{F}(K) = \mathcal{F}^{wa}(K)$. Finally, the non–basic example (1.50) may be treated in exactly the same way as (1.53) above.

2. Quasi–isogenies of p-divisible groups

In this chapter we will define a moduli space for the quasi–isogenies of a given p-divisible group $\mathbf{X}$. This moduli space will be a formal scheme over the Witt vectors.

2.1 Let us review some basic facts on formal schemes in the form needed here. Consider a preadmissible topological ring A. Let $\{\mathcal{I}_\alpha\}$ be a set of ideals of A that form a fundamental system of neighbourhoods of 0. Then we define a contravariant functor $\mathit{Spf}\ A$ on the category of schemes

$$\mathit{Spf}\ A(Z) = \varinjlim_{\alpha} \text{Hom}\ (Z, \mathit{Spec}\ A/\mathcal{I}_\alpha)$$

This is a local functor, i.e. a sheaf for the Zariski topology on the category of quasicompact quasiseparated schemes. If the ring $(A, \mathcal{I}_\alpha)$ is adic, we will call $\mathit{Spf}\ A$ an *affine formal scheme.* A local functor which has a covering by open subfunctors which are affine formal schemes, is called a *formal scheme.* If Λ is a preadmissible ring we may consider the category $Nilp_\Lambda$ of schemes over $\mathit{Spf}\ \Lambda$. Then the category of formal schemes over $\mathit{Spf}\ \Lambda$ is a full subcategory of the category of set valued sheaves on $Nilp_\Lambda$.

2.2 We call a morphism $\mathcal{X} \to \mathcal{Y}$ of formal schemes to be of finite type, étale, lisse, etc., if for any scheme Z and any morphism $Z \to \mathcal{Y}$ the fibre product $\mathcal{X} \times_{\mathcal{Y}} Z$ is a scheme and $\mathcal{X} \times_{\mathcal{Y}} Z \to Z$ is of finite type, étale, smooth etc. in the usual sense.

Let $\mathcal{X}$ be a formal scheme. Then there is a unique morphism $\mathcal{X}_{red} \to \mathcal{X}$, where $\mathcal{X}_{red}$ is a reduced scheme, such that for any reduced scheme Z the following map is bijective

$$\mathrm{Hom}(Z, \mathcal{X}_{red}) \longrightarrow \mathrm{Hom}(Z, \mathcal{X}).$$

$\mathcal{X}$ is called locally noetherian, if it is locally isomorphic to $\mathcal{S}pf\ A$, where A is a noetherian adic ring.

Definition 2.3 *Let $\mathcal{X}$ and $\mathcal{Y}$ be formal schemes that are locally noetherian. A morphism $\mathcal{X} \to \mathcal{Y}$ is called formally of finite type if $\mathcal{X}_{red} \to \mathcal{Y}_{red}$ is of finite type.*

The notion formally locally of finite type is defined in the same way.

2.4 If Λ is an adic noetherian ring and $\mathcal{X}$ is an affine noetherian formal scheme over $\mathcal{S}pf\ \Lambda$, such that $\mathcal{X} \to \mathcal{S}pf\ \Lambda$ is formally of finite type, then there is a Λ-algebra A of finite type with a preadmissible topology $\{\mathcal{I}_\alpha\}_{\alpha \in \mathbf{N}}$ such that

$$\mathcal{X} = \mathcal{S}pf\ (A, \mathcal{I}_\alpha).$$

The ring $(A, \mathcal{I}_\alpha)$ need not to be preadic, but its completion is an adic ring by definition.
The following proposition gives a condition to ensure that the completion of $(A, \mathcal{I}_\alpha)$ is adic. It is a reformulation of [EGA]0_I 7.2.2.

Proposition 2.5 *Let A be a preadmissible ring. Assume a fundamental system of neighbourhoods is given by a chain of ideals*

$$\mathcal{I}_1 \supset \mathcal{I}_2 \supset \cdots \supset \mathcal{I}_r \supset \cdots \qquad r \in \mathbf{N}.$$

Let $\mathbf{I}$ be an ideal of definition of A, such that $\mathbf{I}/\mathbf{I}^2$ is topologically of finite type, i.e. $\mathbf{I}/\mathbf{I}^2 + \mathcal{I}_r$ is a A-module of finite type for all r. Then the completion of A is an adic ring if for each $m \in \mathbf{N}$ the following chain of ideals stabilizes

$$\mathcal{I}_1 + \boldsymbol{I}^m \supset \mathcal{I}_2 + \boldsymbol{I}^m \supset \cdots \supset \mathcal{I}_r + \boldsymbol{I}^m \supset \cdots$$

Proof: Let us denote by $\boldsymbol{a}_m$ the intersection of the ideals in this last chain. Then $\boldsymbol{a}_1 = \mathbf{I}$ and

$$\boldsymbol{a}_m \supset \boldsymbol{a}_{m+1}, \quad \boldsymbol{a}_{m+1} + \boldsymbol{a}_1^m = \boldsymbol{a}_m.$$

Moreover $\mathbf{a}_1/\mathbf{a}_2$ is a A-module of finite type. We conclude by [EGA] loc. cit. that $\varprojlim A/\mathbf{a}_m$ is an adic ring. □

We add a few remarks on isogenies of p-divisible groups. By a p-divisible group X over a scheme S we mean a Barsotti-Tate group in the sense of Grothendieck (see Messing [Me]).

Definition 2.6 *A morphism $f : X \to Y$ of p-divisible groups over S is called an isogeny, iff f is an epimorphism of f.p.p.f. sheaves whose kernel is representable by a finite locally free group scheme.*

If $S \in Nilp_{\mathbf{Z}_p}$, the kernel of an isogeny is of rank a power of p. If the rank is constant and equal to p^h we call h the *height of the isogeny*. We have a converse to this definition.

Proposition 2.7 *Let X be a p-divisible group over a scheme S. Let H be a finite locally free group scheme over S and $H \to X$ a monomorphism over S. Then the f.p.p.f. sheaf X/H is a p-divisible group.*

Proof: Clearly the multiplication by $p : X/H \to X/H$ is an epimorphism and X/H is a p-torsion group. We have to verify that the kernel of the multiplication by p is representable by a finite locally free group scheme over S. Let us denote the kernel of multiplication by p^n on X by $X[n]$. By Oort and Tate [OT] we have that H is a closed subscheme of $X[n]$ for big n (compare EGA IV 8.11.5). Hence for big n we get an exact sequence:

$$0 \to H \to X[n] \to (X/H)[n] \to H \to 0.$$

One knows that the quotient $X[n]/H$ is a finite locally free group scheme and that an extension of finite locally free group schemes in the category of f.p.p.f.-sheaves is again a finite locally free group scheme (see Grothendieck [Gr2]). Hence $(X/H)[n]$ is a finite locally free group scheme. We finish the proof by writing the exact sequence:

$$0 \to (X/H)[1] \to (X/H)[n+1] \to (X/H)[n] \to 0.$$

□

The multiplication by p on a p-divisible group is by definition an isogeny. It follows that the group $\mathrm{Hom}_S(X, Y)$ is a torsion free $\mathbf{Z}_p$-module. Let us denote by $\underline{\mathrm{Hom}}_S(X, Y)$ the Zariski sheaf of germs of homomorphisms.

Definition 2.8 *Let X and Y be p-divisible groups over a scheme S. A quasi–isogeny is a global section f of the sheaf $\underline{Hom}_S(X,Y) \otimes \mathbf{Q}$, such that any point of S has a Zariski neighbourhood, where $p^n f$ is an isogeny for a suitable natural number n.*

Let us denote by $Qisg_S(X,Y)$ the group of quasi–isogenies from X to Y. Quasi–isogenies of p–divisible groups have the following well-known rigidity property. Let $S' \subset S$ be a closed subscheme, such that the defining sheaf of ideals J is locally nilpotent. Assume moreover that p is locally nilpotent on S. Then the canonical homomorphism

$$Qisg_S(X,Y) \longrightarrow Qisg_{S'}(X_{S'},Y_{S'}) \tag{2.1}$$

is bijective (Drinfeld [Dr2]).

Proposition 2.9 *Let $\alpha : X \rightarrow Y$ be a quasi–isogeny of p-divisible groups over a scheme S. Consider the functor:*

$$F(T) = \{\phi \in \mathrm{Hom}(T,S) \mid \quad \phi^*\alpha \quad \textit{is an isogeny}\}.$$

Then F is representable by a closed subscheme of S.

Proof: The question is local for the Zariski topology. Hence we may assume that $p^n\alpha$ is an isogeny for some natural number n. The property that α is an isogeny is equivalent to the property that $p^n\alpha : X[n] \rightarrow Y[n]$ is the zero morphism. To show that this last property is representable by a closed subscheme we prove:

Lemma 2.10 *Let $\alpha : \mathcal{M} \rightarrow \mathcal{L}$ be a morphism of $\mathcal{O}_S$–modules, on a scheme S. Assume that $\mathcal{L}$ is finite and locally free. Then the functor*

$$F(T) = \{\phi \in \mathrm{Hom}(T,S) \mid \quad \phi^*\alpha = 0\}$$

is representable by a closed subscheme of S.

Proof: We have an isomorphism

$$\mathrm{Hom}(\mathcal{M},\mathcal{L}) = \mathrm{Hom}(\mathcal{M} \otimes \mathcal{L}^*, \mathcal{O}_S), \quad \mathcal{L}^* = \mathrm{Hom}(\mathcal{L},\mathcal{O}_S).$$

The morphism α corresponds to $\tilde{\alpha} : \mathcal{M} \otimes \mathcal{L}^* \rightarrow \mathcal{O}_S$. The ideal *Image* $\tilde{\alpha}$ defines the desired closed subscheme.

2.11 Let $\mathbf{X}$ be a p-divisible over $\bar{\mathbf{F}}_p$. Let $W = W(\bar{\mathbf{F}}_p)$ be the ring of Witt vectors. We will define the functor of quasi-isogenies of $\mathbf{X}$, and show that it is representable by a formal scheme. Although we are only interested in the case of the field $\bar{\mathbf{F}}_p$, it will be essential for the proofs to allow other perfect fields L of characteristic p. In this context we set $W = W(L)$. We denote by σ the absolute Frobenius automorphism of W.
The category $Nilp_W$ is the category of schemes S over $\mathcal{S}pec\ W$ such that p is locally nilpotent on S. A scheme $S \in Nilp_W$ may be viewed as a formal scheme with ideal of definition $p\mathcal{O}_S$. We denote by $\bar{S}$ the closed subscheme of S defined by the sheaf of ideals $p\mathcal{O}_S$. By the universal property of Witt vectors (Grothendieck[Gr2]) it is equivalent to give a morphism $S \to \mathcal{S}pf\ W$ or to give a morphism $\bar{S} \to \mathcal{S}pec\ L$.

2.12 We consider isocrystals over L (1.1). Our notation will differ a little from the first chapter. We write $\mathbf{F} = \Phi$ for the Frobenius morphism. We do this because we also need the Verschiebung $\mathbf{V} = p \cdot \Phi^{-1}$. We define the dimension of an isocrystal N by the formula

$$\dim N = \mathrm{ord}_p \det \mathbf{V}.$$

We recall that an isocrystal N is isoclinic of slope $\lambda \in \mathbf{Q}$, if there is a $W(L)$-lattice $M \subset N$, such that $\mathbf{F}^s M = p^r M$, where $s > 0$ and r are integers, such that $\lambda = r/s$. If N is isoclinic of slope λ we have the relation

$$\dim N = (1 - \lambda)\mathrm{height}\ N.$$

We will call a sublattice $M \subset N$ a *crystal* if it is stable under $\mathbf{F}$ and $\mathbf{V}$.

Definition 2.13 *An isocrystal* $(N, \mathbf{F})$ *over* L *is called decent, if the vector space* N *is generated by elements* n *satifying an equation* $\mathbf{F}^s n = p^r n$ *for some integers* r *and* $s > 0$.

Remarks 2.14 Let us write $N = V \otimes W(L)_{\mathbf{Q}}$ for some $\mathbf{Q}_p$-vector space V. Let $G = GL(V)$ considered as an algebraic group over $\mathbf{Q}_p$. Then we get a σ-conjugacy class $\bar{b} \in B(G)$ that is defined by the equation $\mathbf{F} = b(\mathrm{id}_V \otimes \sigma)$, where $b \in G(W(L)_{\mathbf{Q}})$. Then N is decent, iff $\bar{b}$ is decent in the sense of (1.8). An equation of the form $\mathbf{F}^s n = p^r n$ implies that n lies in some isoclinic component of N. Hence N is decent, iff all isoclinic components are decent. Over a finite field decent amounts to saying that on each isoclinic component

$N_\lambda \subset N$ we have $\mathbf{F}^s = p^r$ for suitable integers r and $s > 0$, such that $\lambda = r/s$. Any decent N is obtained by base change from a decent isocrystal over some finite field. Over an algebraically closed field any isocrystal is decent. Hence in the case in which we are interested this definition is empty. We call a p-divisible group over a perfect field L of characteristic p decent, if the corresponding isocrystal is decent.

Definition 2.15 *Let $\mathbf{X}$ be a decent p-divisible group over a perfect field L. We associate to $\mathbf{X}$ the following functor $\mathcal{M}$ on the category $Nilp_W$. For $S \in Nilp_W$ a point of $\mathcal{M}(S)$ is given by the following data:*

1. *A p-divisible group X on S.*
2. *A quasi-isogeny $\varrho : \mathbf{X}_{\bar{S}} \to X_{\bar{S}}$.*

We denote such a point by (X, ϱ). Two points (X_1, ϱ_1) and (X_2, ϱ_2) are identified if $\varrho_1 \circ \varrho_2^{-1}$ lifts to an isomorphism $X_2 \to X_1$, i.e., if they are isomorphic.

Theorem 2.16 *The functor $\mathcal{M}$ is representable by a formal scheme over Spf W, which is locally formally of finite type.*

The proof will depend on several lemmas. Let us start by the remark that $\mathcal{M}$ depends only on the isogeny class of $\mathbf{X}$. Since the isocrystal associated to $\mathbf{X}$ is defined over a finite field, we may assume that the field L is finite. Let P be any perfect field of characteristic p. For the following proposition let $W = W(P)$ be the ring of Witt vectors and τ be some positive power of the Frobenius automorphism σ. The invariants of τ are the Witt vectors $W(L)$ of a finite field L. Let N_0 be a finite dimensional vector space over $W(L)_{\mathbf{Q}}$. Then τ acts via the second factor on the $W_{\mathbf{Q}}$-vector space $N = N_0 \otimes_{W(L)} W$. A finitely generated W-submodule $M \subset N$ is called a lattice, if the natural map $M \bigotimes_W W_{\mathbf{Q}} \to N$ is an isomorphism.

Proposition 2.17 *Let h be the dimension of the $W_{\mathbf{Q}}$-vector space N. Let M be any lattice in N. Then the lattice $M + \tau(M) + \cdots + \tau^{h-1}(M)$ is invariant under τ.*

Proof: We make an induction on h. Clearly for $h = 1$ any lattice is invariant. We choose an invariant vector $e \in N$. We will assume $e \in M$ but $p^{-1}e \notin$

M. Let M' denote the image of M in $N' = N/W_{\mathbf{Q}}e$. We replace M by $M + \cdots + \tau^{h-2}(M)$. Then by induction M' is τ-invariant. Hence it is enough to show that $M + \tau(M)$ is τ -invariant, under the assumption that M' is τ-invariant.
Let $\bar{f}_1, \ldots, \bar{f}_{h-1}$ be a τ-invariant basis of M'. We lift these elements to τ -invariant elements $f_1, \cdots, f_{h-1}$ of N. Hence M has a set of generators of the form

$$e, f_1 - w_1 e, \ldots, f_{h-1} - w_{h-1} e,$$

where $w_i \in W_{\mathbf{Q}}$.
The module $M + \tau(M)$ has the following set of generators

$$e, f_1 - w_1 e, \ldots, f_{h-1} - w_{h-1} e, (\tau(w_1) - w_1)e, \ldots, (\tau(w_{h-1}) - w_{h-1})e.$$

We have to show that $M + \tau(M)$ is invariant under τ. But we have

$$M + \tau(M) = M + \sum_{i=1}^{i=h-1} W p^{ord(\tau(w_i) - w_i)} e.$$

Since the sum is invariant under τ we get the result. □

Proposition 2.18 *Let N be a decent isocrystal over a finite field L. Then there is a natural number c and a finite extension L' of L, such that for any perfect field P containing L', and for any crystal $M \subset N \bigotimes W(P)_{\mathbf{Q}}$ there is a crystal $M' \subset N \bigotimes W(L')_{\mathbf{Q}}$, such that $M \subset M' \bigotimes W(P)$ and has index smaller than c.*

Proof: We assume first that N is isoclinic of slope r/s, $s > 0$. As above we denote the height of N by h. Since N contains a crystal, we have $s \geq r \geq 0$. We may assume that the field L' fixed by σ^s contains L. The operator $\alpha = p^{s-r}\mathbf{V}^{-s}$ acts on $N \bigotimes_{W(L)} W(L') \bigotimes_{W(L')} W(P) = N_P$ by the Frobenius automorphism τ over $W(L')$ on the last factor. The crystal $\hat{M} = M + \tau(M) + \ldots + \tau^{h-1}(M)$ is by the previous proposition of the form $M' \bigotimes_{W(L')} W(P)$. Since $M + \tau(M) \subset \mathbf{V}^{-s}(M)$, we conclude that length $(M + \tau M)/M \leq s \dim N$. Iterating this we get length $\hat{M}/M \leq (h-1)s \dim N$. This number depends only on N (and L).
In the general case let s be a common multiple of the dominators of the slopes of N. Again we assume that the fixed field L' of σ^s contains L.

Let $N = N_0 \oplus \ldots \oplus N_t$ be the isotypic decomposition of N. We order the summands such that the slopes decrease:

$$\frac{r_0}{s} > \frac{r_1}{s} > \cdots > \frac{r_t}{s}.$$

Since N is decent, we way assume that $F^s n = p^{r_i} n$ for each $n \in N$. We prove by induction on the number of isotypic components that there exists an M' as claimed in the proposition, with the additional property that it is a direct sum of isoclinic crystals. Let $N_> = N_1 \oplus \ldots \oplus N_t$ be the direct sum and $M_> \subset N_{>,P} = N_> \otimes_{W(L)} W(P)$ be the image of M by the projection $N_P \longrightarrow N_{>,P}$. We obtain an exact sequence

$$0 \longrightarrow M_0 \longrightarrow M \longrightarrow M_> \longrightarrow 0, \tag{2.2}$$

where the kernel is an isoclinic crystal $M_0 \subset N_{0,P}$. By induction assumption and the isoclinic case we may assume that M_0 and $M_>$ are obtained by the base change from crystals over L' and moreover that $M_>$ is a direct sum of isoclinic crystals. If sequence (2.2) would split as an extension of crystals, we would obtain M by base change from a crystal M' over L'. This would prove the proposition. Therefore it suffices to show that the following is true.

Lemma 2.19 *After push-out by $p^{r_0} : M_0 \rightarrow M_0$ the exact sequence (2.2) splits as a sequence of crystals.*

Proof: Let $W = W(P)$ and $W[F]$ be the non-commutative polynomial ring $(Fw = \sigma(w)F)$. From the exact sequence

$$0 \longrightarrow M_0 \longrightarrow N_0 \longrightarrow N_0/M_0 \longrightarrow 0$$

we obtain an isomorphism

$$\mathrm{Hom}_{W[F]}(M_>, N_0/M_0) \simeq \mathrm{Ext}^1_{W[F]}(M_>, M_0)$$

Hence the extension (2.2) corresponds to a W-linear homomorphism

$$\alpha : M_> \longrightarrow N_0/M_0,$$

which commutes with F. By assumption $M_>$ is generated as a W-module by elements m satisfying some equation $F^s m = p^{r_i} m$, $i = 1, \ldots, t$. Let

$\alpha(m) = \bar{z}$ be the image and $z \in N_0$ a representative. We assume that $k \in \mathbf{Z}$ is chosen such that $p^k z = w \in M_0$, but $p^{k-1} z \notin M_0$. By the equation $F^s m = p^{r_i} m$ we obtain that

$$F^s w \equiv p^{r_i} w \; mod \; p^k M_0.$$

On the other hand we have $F^s M_0 = p^{r_0} M_0$. For $k \geq r_0$ we would get $p^{r_i} w \equiv 0 \; mod \; p^{r_0} M_0$ which contradicts $r_0 > r_i$. Hence we have $k \leq r_0$, i.e. $p^{r_0} \alpha(m) = 0$.
This proves the lemma and the proposition.

Remarks 2.20 (i) The hypothesis that N be decent is indeed necessary for the conclusion of (2.18), as the following example shows.
Let $N = \mathbf{Q}_p^2$ with the standard basis e_1, e_2. We define the Verschiebung by the requirement $\mathbf{V} e_1 = e_1, \mathbf{V} e_2 = a e_2$, where a is a unit in $\mathbf{Z}_p$. Then $(N, \mathbf{F})$ is decent, iff a is a root of unity. Let n be a positive integer and $\epsilon \in W(\bar{\mathbf{F}}_p)$ a unit. Consider the lattice in $N \otimes W(\bar{\mathbf{F}}_p)$:

$$M = W(\bar{\mathbf{F}}_p)(e_1 + p^{-n} \epsilon e_2) + W(\bar{\mathbf{F}}_p) e_2.$$

Then M is a crystal, iff $\mathbf{V} M = M$, i.e.

$$\sigma(\epsilon) a = \epsilon \, (mod \, p^n).$$

An element satisfying this equality always exists.
Let $s > 0$. It is easy to see that the smallest lattice M_s of the form $M_s' \otimes_{\mathbf{F}_{p^s}} W(\bar{\mathbf{F}}_p)$ containing M is

$$M_s = W(\bar{\mathbf{F}}_p)(e_1 + p^{-n} \epsilon e_2) + W(\bar{\mathbf{F}}_p) p^{-n} (\epsilon - \sigma^s(\epsilon)) e_2 + W(\bar{\mathbf{F}}_p) e_2.$$

We note that $\epsilon - \sigma^s(\epsilon) = \sigma^s(\epsilon)(a^s - 1)$. If N is not decent, we have for any s that

$$m_s = \mathrm{ord}_p(a^s - 1) < \infty.$$

Hence we obtain for $n > m_s$

$$M_s = W(\bar{\mathbf{F}}_p)(e_1 + p^{-n} \epsilon e_2) + W(\bar{\mathbf{F}}_p) p^{-n+m_s} e_2.$$

We get that for any s the indices of M in M_s can become arbitrarily large for a suitable choice of M, i.e. n.

(ii) The content of proposition (2.18) may be interpreted as a statement about the Bruhat-Tits building of the general linear group over a p-adic field. As such it can be generalized to any connected reductive group over a p-adic field, comp. [RZ], [Rou].

2.21 Having done these preparations we will write $\mathcal{M}$ as a union of representable subfunctors. The last step is to exhibit Zariski open sets, which remain stable in this union. Let us start by giving an alternative definition of the functor $\mathcal{M}$.
Let $\tilde{\mathbf{X}}$ be a lifting of the p-divisible group $\mathbf{X}$ to $\mathcal{S}pf\ W(L)$. Then a point of $\mathcal{M}$ with values in $S \in Nilp_{W(L)}$ is given by the following data:

1. A p-divisible group X on S.
2. A quasi–isogeny $\tilde{\varrho} : \tilde{\mathbf{X}}_S \to X$ of p-divisible groups on S.

2.22 We define the closed subfunctor $\mathcal{M}^n$ of $\mathcal{M}$ by the condition that $p^n \tilde{\varrho}$ is an isogeny. The functor $\mathcal{M}^n$ is representable by the p-adic completion of a scheme locally of finite type over S. Indeed, $\mathcal{M}^n$ is a union of open and closed subfunctors $\mathcal{M}^{n,m}$, which are given by the condition that $p^n \tilde{\varrho}$ is an isogeny of height m. To give such an isogeny is the same thing as to give a finite locally free group scheme $G \subset \tilde{\mathbf{X}}[m]_S$, which is of height m. Hence we see that the functor $\mathcal{M}^{n,m}$ is representable by the p-adic completion of a closed subscheme of a Grassmanian variety associated to the algebra of functions on $\tilde{\mathbf{X}}(m)$. This proves the representability of $\mathcal{M}^n$. In the sense of Zariski sheaves we have $\mathcal{M} = \varinjlim \mathcal{M}^n$.
To prove the theorem we need still another representation of $\mathcal{M}$ as a union of representable subfunctors. To do this we define for any field extension P of L a quasi-metric on the set $\mathcal{M}(P)$.

Definition 2.23 *Let $\alpha : X \to Y$ be a quasi–isogeny of p-divisible groups over P. We define $q(\alpha) = \text{height}\, p^n \alpha$, where n is the smallest integer such that $p^n \alpha$ is an isogeny.*

If P' is a field extension of P we have $q(\alpha) = q(\alpha_{P'})$.

Lemma 2.24 *Let $\alpha : X \to Y$ be an isogeny of p-divisible groups over a scheme S. For any integer c the set of points $s \in S$ such that $q(\alpha_s) \leq c$ is closed.*

Proof: We prove that the set of points $s \in S$, such that $q(\alpha_s) > c$ is open. The function q does not change if we multiply α by a power of p. Therefore we may assume that α_s is an isogeny, but $p^{-1}\alpha$ is not an isogeny. Then there is a neighbourhood U of s, such that $p^{-1}\alpha_t$ is not an isogeny for $t \in U$. Let n_t be the smallest integer, such that $p^{n_t}\alpha_t$ is an isogeny. Hence n_t is nonnegative for $t \in U$. Assuming that height α_t is a constant function on U we find the result

$$c < q(\alpha_s) = \text{height } \alpha_s = \text{height } \alpha_t \leq \text{height } p^{n_t}\alpha_t = q(\alpha_t).$$

Definition 2.25 *Let $\alpha : X \to Y$ be a quasi-isogeny of p-divisible groups over P. We define $d(\alpha) = q(\alpha)+q(\alpha^{-1})$. For two points of $\mathcal{M}(P)$ we define $d((X,\varrho),(X',\varrho')) = d(\varrho'\varrho^{-1})$.*

If m_+ is the smallest integer such that $p^{m_+}\alpha$ is an isogeny and m_- is the smallest integer such that $p^{m_-}\alpha^{-1}$ is an isogeny, we have $d(\alpha) = (m_+ + m_-)\text{height} X$.

Corollary 2.26 *Lemma 2.24 holds with q replaced by d.*

Because $d((X,\varrho),(X,p\varrho)) = 0$ the function d is not quite a metric on $\mathcal{M}(P)$. To get a metric, we consider for $k \in \mathbf{Z}$ the subfunctor $\mathcal{M}(k) \subset \mathcal{M}$ of quasi-isogenies of height k. We set:

$$\tilde{\mathcal{M}} = \coprod_{h=0}^{\text{height } \mathbf{X}-1} \mathcal{M}(h).$$

It is easily checked that the function d of definition (2.25) is a metric on $\tilde{\mathcal{M}}(P)$.
The proposition 2.18 may be reformulated as follows:

Proposition 2.27 *There is a natural number c and a finite extension L' of L, such that for any perfect field P containing L', and any point $X \in \mathcal{M}(P)$ there is a point $Y \in \mathcal{M}(L')$, such that $d(X, Y_P) \leq c$.*

We define for a natural number c a subfunctor $\mathcal{M}_c$ of $\mathcal{M}$ consider the subfunctor $\mathcal{M}(h)$ of quasi-isogenies of height h. We define

$$\tilde{\mathcal{M}} = \coprod_{h=0}^{\text{height } \mathbf{X}-1} \mathcal{M}(h).$$

It is enough to show that this functor is representable. Since multiplication of an isogeny by p does not change the value of the function d,

$$\mathcal{M}_c(S) = \{(X,\varrho) \in \mathcal{M}(S);\ d(\varrho_s) \leq c \quad \text{for} \quad s \in S\}.$$

Lemma 2.28 *The functor $\mathcal{M}_c$ is representable by a formal scheme, which is locally formally of finite type over Spf $W(L)$.*

Proof: Let $\mathcal{M}_c(h)$ be the open and closed subfunctor of $\mathcal{M}_c$ that consists of points (X,ϱ) such that height $\varrho = h$. Then $\mathcal{M}_c$ is a disjoint union

$$\mathcal{M}_c = \coprod_{h \in \mathbf{Z}} \mathcal{M}_c(h).$$

The multiplication of ϱ by p defines an isomorphism

$$\mathcal{M}_c(h) \to \mathcal{M}_c(h + \text{height } \mathbf{X}).$$

Therefore it is enough to show that the following functor is representable by a formal scheme formally of finite type

$$\tilde{\mathcal{M}}_c = \coprod_{h=0}^{\text{height } \mathbf{X}-1} \mathcal{M}_c(h).$$

We consider the functor $\tilde{\mathcal{M}}_c^n = \mathcal{M}^n \bigcap \tilde{\mathcal{M}}_c$. The functor $\tilde{\mathcal{M}}_c^n$ is represented by the completion of the scheme $\mathcal{M}^n$ along the closed set of points $s \in \mathcal{M}^n$ given by the conditions $d(\varrho_s) \leq c$ and $0 \leq$ height $\varrho_s <$ height $\mathbf{X}$. Hence it is represented by a formal scheme formally of finite type over Spf $W(L)$. Let (X,ϱ) be a point of $\tilde{\mathcal{M}}_c$ with values in a field P. Then $p^{-1}\varrho$ is not an isogeny, because otherwise we would have height $\varrho =$ height $p +$ height $p^{-1}\varrho \geq$ height $\mathbf{X}$. Hence the smallest integer m_+, such that $p^{m_+}\varrho$ is an isogeny must be nonnegative. Since height $\varrho^{-1} = -$ height ϱ, we have the inequalities

$$-\text{height } \mathbf{X} < \text{height } \varrho^{-1} \leq 0.$$

Again we conclude that the smallest integer m_-, such that $p^{m_-}\varrho^{-1}$ is an isogeny, is nonnegative. By the remark after definition (2.25) we conclude that

$$m_+ + m_- \leq c/\text{height } \mathbf{X}.$$

Hence m_+ is bounded by $c/$height $\mathbf{X}$. This implies that we have an equality for $n \geq c/$height $\mathbf{X}$

$$(\tilde{\mathcal{M}}_c^n)_{red} = (\tilde{\mathcal{M}}_c^{n+1})_{red}.$$

The equality follows because a quasi–isogeny $\alpha : X \to Y$ of p-divisible groups over a reduced scheme S is an isogeny, iff α_s is an isogeny for each point $s \in S$ (see 2.9).
We fix an affine open subscheme $U \subset (\tilde{\mathcal{M}}_c^n)_{red}$ for large n. For n big we get an affine open formal subscheme $\mathcal{S}pf\ R_n$ of $\tilde{\mathcal{M}}_c^n$, whose underlying set is U. Hence we have a projective system of surjective maps of adic rings

$$R_{n+1} \longrightarrow R_n.$$

Let R be the projective limit. We write $R_n = R/\mathbf{a}_n$. Let $\mathbf{J}$ be the inverse image of the ideal of definition in some R_n.
In order to show that $\tilde{\mathcal{M}}_c = \varinjlim \tilde{\mathcal{M}}_c^n$ is a formal scheme, we have to prove that the ring R is $\mathbf{J}$ -adic. Since R_n is $\mathbf{J}$ -adic we may write

$$R = \varprojlim R/\mathbf{a}_n + \mathbf{J}^m.$$

The limit is taken independently over all n, m. We claim that for fixed m the following descending sequence stabilizes

$$\ldots \mathbf{a}_n + \mathbf{J}^m \supset \mathbf{a}_{n+1} + \mathbf{J}^m \supset \ldots$$

Indeed, let X_n be the universal p-divisible group on $\mathcal{S}pf\ R_n$. Then $X = \varinjlim X_n$ defines a p-divisible group on $R/\mathbf{J}^m$ for each m. We get $\varrho : \tilde{\mathbf{X}}_{R/\mathbf{J}^m} \to X$ by lifting the existing quasi–isogeny for $m = 1$. By the definition of representable we get for a suitable N a unique map $R_N \to R/\mathbf{J}^m$ that induces the point (X, ϱ). For any $n \geq N$ the composite map

$$R_n \to R_N \to R/\mathbf{J}^m \to R_n/\mathbf{J}^m R_n$$

has to be the canonical one. This implies that the first arrow induces an isomorphism $R_n/\mathbf{J}^m R_n \to R_N/\mathbf{J}^m R_N$. We conclude that the descending sequence of ideals stabilizes.

By proposition (2.5) we conclude that R is an adic ring. This completes the proof of lemma (2.28).

Corollary 2.29 *The functor $\tilde{\mathcal{M}}_c$ is representable by a formal scheme formally of finite type over Spf $W(L)$. The associated reduced scheme $(\tilde{\mathcal{M}}_c)_{red}$, which is the subscheme defined by the largest ideal of definition, is projective.*

Proof: Indeed, for $n \geq c/\text{height}\ \mathbf{X}$ we have

$$(\tilde{\mathcal{M}}_c)_{red} = (\tilde{\mathcal{M}}_c^n)_{red}.$$

But the right hand side is a closed subscheme of $\coprod \mathcal{M}^{n,m}$, where $nh \leq m \leq (n+1)h$ with $h = \text{height}\ \mathbf{X}$. This follows because for each geometric point (X, ϱ) of the right hand side

$$\text{height}\ p^n \varrho = n \,\text{height}\ \mathbf{X}\ +\ \text{height}\ \varrho \leq (n+1)\,\text{height}\ \mathbf{X}.$$

□

Proof of theorem 2.16: Let c and L' as in proposition (2.27). It is enough to show that $\mathcal{M}$ is representable over L'. As in the proof of lemma (2.28), we see that it is equivalent to show, that the subfunctor $\tilde{\mathcal{M}}$ is representable. Obviously the proposition (2.27) remains valid for the functor $\tilde{\mathcal{M}}$.
Let a be an integer. For a point $(Y, y : \mathbf{X}_{L'} \to Y)$ of $\tilde{\mathcal{M}}(L')$ we denote by $\tilde{\mathcal{M}}_a(Y) \subset \tilde{\mathcal{M}}_a$ the closed subset of points $s \in \tilde{\mathcal{M}}_a$, such that $d(X_s, Y_s) \leq c$, where X denotes the universal p-divisible group over $\tilde{\mathcal{M}}_a$. It is easily seen by the triangular inequality, that $\tilde{\mathcal{M}}_a(Y) = \emptyset$, if $d(\mathbf{X}_{L'}, Y) > a + c$.
Let U_a^f be the open formal subscheme of $\tilde{\mathcal{M}}_a$, whose underlying set is the complement of

$$\bigcup_{Y \in \tilde{\mathcal{M}}(L'), d(\mathbf{X}_{L'}, Y) \geq f} \tilde{\mathcal{M}}_a(Y).$$

Note that the last union is finite, because $\tilde{\mathcal{M}}_{a+c}(L')$ is finite by the last corollary.

Claim: If $a \geq f + c$ we have $U_a^f = U_{a+1}^f$.

First we show this equality for the underlying sets. Let $Z \in U_{a+1}^f(P)$ a point with values in some field P. We have to show $d(\mathbf{X}_P, Z) \leq a$. By proposition (2.18) there exists a point $Y \in \tilde{\mathcal{M}}(L')$ such that $d(Y_P, Z_P) \leq c$. But by the

definition of U^f_{a+1} it follows that $d(\mathbf{X}_P, Y_P) < f$. Hence $d(\mathbf{X}_P, Z) < f + c \leq a$.
The equality of formal schemes follows because $\tilde{\mathcal{M}}_a$ is the completion of $\tilde{\mathcal{M}}_{a+1}$ along the closed subset $Image\ \tilde{\mathcal{M}}_a \to \tilde{\mathcal{M}}_{a+1}$. Indeed, this implies that U^f_a is the completion of U^f_{a+1} along the closed subset U^f_{a+1}. Hence the claim follows.
We set $U^f = U^f_a$ for any $a \geq f + c$. Clearly $U^f \to U^{f+1}$ is an open immersion of formal schemes of finite type. We have $\tilde{\mathcal{M}} = \bigcup_f U^f$, because any point s of $\tilde{\mathcal{M}}$ such that $d(\mathbf{X}_s, X_s) < f - c$ is contained in the open set U^f. Indeed, if s is in the complement of U^f, there is a $Y \in \tilde{\mathcal{M}}(L')$, such that $d(X_s, Y_s) \leq c$ and $d(\mathbf{X}_s, Y_s) \geq f$. Hence we get the contradiction $d(\mathbf{X}_s, X_s) \geq f - c$. The theorem is proved. □

2.30 We call a subset $T \subset \tilde{\mathcal{M}}$ *bounded* if there is a natural number N such that $d(\mathbf{X}_t, X_t) \leq N$ for each point $t \in T$. We call a subset $T \subset \mathcal{M}$ bounded if there is an N, such that for each point $t \in T$ represented by (X_t, ϱ_t) we have

$$\text{height } \varrho_t \leq N, \qquad d(\mathbf{X}_t, X_t) \leq N.$$

By the proof of the theorem (2.16) we see that a subset of $\tilde{\mathcal{M}}$ is bounded, iff it is contained in one of the sets U^f. Since $(U^f)_{red}$ is contained in the projective scheme $(\tilde{\mathcal{M}}_a)_{red}$ for a suitable number a, we see that $(U^f)_{red}$ is quasiprojective. We obtain the following:

Corollary 2.31 *For a locally closed subscheme T of $\mathcal{M}$, the following conditions are equivalent:*

(i) T is bounded.

(ii) T is quasicompact.

(iii) T is quasiprojective.

Indeed for the proof it is enough to note that $\tilde{\mathcal{M}}$ is the union of the quasicompact open subsets U^f.

Proposition 2.32 *Any irreducible component of the scheme $\mathcal{M}_{red}$ is projective.*

Proof: Let us first show that an irreducible component is quasiprojective. By the last corollary it is enough to verify that an irreducible component is bounded. Let η be the general point of the component. By corollary (2.26) we have for each point of the irreducible component the inequality $d(\mathbf{X}_t, X_t) \leq d(\mathbf{X}_\eta, X_\eta)$. This shows quasiprojectivity.
To finish the proof it suffices to show that the valuative criterion for properness is satisfied for our irreducible component C of $\mathcal{M}_{red}$. Let R be a discrete valuation ring over L with field of fractions Q. Suppose we have a Q-valued point (X, ϱ) of C. To extend this point to $\mathcal{S}pec\, R$ we may replace ϱ by $p^n \varrho$ and hence assume that ϱ is an isogeny. Then Ker $\varrho \subset \mathbf{X}_Q[n]$ for some n. Taking the scheme theoretic closure we extend Ker ϱ to a finite flat group scheme $H \subset \mathbf{X}_R[n]$. Then $\varrho_R : \mathbf{X}_R \to \mathbf{X}_R/H$ is the desired extension of (X, ϱ). □

2.33 Let us denote by $J(\mathbf{Q}_p)$ the group of quasi–isogenies of $\mathbf{X}$. There is a natural right action of $J(\mathbf{Q}_p)$ on the formal scheme $\mathcal{M}$,

$$(X, \varrho) \longmapsto (X, \varrho \circ \gamma), \qquad \gamma \in J(\mathbf{Q}_p).$$

We will give conditions for the existence of the quotient of $\mathcal{M}$ by the action of a discrete subgroup of $J(\mathbf{Q}_p)$.

Proposition 2.34 *Let $\Gamma \subset J(\mathbf{Q}_p)$ be a discrete subgroup. Let $U \subset \mathcal{M}$ be a quasicompact open formal subscheme. Then the set*

$$\{\gamma \in \Gamma\, ;\, U\gamma \cap U \neq \emptyset\} \tag{2.3}$$

is finite.

Proof: There is a finite field $L_0 \subset L$ such that $\mathbf{X}$ is obtained by base change from a decent p–divisible group $\mathbf{X}_0$ over L_0. After a finite extension of L_0 we may assume

$$\operatorname{End} \mathbf{X}_0 = \operatorname{End} \mathbf{X}.$$

Indeed, to see this we may assume that $\mathbf{X}$ is isotypical. Then there are integers r and $s > 0$, such that $p^{-r}\mathbf{F}^s$ acts identically on the Cartier module $\mathbf{M}_0$ of $\mathbf{X}_0$. We may assume $L_0 = \mathbf{F}_{p^s}$. Let $\mathbf{M}$ be the Cartier module of $\mathbf{X}$. Then $\mathbf{M}_0$ are the invariants of the operator $p^{-r}\mathbf{F}^s$. Hence any endomorphism of $\mathbf{X}$ maps $\mathbf{M}_0$ to itself.

This remark shows that we may suppose that L is a finite field. Since $\mathcal{M}_{red}$ is locally of finite type the conditions

$$U\gamma \cap U = \emptyset \quad \text{and} \quad (U(\bar{L}))\gamma \cap U(\bar{L}) = \emptyset$$

are equivalent.
The points of $U(\bar{L})$ correspond to subcrystals M of the isocrystal $\mathbf{M} \otimes_{W(L)} W(\bar{L})_{\mathbf{Q}}$. Since U is quasicompact and therefore bounded there are integers $a, b \in \mathbf{Z}$ such that

$$p^a\mathbf{M} \subset M \subset p^b\mathbf{M}$$

for each $M \in U(\bar{L})$. If $\gamma M \in U(\bar{L})$ for some $\gamma \in \Gamma$, we get the inequalities

$$p^{a-b}\mathbf{M} \subset \gamma^{-1}\mathbf{M} \subset p^{b-a}\mathbf{M}.$$

Hence if γ runs through (2.3) the set of $W(L)$-lattices $\{\gamma^{-1}\mathbf{M}\}$ is finite. Since the set of elements of $J(\mathbf{Q}_p) \subset \operatorname{End}_{W(L)} \mathbf{M} \otimes \mathbf{Q}$ that fix the lattice $\mathbf{M}$ is compact, there are for a given $\gamma \in \Gamma$ only finitely many elements $\gamma' \in \Gamma$ such that

$$\gamma^{-1}\mathbf{M} = \gamma'^{-1}\mathbf{M}.$$

This proves the proposition. □

Corollary 2.35 *If the group Γ is torsionfree, it acts without fixed points on $\mathcal{M}$.*

Proof: If x is a fixed point of $\gamma \in \Gamma, \gamma \neq 1$, we choose a bounded open neighbourhood U of x. Then we have

$$U\gamma^n \cap U \neq \emptyset \qquad \text{for} \quad n \in \mathbf{Z}.$$

Hence the proposition implies that γ is a torsion element.

Lemma 2.36 *Let $\Gamma \subset J(\mathbf{Q}_p)$ be a discrete subgroup. There is a family of quasicompact open subsets $\{U_i\}_{i\in I}$ such that for any quasicompact open subset V of $\mathcal{M}$, there are only finitely many pairs (γ, i) such that*

$$V \cap U_i\gamma \neq \emptyset$$

and such that $\bigcup_{\Gamma\times I} U_i\gamma$ is a covering of $\mathcal{M}$.

Proof: We may assume that the ground field L is finite. Then there is a constant c such that any point of $\mathcal{M}$ has distance less than c to a point of $\mathcal{M}(L)$.
For any point $x \in \mathcal{M}(L)$ we define an open neighborhood

$$U_x = \mathcal{M} \setminus \bigcup_{\substack{y \in \mathcal{M}(L) \\ d(y,x) > 2c}} B_c(y),$$

where $B_c(y)$ denotes the closed ball around y of radius c. One checks that the open set U_c has the properties:

$$B_c(x) \subset U_x \subset B_{3c}(x), \qquad U_x\gamma = U_{x\gamma} \qquad \text{for} \quad \gamma \in J(\mathbf{Q}_p).$$

Then $\bigcup_{x \in \mathcal{M}(L)} U_x = \mathcal{M}$ and only finitely many U_x meet a given bounded subset of $\mathcal{M}$. We get the desired system $\{U_i\}_{i \in I}$, if we take for $I \subset \mathcal{M}(L)$ a set of representatives of $\mathcal{M}(L)/\Gamma$. □

Let us call the group Γ *separated,* if it is separated in the profinite topology. This means that for any $\gamma \in \Gamma$, $\gamma \neq 1$, there is a normal subgroup $\Gamma' \subset \Gamma$ of finite index that does not contain γ.

Proposition 2.37 *Let $\Gamma \subset J(\mathbf{Q}_p)$ be a torsionfree separated discrete subgroup. Then the quotient $\mathcal{M}/\Gamma$ as a sheaf for the étale topology is a formal algebraic space locally formally of finite type over Spf $W(L)$.*

Proof: We define open subfunctors of $\mathcal{M}/\Gamma$ which are formal algebraic spaces. Let $\{U_i\}_{i \in I}$ be as in the previous lemma. We choose for a fixed i a normal subgroup $\Gamma' \subset \Gamma$ such that $U_i \cap U_i\gamma' = \emptyset$ for $\gamma' \in \Gamma'$. For each $\gamma \in \Gamma$ we have an open immersion of locally ringed spaces

$$U_i\gamma \longrightarrow (\mathcal{M}/\Gamma')_{\text{ringed space}}.$$

Hence the image of the morphism

$$\coprod_{\gamma \in \Gamma} U_i\gamma \longrightarrow (\mathcal{M}/\Gamma')_{\text{ringed space}}$$

is a scheme $\tilde{V}_i$, where the finite group Γ/Γ' acts without fixed points.

Then $\tilde{V}_i \rightarrow \mathcal{M}/\Gamma'$ is an open immersion because it is the union of the open subfunctors $U_i\gamma$. The quotient of this open immersion by the group $G = \Gamma/\Gamma'$ is an open immersion of an algebraic space V_i:

$$V_i \longrightarrow \mathcal{M}/\Gamma.$$

Hence we have a cartesian diagram

$$\begin{array}{ccc} \tilde{V}_i & \longrightarrow & \mathcal{M}/\Gamma' \\ \downarrow & & \downarrow \\ V_i & \longrightarrow & \mathcal{M}/\Gamma \end{array}$$

Since the upper horizontal arrow is an open immersion so is the lower one, which proves that V_i is an algebraic space open in $\mathcal{M}/\Gamma$.
To prove the proposition we have to show that the map

$$V = \coprod_{i \in I} V_i \longrightarrow \mathcal{M}/\Gamma \tag{2.4}$$

is étale and surjective in the sense of sheaves. To see this we note that $V_i \times_{\mathcal{M}/\Gamma} \mathcal{M}$ is the open subset $\bigcup_{\gamma \in \Gamma} U_i\gamma \subset \mathcal{M}$. This implies surjectivity of (2.4) by the previous lemma. To show that (2.4) is étale it is enough to verify that it is of finite type, because we already know that $V_i \rightarrow \mathcal{M}/\Gamma$ is an open immersion. Since the map is locally of finite type it suffices to show that for any quasicompact scheme T the scheme $V \times_{\mathcal{M}/\Gamma} T$ is quasicompact. Again we may assume that there is a section $T \rightarrow \mathcal{M}$. Then there are only finitely many indices $i \in I$ such that $U_i\gamma$ meets the image of T for some $\gamma \in \Gamma$, by lemma (2.36). For the remaining indices $i \in I$ we have $V_i \times_{\mathcal{M}/\Gamma} T = \emptyset$. Since each V_i is quasicompact the proposition is proved.

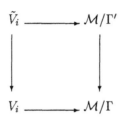

3. Moduli spaces of p–divisible groups

In this chapter we formulate a moduli problem of p–divisible groups *with a level structure of parahoric type* and show that it is representable by a formal scheme.

We fix a prime number p. Let B be a finite–dimensional semisimple algebra over $\mathbf{Q}_p$ and V be a finite left B-module. We fix a maximal order O_B of B. We are going to define the notion of a multichain of lattices in V. We consider first the case, where B is a simple algebra.

Definition 3.1 *A chain of lattices is a subset $\mathcal{L}$ of the set of O_B-lattices of V, that satisfies the following conditions:*

1. *If Λ and Λ' are two lattices of $\mathcal{L}$ then either*

$$\Lambda \subset \Lambda' \quad or \quad \Lambda' \subset \Lambda .$$

2. *Let $x \in B^{\times}$ be a unit of B, which normalizes O_B. Then $x\Lambda \in \mathcal{L}$, if $\Lambda \in \mathcal{L}$.*

3.2 We may make this definition a little more explicit. There is a division algebra D and an isomorphism $B \simeq M_n(D)$, that takes O_B to $M_n(O_D)$, where O_D is the unique maximal order of D. We consider D as a subalgebra of $M_n(D)$ via the diagonal embedding.
The normalizer of $M_n(O_D)$ in $M_n(D)^{\times}$ is $D^{\times} \cdot M_n(O_D)^{\times}$. Indeed, consider an $x \in M_n(D)$ of that normalizer. Then $xO_D^n \subset D^n$ is an $M_n(O_D)$-lattice. Hence it is enough to show that any $M_n(O_D)$-lattice in D^n has the form dO_D^n

for some $d \in D^\times$. This is well-known for $n = 1$. For general n it follows from the so called Morita equivalence (easy exercise), that asserts that the functor from the category of O_D-modules to the category of $M_n(O_D)$-modules given by

$$W \longmapsto O_D^n \otimes_{O_D} W$$

is an equivalence of categories.
If we fix a prime element Π of O_D, we may reformulate our second condition as follows:

2') *For any lattice $\Lambda \in \mathcal{L}$ the lattices $\Pi^{\pm 1}\Lambda$ belong to $\mathcal{L}$.*

Indeed any element b of the normalizer of O_B has the form $b = \Pi^k u$, where k is some integer and u is a unit in O_B. Hence $b\Lambda = \Pi^k \Lambda$.
We call b a *maximal element* of the normalizer, iff $k = 1$. It is equivalent to say that $b \in O_B$ and that bO_B is a maximal two-sided ideal in O_B.
Fix some lattice $\Lambda_0 \in \mathcal{L}$. By the property 2') it is enough to know the lattices between $\Pi^{-1}\Lambda_0$ and Λ_0 that lie in $\mathcal{L}$ to recover the whole chain. Hence to give a chain $\mathcal{L}$ of lattices is equivalent to giving a finite set $\{\Lambda_0, \ldots, \Lambda_{r-1}\}$ of $M_n(O_D)$-lattices in V, such that

$$\Lambda_0 \subsetneqq \Lambda_1 \subsetneqq \cdots \subsetneqq \Lambda_{r-1} \subsetneqq \Pi^{-1}\Lambda_0$$

The number r that appears here is called the *period of the chain.* We see that $\mathcal{L} = \{\Lambda_i\}_{i \in \mathbf{Z}}$, where the Λ_i for i not in the intervall $[0, r-1]$ are defined by the condition

$$\Lambda_{i-r} = \Pi\ \Lambda_i\ .$$

3.3 Next we consider the case of a semisimple algebra B. It is a product of simple algebras

$$B = B_1 \times \ldots \times B_m\ .$$

There are maximal orders O_{B_i} of $B_i, i = 1, \ldots, m$, such that

$$O_B = O_{B_1} \times \ldots \times O_{B_m}\ .$$

We get a corresponding decomposition of V,

$$V = V_1 \oplus \cdots \oplus V_m .$$

Moreover, each O_B-lattice $\Lambda \subset V$ may be written in a unique way:

$$\Lambda = \Lambda_1 \oplus \cdots \oplus \Lambda_m ,$$

where $\Lambda_i \subset V_i$ is an O_{B_i}-lattice. Let us call Λ_i the ith *projection of* Λ and denote it by $pr_i \Lambda$.

Definition 3.4 *A set $\mathcal{L}$ of O_B-lattices $\Lambda \subset V$, is called a multichain, iff there exists for each $i = 1, \ldots, m$ a chain of O_{B_i}-lattices $\mathcal{L}_i$ in V_i, such that $\mathcal{L}$ consists of the O_B-lattices for which $pr_i\ \Lambda \in \mathcal{L}_i$ for $i = 1, \ldots, m$.*

3.5 Let T be a $\mathbf{Z}_p$-scheme, such that p is locally nilpotent on T. We are going to define the notion of a multichain of $O_B \otimes_{\mathbf{Z}_p} \mathcal{O}_T$-modules of type $(\mathcal{L})$. A typical multichain on T will be $\{\Lambda \otimes_{\mathbf{Z}_p} \mathcal{O}_T\}$ where $\Lambda \in \mathcal{L}$.
Let us fix a notation. Assume that $b \in B^\times$ is in the normalizer of O_B. Then conjugation by b^{-1} defines an isomorphism

$$O_B \longrightarrow O_B \qquad x \longmapsto b^{-1} x b$$

Let M be a $O_B \otimes \mathcal{O}_T$-module. We denote by M^b the module obtained via restriction of scalars with respect to this isomorphism. Then multiplication by b induces a homomorphism

$$b : M^b \longrightarrow M .$$

Let us begin with the case, where B is simple. We consider the chain $\mathcal{L}$ as a category with inclusions as morphisms.

Definition 3.6 *A chain of $O_B \otimes_{\mathbf{Z}_p} \mathcal{O}_T$-modules of type $(\mathcal{L})$ on T is a functor*

$$\Lambda \longmapsto M_\Lambda$$

from the category $\mathcal{L}$ to that of $O_B \otimes \mathcal{O}_T$-modules. Moreover, for each $b \in B^\times$ in the normalizer of O_B, a periodicity isomorphism

$$\theta_b : M_\Lambda^b \xrightarrow{\sim} M_{b\Lambda}$$

is given.
We require that the following conditions are satisfied:

1. *Locally on T there exist isomorphisms of $O_B \otimes \mathcal{O}_T$-modules*

$$M_\Lambda \simeq \Lambda \otimes_{\mathbf{Z}_p} \mathcal{O}_T \ .$$

2. *If b is maximal (cf.(3.2)) and $\Lambda, \Lambda' \in \mathcal{L}$ are such that $b\Lambda \subset \Lambda' \subset \Lambda$, we have an isomorphism of $O_B/bO_B \otimes \mathcal{O}_T$-modules locally on T*

$$M_\Lambda / \varrho_{\Lambda,\Lambda'}(M_{\Lambda'}) \simeq \Lambda/\Lambda' \otimes \mathcal{O}_T \ .$$

Here $\varrho_{\Lambda,\Lambda'} : M_{\Lambda'} \to M_\Lambda$ denotes the homomorphism that corresponds by functoriality to the inclusion $\Lambda' \subset \Lambda$.

3. *The periodicity isomorphisms are functorial, i.e. for any inclusion $\Lambda' \subset \Lambda$ the following diagram is commutative:*

$$\begin{array}{ccc} M_{\Lambda'}^b & \xrightarrow{\ \theta_b\ } & M_{b\Lambda'} \\ {\scriptstyle \varrho_{\Lambda,\Lambda'}} \downarrow & & \downarrow {\scriptstyle \varrho_{b\Lambda,b\Lambda'}} \\ M_\Lambda^b & \xrightarrow{\ \theta_b\ } & M_{b\Lambda} \ . \end{array}$$

The θ_b satisfy the cocycle condition:

$$\begin{array}{ccc} M_\Lambda^{b_1 b_2} & \xrightarrow{\ \theta_{b_1 b_2}\ } & M_{b_1 b_2 \Lambda} \\ \uparrow & & \uparrow {\scriptstyle \theta_{b_1}} \\ (M_\Lambda^{b_2})^{b_1} & \xrightarrow{\ \theta_{b_2}^{b_1}\ } & M_{b_2\Lambda}^{b_1} \ . \end{array}$$

4. *For each $b \in B^\times$, which is in the normalizer of O_B the composition*

$$M_\Lambda^b \xrightarrow{\theta_b} M_{b\Lambda} \xrightarrow{\varrho_{\Lambda,b\Lambda}} M_\Lambda$$

is multiplication by b.

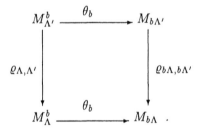

Let us reformulate this definition more explicitly. As above we represent B as a matrix algebra over a division algebra D in such a way that $O_B = M_n(O_D)$. We fix a prime element $\Pi \in O_D$. Then we may represent $\mathcal{L}$ as a chain of O_B-lattices in V

$$\cdots \subset \Lambda_i \subset \Lambda_{i+1} \subset \cdots \qquad , i \in \mathbf{Z},$$

such that $\Lambda_{i-r} = \Pi\Lambda_i$ for some fixed natural number r and any $i \in \mathbf{Z}$. We may reformulate our definition as follows:

Corollary 3.7 *A chain of $O_B \otimes \mathcal{O}_T$-modules of type $(\mathcal{L})$ on T is an indexed set of $O_B \otimes \mathcal{O}_T$-modules $\{M_i\}_{i \in \mathbf{Z}}$, such that*

$$M_{i-r} = M_i^{\Pi}, \qquad i \in \mathbf{Z}.$$

Moreover there is a $O_B \otimes \mathcal{O}_T$-homomorphism of degree one

$$\varrho : M_i \longrightarrow M_{i+1}$$

such that the following conditions are satisfied.

1. *We have isomorphisms of $O_B \otimes \mathcal{O}_T$-modules locally on T:*

$$M_i \simeq \Lambda_i \otimes \mathcal{O}_T, \qquad M_i/\varrho(M_{i-1}) \simeq \Lambda_i/\Lambda_{i-1} \otimes \mathcal{O}_T.$$

2. *The map*

$$\varrho^r : M_{i-r} = M_i^{\Pi} \longrightarrow M_i$$

is the multiplication by Π.

We note that the condition 1) does not claim any functoriality in Λ_i. It just says that M_i is locally on T a free $O_D \otimes \mathcal{O}_T$-module of the same rank as the O_D-module Λ_i (i.e. $\dim_D V$) and $M_i/\varrho(M_{i-1})$ is a free $O_D/\Pi O_D \otimes \mathcal{O}_T$–module of the same rank as the $O_D/\Pi O_D$-vector space Λ_i/Λ_{i-1}.

3.8 We will also consider chains $\{M_i\}_{i \in \mathbf{Z}}$ where we replace the condition 1) by the weaker conditions, that M_i is locally on T a free $O_D \otimes \mathcal{O}_T$-module, and that $M_i/\varrho(M_{i-1})$ is locally on T a free $O_D/\Pi O_D \otimes \mathcal{O}_T$-module. Then we speak just of a *chain of $O_D \otimes \mathcal{O}_T$-modules on T* without fixing a type. The type $(\mathcal{L})$ enters in the definition 3.6 only via the ranks of the modules above.

We note that the Morita equivalence induces a bijection between chains of $O_B \otimes \mathcal{O}_T$-modules and chains of $O_D \otimes \mathcal{O}_T$-modules.

3.9 Let us return to the general case, where B need not to be simple. We consider the decomposition into simple algebras. Let $\mathcal{L}$ be a multichain of O_B-lattices in V. We denote by $\mathcal{L}_i$ the chain of O_{B_i}-lattices in V_i, which is the projection of $\mathcal{L}$,

$$\mathcal{L}_i = \{pr_i\Lambda \mid \Lambda \in \mathcal{L}\}.$$

Definition 3.10 *A multichain of $O_B \otimes_{\mathbf{Z}_p} \mathcal{O}_T$-modules on T of type $(\mathcal{L})$ is a set $\{\mathcal{M}_1, \ldots, \mathcal{M}_m\}$, where $\mathcal{M}_i$ is a chain of $O_{B_i} \otimes_{\mathbf{Z}_p} \mathcal{O}_T$-modules of type $(\mathcal{L}_i)$.*

If $\Lambda \in \mathcal{L}$ has the decomposition

$$\Lambda = \Lambda_1 \oplus \cdots \oplus \Lambda_m\ , \quad \Lambda_i \in \mathcal{L}_i$$

we write $M_\Lambda = \bigoplus_{i=1}^{m} M_{\Lambda_i}$, where M_{Λ_i} is defined by the chain $\mathcal{M}_i$. Again $\Lambda \mapsto M_\Lambda$ is a functor from $\mathcal{L}$ to the category of $O_B \otimes_{\mathbf{Z}_p} \mathcal{O}_T$-modules. Moreover for any $b \in B^\times$ that normalizes O_B we have a periodicity isomorphism

$$\theta_b : M_\Lambda^b \longrightarrow M_{b\Lambda},$$

such that the following diagram is commutative for $b\Lambda \subset \Lambda$:

$$\begin{array}{ccc} M_\Lambda^b & \longrightarrow & M_{b\Lambda} \\ & \searrow^{\,b} & \downarrow{\scriptstyle \varrho_{\Lambda,b\Lambda}} \\ & & M_\Lambda \end{array}$$

Theorem 3.11 *Let $\{M_\Lambda\}$ be a multichain of $O_B \otimes \mathcal{O}_T$-modules of type $(\mathcal{L})$ on a $\mathbf{Z}_p$-scheme T, where p is locally nilpotent. Then locally for the Zariski topology on T the multichain $\{M_\Lambda\}$ is isomorphic to $\{\Lambda \otimes \mathcal{O}_T\}_{\Lambda \in \mathcal{L}}$.*

If $\{M'_\Lambda\}$ is a second multichain of type $(\mathcal{L})$ on T, then the following functor on the category of T-schemes

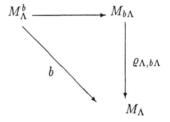

$$T' \longmapsto \text{Isom}\ (\{M_\Lambda \otimes_{O_T} O_{T'}\},\ \{M'_\Lambda \otimes_{O_T} O_{T'}\})$$

is representable by a smooth affine scheme over T.

We remark that by the Morita equivalence it is enough to prove this theorem in the case, where $B = D$ is a division algebra. We refer to the appendix to this chapter for the proof and first formulate a similar theorem in the presence of a polarization. *Whenever we consider the polarized case we will make the blanket assumption that $p \neq 2$.*

3.12 We fix data $(F, B, V, (\ ,\ ))$ of type (PEL), cf. (1.36). Let $*$ denote the involution on B. We let O_B be a maximal order of B invariant under $*$. If W is a right B-module, we define a left B-module by restriction of scalars

$$* : B \longrightarrow B^{opp}\ .$$

With this convention the dual vector space $V^* = \text{Hom}_{\mathbf{Q}_p}(V, \mathbf{Q}_p)$ is a left B-module and $(\ ,\ )$ induces an isomorphism of B-modules

$$\psi : V \longrightarrow V^*.$$

In the same way for an O_B-lattice Λ in V, $\Lambda^* = \text{Hom}_{\mathbf{Z}_p}(\Lambda, \mathbf{Z}_p)$ is a left O_B-module. The image of Λ^* by the map

$$\Lambda^* \longrightarrow V^* \xrightarrow[\sim]{\psi^{-1}} V$$

is the dual lattice with respect to $(\ ,\ )$. We will denote it by Λ^* as well.

Definition 3.13 *A multichain $\mathcal{L}$ of lattices in V is called selfdual, if $\Lambda \in \mathcal{L}$ implies $\Lambda^* \in \mathcal{L}$.*

Definition 3.14 *Let $\mathcal{L}$ be a selfdual multichain of lattices in V. A polarized multichain of $O_B \otimes_{\mathbf{Z}_p} O_T$-modules on the scheme T of type $(\mathcal{L})$ is a multichain of $O_B \otimes_{\mathbf{Z}_p} O_T$-modules $\{M_\Lambda\}$ of type $(\mathcal{L})$ together with perfect O_T-linear pairings*

$$\mathcal{E}_\Lambda : M_\Lambda \times M_{\Lambda^*} \longrightarrow O_T$$

such that the following conditions are satisfied

1. $\mathcal{E}_\Lambda(am, m') = \mathcal{E}_\Lambda(m, a^*m'),\ m \in M_\Lambda, m' \in M_{\Lambda^*},\ a \in O_B$

2. $\mathcal{E}_\Lambda(m, m') = -\mathcal{E}_{\Lambda^*}(m', m),\ m \in M_\Lambda, m' \in M_{\Lambda^*}$.

3. *Let $\Lambda_1 \subset \Lambda_2$ be lattices in $\mathcal{L}$. Then*

$$\mathcal{E}_{\Lambda_1}(m, \varrho_{\Lambda_1^*,\Lambda_2^*} n) = \mathcal{E}_{\Lambda_2}(\varrho_{\Lambda_2,\Lambda_1} m, n),\ m \in M_{\Lambda_1}, n \in M_{\Lambda_2^*}.$$

4. *Let $b \in B^\times$ be in the normalizer of O_B. We set $\tilde{b} = (b^{-1})^*$ so that for a lattice Λ we have the relation $(b\Lambda)^* = \tilde{b}\Lambda^*$. We consider for a lattice $\Lambda \in \mathcal{L}$ the periodicity isomorphisms*

$$\theta_b : M_\Lambda^b \xrightarrow{\sim} M_{b\Lambda} \quad \theta_{\tilde{b}} : M_{\Lambda^*}^{\tilde{b}} \longrightarrow M_{\tilde{b}\Lambda^*} = M_{(b\Lambda)^*}.$$

Then we have the relation

$$\mathcal{E}_\Lambda(m_1, m_2) = \mathcal{E}_{b\Lambda}(\theta_b m_1, \theta_{\tilde{b}} m_2),\ m_1 \in M_\Lambda,\ m_2 \in M_{\Lambda^*}.$$

3.15 On the selfdual chain $\mathcal{L}$ we may consider the functor $\Lambda \longmapsto \hat{M}_\Lambda = M_{\Lambda^*}^* = \operatorname{Hom}_{\mathbf{Z}_p}(M_{\Lambda^*}, \mathcal{O}_T)$. We have a periodicity map on this functor defined by the diagram

$$\left(\hat{M}_\Lambda\right)^b = \left(M_{\Lambda^*}^{\tilde{b}}\right)^* \xleftarrow{\sim} (M_{\tilde{b}\Lambda^*})^* = M_{(b\Lambda)^*}^* = \hat{M}_{b\Lambda}.$$

One verifies with little pain that $\{\hat{M}_\Lambda\}$ is a multichain of $O_B \otimes \mathcal{O}_T$-modules of type $(\mathcal{L})$. Let us call $\hat{M}_\Lambda$ the *dual chain*. We may restate the definition of a polarized multichain in this set-up more elegantly:

A polarized multichain over the scheme T of type $(\mathcal{L})$ is a multichain $\{M_\Lambda\}$ of type $(\mathcal{L})$ together with an antisymmetric isomorphism of multichains

$$\{M_\Lambda\} \longrightarrow \{\hat{M}_\Lambda\}.$$

The analogue of theorem (3.11) in the polarized case is the following theorem. For the proof we again refer to the appendix to this chapter.

Theorem 3.16 *Let $\mathcal{L}$ be a selfdual multichain of O_B-lattices in V. Let T be a $\mathbf{Z}_p$-scheme, where p is locally nilpotent. Let $\{M_\Lambda\}$ be a polarized multichain of $O_B \otimes_{\mathbf{Z}_p} \mathcal{O}_T$-modules of type $(\mathcal{L})$. Then locally for the étale* *topology on T the polarized multichain $\{M_\Lambda\}$ is isomorphic to the polarized multichain $\{\Lambda \otimes \mathcal{O}_T\}$.*

Moreover, if $\{M'_\Lambda\}$ is a second polarized multichain of type $(\mathcal{L})$ on T then the functor of isomorphisms of polarized multichains on the category of T-schemes

$$T' \longmapsto \text{Isom}\left(\{M_\Lambda \otimes \mathcal{O}_{T'}\}, \{M_{\Lambda'} \otimes \mathcal{O}_{T'}\}\right)$$

is representable by a smooth affine scheme over T.

3.17 We will now define moduli problems of p-divisible groups, that are variants of the problem in chapter 2. Our starting point is one of the following two situations:

Case (EL):

> We fix (F, B, V) as in (1.38), and a maximal order O_B in B. Let G be the corresponding algebraic group over $\mathbf{Q}_p$.

Case (PEL):

> We fix $(F, B, V, (\ ,\ ))$ as in (1.38), and a maximal order O_B in B fixed by the involution $*$. Let G be the corresponding algebraic group over $\mathbf{Q}_p$.

To define the variants of our functor $\mathcal{M}$ we need a replacement for $\mathbf{X}$. In terms of the group G this is given by an admissible pair, cf. (1.18).

Definition 3.18 *A set of data for moduli of p-divisible groups in the case (EL) relative to an algebraically closed field L of characteristic p is a tuple:*

$$(F, B, O_B, V, b, \mu, \mathcal{L}).$$

Here (F, B, O_B, V) are the data of case (EL). We denote by G the associated reductive algebraic group over $\mathbf{Q}_p$. Let us denote by K_0 the quotient field of $W(L)$. The datum b is an element of $G(K_0)$. The next datum is a cocharacter

$$\mu : \mathbf{G}_m \longrightarrow G$$

that is defined over a finite field extension K of K_0. Finally $\mathcal{L}$ is a multichain of O_B-lattices in V.
We require, that the following conditions are fullfilled.

(i) *The pair (b, μ) is admissible, cf.(1.18).*

(ii) *The isocrystal $(V \otimes K_0, b\sigma)$ has slopes in the interval $[0, 1]$.*

(iii) *The weight decomposition of $V \otimes K$ with respect to the cocharacter μ contains only the weights 0 and 1:*

$$V \otimes K = V_0 \oplus V_1.$$

In the case (PEL) we have in addition to the data above the nondegenerate antisymmetric pairing $(\, , \,)$ on V that induces an involution $*$ on O_B. We require that the multichain $\mathcal{L}$ is selfdual. The multiplier of the corresponding group G is denoted by c. Let us denote by ν the slope morphism associated to b. In addition to the conditions above we require:

(iv) *The character $c\nu : \mathbf{D} \to \mathbf{G}_m$ is the character χ_1 that corresponds to the rational number 1.*

Let us fix the set of data (EL) respectively (PEL). We consider two sets of data $(b, \mu, \mathcal{L})$ and $(b', \mu', \mathcal{L}')$ to be equivalent, iff b and b' are in the same σ-conjugacy class, μ and μ' are conjugate over a suitable finite extension K'' of K_0, and there exists a bijection $\Lambda \mapsto \Lambda'$ between the chains $\mathcal{L}$ and $\mathcal{L}'$, such that for any pair Λ_1 and Λ_2

$$length_{O_F} \Lambda_1/\Lambda_2 = length_{O_F} \Lambda'_1/\Lambda'_2.$$

Moreover in the case (PEL) we require that the bijection $\Lambda \mapsto \Lambda'$ commutes with taking the dual lattice. Note that we do not require the equivalence of the pairs (b, μ) and $(b', /mu')$ in the sense of definition (1.23).

3.19 Let us make a few comments on this definition.
a) By the crystal associated to a p-divisible group X over L we mean the Lie algebra of the universal extension of some lifting of X to $W(L)$. It is canonically isomorphic to the Cartier module of X.
The condition (ii) above says that $(N, \mathbf{F}) = (V \otimes K_0, b\sigma)$ is the isocrystal of some p-divisible group $\mathbf{X}$ over L.

The conditions (i) – (iii) are satisfied if there is a p-divisible group X over the ring of integers O_K of K, such that its reduction X_L modulo the maximal ideal is equipped with a quasi-isogeny $\mathbf{X} \rightarrow X_L$, and such that the following condition is satisfied.

In general let X be a p-divisible group over a base scheme S, where p is locally nilpotent. Then we denote by $M(X)$ the Lie algebra of the universal extension of X.

In our case where $S = Spec\ O_K$ the definition of $M(X)$ makes sense because O_K is a p-adic ring. The given quasi-isogeny allows us to identify $M(X)\otimes\mathbf{Q}$ with the K-vector space $N \otimes_{K_0} K$. Indeed we have a quasi-isogeny

$$\mathbf{X} \times_{Spec\ L} Spec\ O_K/p \longrightarrow X \times_{Spf\ O_K} Spec\ O_K/p$$

that lifts the quasi-isogeny $\mathbf{X} \rightarrow X_L$. This induces a quasi-isogeny between the values of the crystals associated to the p-divisible groups at the divided power thickening $Spec\ O_K/p \rightarrow Spf\ O_K$. We get the desired identification (comp. also (5.15)).

The condition is that under this identification the canonical filtration on the universal extension

$$0 \rightarrow Fil^1 \rightarrow M(X) \otimes \mathbf{Q} \rightarrow LieX \otimes \mathbf{Q} \rightarrow 0$$

coincides with the filtration given by μ

$$0 \rightarrow V_1 \rightarrow V \otimes K \rightarrow V/V_1 \rightarrow 0.$$

Conversely one expects that the existence of an X with the properties above follows from the conditions (i) – (iii). If this is false it could happen that the moduli functors we are going to define are empty for some of the data of definition (3.18).

b) The condition that (b, μ) is admissible implies that for each character χ of G that is defined over $\mathbf{Q}_p$, we have

$$< \mu, \chi >= \operatorname{ord}_p \chi(b),$$

(cf.(1.21)). If we take for χ the determinant of an element $g \in G(\mathbf{Q}_p)$ acting on the $\mathbf{Q}_p$–vector space V

$$\det_{\mathbf{Q}_p} : G \longrightarrow \mathbf{G}_m,$$

we get the equality:

$$\dim V_1 = \mathrm{ord}_p \det_{K_0}(b; V \otimes K_0).$$

If we take in the case (PEL) for χ the multiplier c, we get from the condition (iv) that

$$< \mu, c >= 1.$$

This implies that the subspaces V_0 and V_1 are isotropic with respect to the pairing obtained on $V \otimes K$ by extension of scalars.
We also note that conditions (i) and (iii) imply condition (ii).

3.20 We recall (cf. (1.38)) that the isocrystal N is equipped with an action of B. In the case (PEL) it is also equipped with an alternating bilinear form of isocrystals,

$$\psi : N \otimes N \longrightarrow \mathbf{1}(1).$$

Indeed since L is algebraically closed and since $\mathrm{ord}_p\, c(b) = 1$, we find $u \in W(L)^\times$ such that $c(b) = p\, u^{-1}\sigma(u)$ and put $\psi(v, v') = u\,(v, v')$. If we choose another u, we change ψ by a factor from $\mathbf{Q}_p^\times$. We call the set $\mathbf{Q}_p^\times \psi$ of bilinear forms on N a *$\mathbf{Q}_p$-homogeneous formal polarization.*
The form ψ defines a polarization on the p-divisible group $\mathbf{X}$, i.e. an anti-symmetric quasi-isogeny $\lambda : \mathbf{X} \rightarrow \hat{\mathbf{X}}$. The isogeny class of the pair $(\mathbf{X}, \mathbf{Q}_p^\times \lambda)$ is well defined by the data of definition (3.18).
Let E denote the Shimura field, i.e. the field of definition of the conjugacy class of μ, cf. (1.31). We denote by $\breve{E}$ the complete unramified extension of E with residue class field L, which is contained in K. We define a functor on the category $Nilp_{O_{\breve{E}}}$ that is associated to the data of definition (3.18). For a scheme S in $Nilp_{O_{\breve{E}}}$ we will denote by $\bar{S}$ the closed subscheme of S defined by the sheaf of ideals $p\mathcal{O}_S$. The structure morphism $\phi : S \rightarrow Spec\ O_{\breve{E}}$ induces a morphism

$$\bar{\phi} : \bar{S} \rightarrow Spec\, O_{\breve{E}}/pO_{\breve{E}} \rightarrow Spec\ L,$$

which allows us to consider $\bar{S}$ as a scheme over L.
Let X be a p-divisible group over S with an action

$$\iota : O_B \longrightarrow \mathrm{End}\, X.$$

For an element $a \in B^{\times}$ that normalizes O_B, we define $\iota^a(x) = \iota(a^{-1}xa)$. By abuse of notation we write X^a for the pair (X, ι^a). The multiplication by $\iota(a)$ induces a morphism of O_B-modules:

$$X^a \longrightarrow X.$$

Definition 3.21 *Let $\breve{\mathcal{M}}$ be the contravariant set-valued functor on the category $Nilp_{O_{\breve{E}}}$, such that a point with values in $S \in Nilp_{O_{\breve{E}}}$ is given by the following data up to isomorphism.*

(1) For each lattice $\Lambda \in \mathcal{L}$ a p-divisible group X_Λ over S, with an action of the algebra O_B:

$$O_B \longrightarrow End\, X_\Lambda$$

(2) For each lattice $\Lambda \in \mathcal{L}$ a quasi–isogeny

$$\varrho_\Lambda : \mathbf{X} \times_{Spec\, L} \bar{S} \longrightarrow X_\Lambda \times_S \bar{S},$$

which commutes with the action of O_B.

We require that the following conditions are satisfied:

Let us denote by M_Λ the Lie algebra of the universal extension of X_Λ. It is a locally free $\mathcal{O}_S$-module. We will write

$$\tilde{\varrho}_{\Lambda',\Lambda} : X_\Lambda \longrightarrow X_{\Lambda'}$$

for the quasi–isogeny that lifts $\varrho_{\Lambda'}\varrho_\Lambda^{-1}$.

(i) Locally on S the $O_B \otimes \mathcal{O}_S$-module M_Λ is isomorphic to $\Lambda \otimes \mathcal{O}_S$.

(ii) Let $\Lambda \subset \Lambda'$ be two neighbours in the multichain $\mathcal{L}$. Then the quasi–isogeny $\tilde{\varrho}_{\Lambda',\Lambda}$ is an isogeny. The cokernel of the induced map $M_\Lambda \rightarrow M_{\Lambda'}$ is locally on S isomorphic to $\Lambda'/\Lambda \otimes_{\mathbf{Z}_p} \mathcal{O}_S$ as an $O_B \otimes_{\mathbf{Z}_p} \mathcal{O}_S$-module.

(iii) For any $a \in B^{\times}$ that normalizes O_B the map $\mathbf{X}^a \rightarrow \mathbf{X}$ defined above induces an isomorphism

$$\theta_a : X_\Lambda^a \longrightarrow X_{a\Lambda}.$$

(iv) For each $\Lambda \in \mathcal{L}$, we have an equality of polynomial functions on O_B:

$$\det_{\mathcal{O}_S}(a; Lie\ X_\Lambda) = \det_K(a; V_0), \quad a \in O_B.$$

(v) In the case (PEL) there exists an isomorphism p_Λ for each lattice $\Lambda \in \mathcal{L}$

$$p_\Lambda : X_\Lambda \longrightarrow X^{\wedge}_{\Lambda^{\wedge}},$$

such that the following diagrams are commutative up to a constant in $\mathbf{Q}_p^\times$ which is independent of Λ,

$$\begin{array}{ccc} (X_\Lambda)_{\bar{S}} & \longrightarrow & (X^{\wedge}_{\Lambda^{\wedge}})_{\bar{S}} \\ \varrho_\Lambda \big\uparrow & & \big\downarrow \varrho^{\wedge}_{\Lambda^{\wedge}} \\ \mathbf{X}_{\bar{S}} & \underset{\bar{\lambda}}{\longrightarrow} & \mathbf{X}^{\wedge}_{\bar{S}}. \end{array}$$

We often write $(X_{\mathcal{L}}, \varrho)$ for a point of the moduli problem $\breve{\mathcal{M}}$. We remark that the functor $\breve{\mathcal{M}}$ does not depend on the choice of the p-divisible group $\mathbf{X}$ but only on the polarized isocrystal $(N, \psi\mathbf{Q}_p^\times)$. Indeed, for another choice $\mathbf{X}'$ we have a canonical quasi–isogeny $\alpha : \mathbf{X}' \to \mathbf{X}$ that respects the homogeneous polarization. The map $(X, \varrho) \mapsto (X, \varrho\alpha_{\bar{S}})$ gives the canonical isomorphism between the functors defined by $\mathbf{X}$ respectively $\mathbf{X}'$.
The automorphism group $J(\mathbf{Q}_p)$ of the homogeneously polarized isocrystal $(N, \psi\mathbf{Q}_p^\times)$ acts by quasi–isogenies on $\mathbf{X}$.

Definition 3.22 *Let $g \in J(\mathbf{Q}_p)$ and let $(X_{\mathcal{L}}, \varrho)$ be a point of $\breve{\mathcal{M}}(S)$. Let $g(X_{\mathcal{L}}, \varrho) \in \breve{\mathcal{M}}(S)$ be the point $(X_{\mathcal{L}}, \varrho g^{-1})$. This is an action from the left of $J(\mathbf{Q}_p)$ on the functor $\breve{\mathcal{M}}$.*

This action is independent of the choice of $\mathbf{X}$.

3.23 Let us explain the data and conditions of this definition.
a) The condition (iv) on the determinants is taken from Kottwitz [Ko3]. The precise formulation is as follows. Let $\mathbf{V}$ be the scheme over $\mathbf{Z}_p$, whose set of points $\mathbf{V}(R)$ with values in a $\mathbf{Z}_p$-algebra R is $O_B \otimes_{\mathbf{Z}_p} R$. We choose a O_B-invariant O_K-lattice $\Gamma \subset V_0$. For an O_K-algebra R we define a map:

$$\begin{array}{ccc} \mathbf{V}(R) & \longrightarrow & \mathbf{A}^1(R) \\ b \in O_B \otimes R & \mapsto & \det(b; \Gamma \otimes_{O_K} R) \end{array}$$

This defines a map of schemes $\mathbf{V}_{O_K} \to \mathbf{A}^1_{O_K}$, which is easily seen to be defined over O_E, where E is the Shimura field. Since it is determined by its restriction to the general fibre, it does not depend on the choice of Γ. In the definition we denote by $\det(b; V_0)$ this morphism. In the same way we view $\det(b; \mathit{Lie}\, X_\Lambda)$ as a morphism $\mathbf{V}_S \to \mathbf{A}^1_S$. What we mean by condition (iv) is an equality of morphisms of schemes over S. We cannot interpret condition (iv) in the naive way, because in general different polynomial functions may have the same value on the finite set $O_B/p^n O_B$.

b) Let us point out a consequence of the condition (iv). We write B as a product of simple algebras,

$$B = \prod_{i=1} B_i.$$

The algebras B_i are matrix algebras over divison algebras D_i,

$$B_i \cong M_{n_i}(D_i).$$

We may choose the isomorphism in such a way that

$$O_B \cong \prod_{i=1} M_{n_i}(O_{D_i}).$$

This induces decompositions of the lattices in our chain $\mathcal{L}$ and of the corresponding p-divisible groups,

$$\Lambda = \bigoplus \Lambda_i$$

$$X_\Lambda = \prod X_{\Lambda_i}.$$

The condition (iv) is a condition on each X_{Λ_i} separately. Therefore we may restrict to the case $i = 1$ in the discussion of that condition. Hence we assume that $O_B = M_n(O_D)$ is a matrix algebra over the ring of integers in a division algebra D with center F. We denote by $\tilde{F}$ an unramified extension of F that is contained in D and splits D. Let F^t respectively $\tilde{F}^t$ the maximal unramified extensions of $\mathbf{Q}_p$ contained in F respectively $\tilde{F}$. Assume that $\tilde{F}^t$ embeds into K. Then we get a decomposition:

$$V_0 = \bigoplus_{\phi:\tilde{F}^t \to K} V_0^{\phi}.$$

Here ϕ runs through all possible embeddings and

$$V_0^{\phi} = \{v \in V_0; fv = \phi(f)v, \quad f \in \tilde{F}^t\}.$$

Let Π be a prime element of D that normalizes $\tilde{F}$. Let us denote by τ the automorphism of $\tilde{F}^t$ induced by conjugation with Π. Then Π induces an isomorphism:

$$\Pi : V_0^{\phi\tau} \longrightarrow V_0^{\phi}.$$

It follows that the rank of the K-vectorspaces V_0^{ϕ} and $V_0^{\phi'}$ agree, if the restrictions of ϕ and ϕ' to F^t are the same. The restriction of the polynomial function $\det_K(a; V_0)$ to $O_{\tilde{F}^t} \otimes O_K$ is uniquely determined by the ranks of the K-vectorspaces V_0^{ϕ}.
Assume that we are given a point of the functor $\breve{\mathcal{M}}$ with values in S. We have a decomposition,

$$O_{\tilde{F}^t} \otimes \mathcal{O}_S = \prod_{\phi:\tilde{F}^t \to K} \mathcal{O}_S.$$

Note that $\mathcal{O}_S$ is a $O_{\breve{E}}$-algebra and that ϕ maps $O_{\tilde{F}^t}$ to the subring $O_{\breve{E}}$ of K. Since the Lie algebra of X_Λ is a $O_{\tilde{F}^t} \otimes \mathcal{O}_S$-module, we get a decomposition

$$Lie\ X_\Lambda = \bigoplus_{\phi} Lie^{\phi} X_\Lambda.$$

The restriction of the condition (iv) to the subalgebra $O_{\tilde{F}^t}$ of O_B says exactly that the rank of the locally free $\mathcal{O}_S$-module $Lie^{\phi} X_\Lambda$ coincides with the rank of the K-vectorspace V_0^{ϕ}. This condition on the ranks is weaker than condition (iv), if F is not an unramified extension of $\mathbf{Q}_p$.

c) Let us show that the conditon (iv) on the Lie algebra implies the condition (i) of the definition (3.21). Let M_{Λ_i} be the Lie algebra of the universal extension of X_{Λ_i}. Then the condition (i) of definition (3.21) is equivalent to the condition that for each $i = 1, \ldots, m$ the $O_{D_i} \otimes \mathcal{O}_S$-module M_{Λ_i} is locally on S free of the same rank as the O_{D_i}-module Λ_i. We note that the condition on the rank is automatic by the existence of the quasi–isogeny ϱ_Λ of definition (3.21) .

Lemma 3.24 *Let D be a finite–dimensional division algebra over $\mathbf{Q}_p$. Let F be the center of D. We denote by O_D the maximal order in D. We denote by k respectively $\tilde{k}$ the residue class fields of F respectively D. We note that k and $\tilde{k}$ are in a canonical way subalgebras of O_D/pO_D.*
Let X be a p-divisible group over an algebraically closed field P of characteristic p, with a faithful action of O_D. Then $Lie\, X$ is a $\tilde{k}\otimes P$-module. We have a decomposition

$$k \otimes P = \prod_{\alpha: k \rightarrow P} P$$

which induces a decomposition of the Liealgebra:

$$Lie\, X = \oplus_{\alpha} Lie_{\alpha}\, X.$$

The following conditions are equivalent:

(i) For each α the $\tilde{k}\otimes_{k,\alpha} P$-module $Lie_{\alpha}\, X$ is free.

(ii) The Cartier module M of X is a free $O_D \otimes W(P)$-module.

Proof: Let F be the center of D. Let d^2 be the degree of D over F. We fix an unramified extension $\tilde{F}$ of F of degree d that is contained in D. We write τ for the Frobenius of $\tilde{F}$ over F. We denote by F^t respectively $\tilde{F}^t$ the maximal unramified extensions of $\mathbf{Q}_p$, which are contained in F respectively $\tilde{F}$. We fix a prime element Π of D that normalizes $\tilde{F}$. Then O_D is the algebra,

$$O_D = O_{\tilde{F}}[\Pi], \quad \Pi a = \tau^s(a)\Pi, \quad a \in \tilde{F}.$$

Here s is some natural number prime to d. It follows that Π^d is a prime element π of F.
We have a decomposition:

$$O_{\tilde{F}^t} \otimes W(P) = \bigoplus_{\phi: \tilde{F}^t \rightarrow W(P)_{\mathbf{Q}}} W(P).$$

The sum ranges over all possible embeddings ϕ. This decomposition induces decompositions of the Lie algebra and the Cartier module of X,

$$Lie\ X = \oplus_{\phi} Lie^{\phi} X$$

$$M = \oplus_{\phi} M^{\phi}.$$

Explicitly the $W(P)$-submodule M^{ϕ} is given by the condition:

$$M^{\phi} = \{m \in M; (f \otimes 1)m = (1 \otimes \phi(f))m\}.$$

One easily checks that the action of Π on X induces on the Cartier module a map

$$\Pi : M^{\phi} \longrightarrow M^{\phi\tau^{-s}}.$$

An element $m \in M^{\phi}$, which is not in $\Pi M^{\phi\tau^{s}}$ generates a direct summand of M as an $O_D \otimes W(P)$-module, which is isomorphic to $O_D \otimes_{\tilde{F}^t,\phi} W(P)$. Hence the condition (ii) of the lemma says that the dimension as a P-vector space of the cokernel of the above map is independent of ϕ.
The Verschiebung $\mathbf{V}$ induces a map

$$\mathbf{V} : M^{\phi} \longrightarrow M^{\sigma^{-1}\phi},$$

which becomes an isomorphism when tensored with $\mathbf{Q}_p$. We consider the commutative diagram:

$$\begin{array}{ccccccccc}
0 & \longrightarrow & M^{\sigma\phi\tau^{s}} & \xrightarrow{\ \Pi\ } & M^{\sigma\phi} & \longrightarrow & C^{\sigma\phi} & \longrightarrow & 0 \\
 & & \downarrow \mathbf{V} & & \downarrow \mathbf{V} & & \downarrow & & \\
0 & \longrightarrow & M^{\phi\tau^{s}} & \xrightarrow{\ \Pi\ } & M^{\phi} & \longrightarrow & C^{\phi} & \longrightarrow & 0 \\
 & & \downarrow & & \downarrow & & & & \\
 & & Lie^{\phi\tau^{s}} & \longrightarrow & Lie^{\phi} & & & &
\end{array}$$

It follows that the cokernels C^{ϕ} have all the same dimension, iff Lie^{ϕ} and $Lie^{\phi\tau}$ have the same dimension for all ϕ. But this is exactly the condition (i) we put on the Lie algebra in the lemma. □

The lemma shows, that the determinantal condition in our moduli problem implies the condition, that the value M_{Λ} of the crystal of X_{Λ} on the scheme S is locally for the Zariski topology on S isomorphic to $\Lambda \otimes \mathcal{O}_S$. Indeed, to show this we may restrict to the case, where B is simple. By the Morita equivalence we may further assume that $B = D$ is a division algebra over

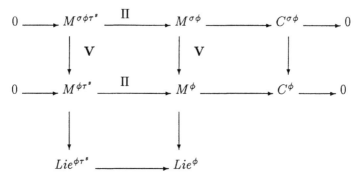

$\mathbf{Q}_p$. In this case the condition (i) of definition (3.21) simply says that locally on S the $O_D \otimes \mathcal{O}_S$-module M_Λ is free. By the remarks made on condition (iv) the assumptions of the lemma above are satisfied. Hence for any geometric point $Spec\ P \to S$ the crystal induced by X_Λ over $Spec\ P$ is a free $O_D \otimes W(P)$-module and a fortiori the value of this crystal at $Spec\ P$ is a free $O_D \otimes P$-module. Since M_Λ is known to be a locally free $\mathcal{O}_S$-module this suffices to prove our assertion.

d) Next we claim that the condition (ii) of the definition in view of condition (iv) is equivalent to the following:

(ii bis) For any neighbours $\Lambda \subset \Lambda'$ of the chain $\mathcal{L}$ the height of the quasi–isogeny $X_{\Lambda_i} \longrightarrow X_{\Lambda'_i}$ is equal to $\log_p |\Lambda'_i/\Lambda_i|$, $i = 1, \ldots, m$.

Indeed, to show this one reduces as above to the case where $O_B = O_D$ is the ring of integers in a division algebra D, with residue class field $\tilde{k} = O_D/\Pi O_D$. The equivalence of the conditions follows if we can show that the cokernel N of the map $M_\Lambda \longrightarrow M_{\Lambda'}$ is a locally free $\tilde{k} \otimes \mathcal{O}_S/p\mathcal{O}_S$-module. By Breen-Berthelot-Messing [BBM] Prop.4.3.1 (compare also de Jong [dJ1] 2.3) we know that N is a locally free $\mathcal{O}_S/p\mathcal{O}_S$-module. Hence it is enough to consider the case, where S is the spectrum of an algebraically closed field P. As before we have a decomposition,

$$N = \bigoplus_{\phi:\tilde{k} \to P} N^\phi.$$

We get a decomposition of the Cartier module M of X_Λ:

$$M = \oplus_\phi M^\phi.$$

A similar decomposition holds for the Cartier module M' of $X_{\Lambda'}$. We see that N is a free $\tilde{k} \otimes P$-module, iff the dimension of the P-vectorspace N^ϕ is independent of ϕ. This follows in the same way as in the proof of lemma (3.24) from the following commutative diagram:

$$\begin{array}{ccccccc}
0 & \to & M^{\sigma\phi} & \xrightarrow{V} & M^{\phi} & \longrightarrow & Lie^{\phi} X_{\Lambda} & \to 0 \\
 & & \downarrow & & \downarrow & & \downarrow & \\
0 & \to & M'^{\sigma\phi} & \xrightarrow{V} & M'^{\phi} & \longrightarrow & Lie^{\phi} X_{\Lambda'} & \to 0 \\
 & & \downarrow & & \downarrow & & & \\
 & & N^{\sigma\phi} & \longrightarrow & N^{\phi} & & & \\
 & & \downarrow & & \downarrow & & & \\
 & & 0 & & 0 & & &
\end{array}$$

Theorem 3.25 *The functor $\breve{\mathcal{M}}$ is representable by a formal scheme, which is formally locally of finite type over Spf $O_{\breve{E}}$.*

Proof: We start with the representable functor $\mathcal{M}$ of theorem (2.16), for our $\mathbf{X}$ at hand. In fact the theorem is applicable, because over an algebraically closed field any isocrystal is decent. Let $(X, \varrho) \in \mathcal{M}(S)$ be a point. We transport the action of B on $\mathbf{X}$ by quasi–isogenies via ϱ to an action of B on X by quasi–isogenies. Let $\mathcal{M}_O$ be the subfunctor of $\mathcal{M}$, where O_B acts by isogenies. This is clearly a closed subfunctor and therefore representable. We have an obvious morphism of functors:

$$j : \breve{\mathcal{M}} \longrightarrow \prod_{\Lambda \in \mathcal{L}} \mathcal{M}_O.$$

It is enough to show that this morphism is representable. We know (2.9) that the condition, that a quasi–isogeny of p–divisible groups over a scheme S is an isogeny, is representable by a closed subscheme. Hence the condition that O_B acts on X_Λ, the conditions (iii) and (v), and the condition that $\tilde{\varrho}_{\Lambda',\Lambda}$ is an isogeny for any two neighbours $\Lambda \subset \Lambda'$ is relatively representable with respect to j. The condition (iv) is clearly representable. In the presence of condition (iv), condition (i) is automatic and condition (ii) is equivalent to condition
(ii bis) prescribing the degree of certain isogenies. This is obviously representable by an open subscheme.

3.26 Next we will consider the problem of determining the local equations of the formal scheme given by definition (3.21). We reduce this to a problem of linear algebra by constructing a *local model*, comp. [R1].

Let us start with a set of data $(B, F, O_B, V, \mu, \mathcal{L})$ in the case (EL). In the case (PEL) we have in addition a nondegenerate antisymmetric $\mathbf{Q}_p$-pairing $(\ ,\ )$ on V.
The cocharacter μ is given over the field K. Let $E \subset K$ be the Shimura field. Let us define a functor $\mathbf{M}^{loc}$ on the category of O_E-schemes.

Definition 3.27 *A point of $\mathbf{M}^{loc}$ with values in an O_E-scheme S is given by the following data.*

1. *A functor from the category $\mathcal{L}$ to the category of $O_B \otimes \mathcal{O}_S$-modules on S:*

$$\Lambda \longmapsto t_\Lambda, \qquad \Lambda \in \mathcal{L}.$$

2. *A morphism of functors*

$$\varphi_\Lambda : \Lambda \otimes_{\mathbf{Z}_p} \mathcal{O}_S \longrightarrow t_\Lambda.$$

We require that the following conditions are satisfied:

(i) *t_Λ is a finite locally free $\mathcal{O}_S$-module. For the action of O_B on t_Λ we have the following identity of polynomial functions*

$$\det_{\mathcal{O}_S}(a\,;\,t_\Lambda) = \det_K(a\,;\,V_0), \qquad a \in O_B\,.$$

(ii) *The morphisms φ_Λ are surjective.*

(iii) *In the case (PEL) the composite of the following maps is zero for each Λ:*

$$t_\Lambda^* \xrightarrow{\varphi_\Lambda^*} (\Lambda \otimes \mathcal{O}_S)^* \underset{\psi}{\cong} \hat{\Lambda} \otimes \mathcal{O}_S \xrightarrow{\varphi_{\hat{\Lambda}}} t_{\hat{\Lambda}}\,.$$

Clearly the functor $\mathbf{M}^{loc}$ is represented by a closed subscheme in a product of Grassmannians. We write $\breve{\mathbf{M}}^{loc}$ for $\mathbf{M}^{loc} \otimes_{O_E} O_{\breve{E}}$.
Let us introduce a smooth covering of the formal scheme $\breve{\mathcal{M}}$.

Definition 3.28 *Let $\mathcal{N}$ be the contravariant set-valued functor on the category $Nilp_{O_{\breve{E}}}$, such that a point of $\mathcal{N}$ with values in $S \in Nilp_{O_{\breve{E}}}$ is given by the following data*

1. *A point $(X_\Lambda, \varrho_\Lambda)$ of $\breve{\mathcal{M}}(S)$.*

2. An isomorphism of (polarized) multichains

$$\gamma_\Lambda : M_\Lambda \xrightarrow{\sim} \Lambda \otimes_{\mathbf{Z}_p} \mathcal{O}_S .$$

Here M_Λ denotes the value of the crystal of X_Λ on S.

3.29 The smooth formal group

$$\mathcal{P}(S) = \mathrm{Aut}(\{\Lambda \otimes_{\mathbf{Z}_p} \mathcal{O}_S\})$$

acts on $\mathcal{N}$ via the data 2):

$$p \cdot (X_\Lambda, \varrho_\Lambda, \gamma_\Lambda) = (X_\Lambda, \varrho_\Lambda, p\gamma_\Lambda), \qquad p \in \mathcal{P}(S) .$$

We see that $\mathcal{N}$ is a left $\mathcal{P}$-torsor over $\breve{\mathcal{M}}$ and therefore representable by a formal scheme, which is of finite type over $\breve{\mathcal{M}}$. There is a natural morphism

$$\begin{array}{ccc} \mathcal{N} & \longrightarrow & \breve{\mathbf{M}}^{loc} \\ (X_\Lambda, \varrho_\Lambda, \gamma_\Lambda) & \longmapsto & \Lambda \otimes_{\mathbf{Z}_p} \mathcal{O}_S \xrightarrow{\gamma_\Lambda^{-1}} M_\Lambda \longrightarrow \mathrm{Lie}\, X_\Lambda , \end{array}$$

that factors through the p-adic completion $\hat{\mathbf{M}}^{loc}$ of $\breve{\mathbf{M}}^{loc}$.
By Grothendieck and Messing [Me] the map

$$\mathcal{N} \longrightarrow \hat{\mathbf{M}}^{loc}$$

is formally smooth. It is formally locally of finite type since $\mathcal{N}$ and $\hat{\mathbf{M}}^{loc}$ are formally locally of finite type over $Spf\ O_{\breve{E}}$.

3.30 Let us fix a closed point $x \in \breve{\mathcal{M}}$. We identify its residue class field $\kappa(x)$ with the residue class field $\kappa = L$ of $O_{\breve{E}}$. We will consider sections of the smooth morphism

$$\mathcal{N} \longrightarrow \breve{\mathcal{M}}$$

in a pointed étale neighbourhood $(\mathcal{U}, y)$ of x. By definition of $\mathcal{N}$ a section s over $\mathcal{U}$ is given by an isomorphism of (polarized) chains

$$\gamma_\Lambda : M_\Lambda \otimes_{\mathcal{O}_{\breve{\mathcal{M}}}} \mathcal{O}_\mathcal{U} \longrightarrow \Lambda \otimes_{\mathbf{Z}_p} \mathcal{O}_\mathcal{U} .$$

We are going to explain a condition on the section s that ensures that the composition

$$\mathcal{U} \xrightarrow{s} \mathcal{N} \longrightarrow \hat{\mathbf{M}}^{loc}$$

is formally étale.
Consider a local artinian augmented κ-algebra A, such that the square of the maximal ideal of A is zero. Consider a morphism

$$\tilde{a} : Spec\ A \longrightarrow \mathcal{U} \tag{3.1}$$

which is concentrated in x. Let us denote by Y_Λ for $\Lambda \in \mathcal{L}$ the p-divisible groups on Spec$A$ induced by the universal p-divisible groups X_Λ on $\breve{\mathcal{M}}$. Let $\bar{Y}_\Lambda = Y_\Lambda \times_{Spec\ A} Spec\ \kappa$ be its reduction. Let N_Λ (respectively $\bar{N}_\Lambda$) be the Lie-algebra of the universal extension of Y_Λ (respectively $\bar{Y}_\Lambda$). By the crystalline nature of N_Λ we have a canonical isomorphism

$$\tau : \bar{N}_\Lambda \otimes_\kappa A \xrightarrow{\sim} N_\Lambda\,.$$

On the other hand the section s provides an isomorphism

$$\gamma_A : N_\Lambda \xrightarrow{\sim} \Lambda \otimes_{\mathbf{Z}_p} A\,.$$

Definition 3.31 *We call a section s rigid of the first order in x, if for all algebras A as above and morphisms $\tilde{a}$ as above the following diagram is commutative*

$$\begin{array}{ccc} \bar{N}_\Lambda \otimes_\kappa A & \xrightarrow{\tau} & N_\Lambda \\ & {\scriptstyle \gamma_\kappa \otimes_\kappa A} \searrow \quad \swarrow {\scriptstyle \gamma_A} & \\ & \Lambda \otimes_{\mathbf{Z}_p} A & \end{array}$$

where $\gamma_\kappa = \gamma_A \otimes_A \kappa$.

3.32 Any closed point $x \in \breve{\mathcal{M}}$ has an étale neighbourhood, such that there is a section s_Λ in this neighbourhood which is rigid of the first order in x. Indeed, let $\mathcal{I}$ be the maximal ideal of definition of $\mathcal{O}_{\breve{\mathcal{M}}}$. Let $\breve{\mathcal{M}}_2$ be the closed subscheme of $\breve{\mathcal{M}}$ defined by $\mathcal{I}^2$. For any formal scheme $\mathcal{X}$ of finite type over $\breve{\mathcal{M}}$ we will denote by $\mathcal{X}_2$ the scheme $\mathcal{X} \times_{\breve{\mathcal{M}}} \breve{\mathcal{M}}_2$. Since $\mathcal{N} \to \breve{\mathcal{M}}$ is a smooth morphism, it is enough to find an étale neighbourhood $\mathcal{U}_2 \to \breve{\mathcal{M}}_2$

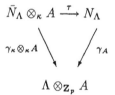

of x and a section $\tilde{s} : \mathcal{U}_2 \to \check{\mathcal{N}}_2$ which is rigid of the first order in x in an obvious sense. Since $\mathcal{N}_2 \to \check{\mathcal{M}}_2$ is a morphism of schemes of finite type it is enough to ask for the existence of a section over Spec O_x^{sh}, where O_x^{sh} is the strict henselization of the local ring of $\check{\mathcal{M}}_2$ at x. By Hensel's lemma it is enough to find a section over Spec O_x^{sh}/m_x^2. Since any morphism (3.1) factors through the spectrum of the artinian ring $A' = O_x^{\mathrm{sh}}/m_x^2 + \pi O_x^{\mathrm{sh}}$, it is enough to construct a section over *Spec* A', such that the diagram of definition (3.31) is commutative. This is obvious.

Proposition 3.33 *Let $x \in \check{\mathcal{M}}$ be a closed point and let $s : (\mathcal{U}, y) \to \mathcal{N}$ be a section in a pointed étale neighbourhood $(\mathcal{U}, y)$ of x which is rigid of the first order. Then the composition*

$$\mathcal{U} \xrightarrow{s} \mathcal{N} \longrightarrow \hat{\mathbf{M}}^{loc}$$

is formally étale in a Zariski open neighbourhood of y in $\mathcal{U}$.

It follows that any point of $\check{\mathcal{M}}$ has an étale neighbourhood, which is formally étale over $\hat{\mathbf{M}}^{loc}$. For the proof we need the following general result on formal smoothness contained in EGA.

Lemma 3.34 *Let $\mathcal{N} \to \mathcal{S}$ be a morphism of locally noetherian formal schemes, which is formally of finite type and formally smooth. Let $\mathcal{M} \to \mathcal{N}$ be a closed subscheme of $\mathcal{N}$ defined by a coherent sheaf of ideals $\mathcal{K} \subset \mathcal{O}_{\mathcal{N}}$. Let x be a point of $\mathcal{M}$, and y be its image in $\mathcal{N}$. Then the composite $\mathcal{M} \to \mathcal{S}$ of the morphisms above is formally smooth in a Zariski open neighbourhood of the point x, iff the map*

$$\mathcal{K}/\mathcal{K}^2 \otimes_{\mathcal{O}_{\mathcal{M}}} \kappa(x) \longrightarrow \Omega^1_{\mathcal{N}/\mathcal{S}} \otimes_{\mathcal{O}_{\mathcal{N}}} \kappa(y),$$

induced by the universal derivation ([EGA] O_{IV} 20.5.11.2), is injective.

Proof: Consider the standard exact sequence

$$\mathcal{K}/\mathcal{K}^2 \xrightarrow{\delta} \Omega^1_{\mathcal{N}/\mathcal{S}} \otimes_{\mathcal{O}_{\mathcal{N}}} \mathcal{O}_{\mathcal{M}} \longrightarrow \Omega^1_{\mathcal{M}/\mathcal{S}} \longrightarrow 0.$$

By [EGA] O_{IV} 20.7.8 the condition that $\mathcal{M} \to \mathcal{S}$ is formally smooth in a neighbourhood of x is equivalent to the condition that δ is formally left invertible in a neighbourhood of x. The topological $\mathcal{O}_{\mathcal{N}}$-module $\Omega^1_{\mathcal{N}/\mathcal{S}}$ is formally projective [EGA] 0_{IV} 20.4.9. The topology on $\Omega^1_{\mathcal{N}/\mathcal{S}}$ is the $\mathcal{J}$-adic topology, where $\mathcal{J} \subset \mathcal{O}_{\mathcal{N}}$ is some ideal of definition [EGA] 0_{IV} 20.4.5. It follows that $\Omega^1_{\mathcal{N}/\mathcal{S}} \otimes_{\mathcal{O}_{\mathcal{N}}} \mathcal{O}_{\mathcal{M}}$ is a formally projective $\mathcal{O}_{\mathcal{M}}$-module, that carries the adic topology induced from $\mathcal{O}_{\mathcal{M}}$. Let us denote by $\mathcal{I}$ the maximal ideal of definition of $\mathcal{O}_{\mathcal{M}}$. By [EGA] 0_{IV} 19.1.9 the condition that δ is formally left invertible in a neighbourhood of x is equivalent to the condition that

$$\mathcal{K}/\mathcal{K}^2 \otimes_{\mathcal{O}_{\mathcal{M}}} \mathcal{O}_{\mathcal{M}}/\mathcal{I} \longrightarrow \Omega^1_{\mathcal{N}/\mathcal{S}} \otimes_{\mathcal{O}_{\mathcal{N}}} \mathcal{O}_{\mathcal{M}}/\mathcal{I}$$

is left invertible. One checks that both modules are coherent modules over $\mathcal{O}_{\mathcal{M}}/\mathcal{I}$. Hence we conclude the proof of the lemma by [EGA] 0_{IV} 19.1.12.

Corollary 3.35 *If the map in (3.34) is an isomorphism, then $\mathcal{M} \to \mathcal{S}$ is formally étale in a neighbourhood of x.*

Proof: This follows from the fact that $\Omega^1_{\breve{\mathcal{M}}/\mathcal{S}} = (0)$ in a neighbourhood of x in this case, [EGA] 0_{IV} 20.1.1.

Proof of Proposition (3.33): First we show that the map is formally smooth. We write $\mathcal{X} = \mathcal{U}, \mathcal{Y} = \mathcal{N} \times_{\breve{\mathcal{M}}} \mathcal{U}, \mathcal{Z} = \hat{\mathbf{M}}^{loc}$. Consider the diagram of formal schemes over $\mathcal{T} = Spf\ O_{\breve{E}}$.

$$\begin{array}{ccccc} & & \mathcal{Y} & & \\ & {\scriptstyle s}\nearrow\swarrow{\scriptstyle p} & & \searrow{\scriptstyle q} & \\ \mathcal{X} & & & & \mathcal{Z} \end{array}$$

By [EGA] 0_{IV} 20.7.18 we have split exact sequences

$$\begin{array}{ccccccccc} 0 & \longrightarrow & q^*\Omega^1_{\mathcal{Z}/\mathcal{T}} & \longrightarrow & \Omega^1_{\mathcal{Y}/\mathcal{T}} & \longrightarrow & \Omega^1_{\mathcal{Y}/\mathcal{Z}} & \longrightarrow & 0 \\ 0 & \longrightarrow & p^*\Omega^1_{\mathcal{X}/\mathcal{T}} & \longrightarrow & \Omega^1_{\mathcal{Y}/\mathcal{T}} & \longrightarrow & \Omega^1_{\mathcal{Y}/\mathcal{X}} & \longrightarrow & 0. \end{array}$$

If we apply s^* to the lower exact sequence, we get a canonical splitting

$$\begin{array}{ccccccccc} 0 & \longrightarrow & s^*p^*\Omega^1_{\mathcal{X}/\mathcal{T}} & \longrightarrow & s^*\Omega_{\mathcal{Y}/\mathcal{T}} & \longrightarrow & s^*\Omega_{\mathcal{Y}/\mathcal{X}} & \longrightarrow & 0 \\ & & \simeq | & & || & & | \simeq & & \\ & & \Omega^1_{\mathcal{X}/\mathcal{T}} & \stackrel{s^*}{\longleftarrow} & s^*\Omega_{\mathcal{Y}/\mathcal{T}} & \stackrel{d}{\longleftarrow} & \mathcal{K}/\mathcal{K}^2. & & \end{array}$$

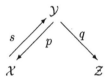

Let $y = s(x)$ and $q(y) = z$. The morphism qs defines an inclusion $\kappa(z) \longrightarrow \kappa(x)$. Let us use the notation $\Omega^1_{\mathcal{X}/T}(x) = \Omega^1_{\mathcal{X}/T} \otimes_{\mathcal{O}_\mathcal{X}} \kappa(x)$ for the geometric fibres.

We claim that the formal smoothness of qs is equivalent to the assertion that the map

$$\Omega^1_{\mathcal{Z}/T}(z) \otimes_{\kappa(z)} \kappa(x) \xrightarrow{q^*} \Omega^1_{\mathcal{Y}/T}(y) \otimes_{\kappa(y)} \kappa(x) \xrightarrow{s^*} \Omega^1_{\mathcal{X}/T}(x)$$

is injective.

Indeed, consider the diagram

$$\begin{array}{ccccc}
 & & s^*q^*\Omega^1_{\mathcal{Z}/T} & & \\
 & \swarrow^{\mu} & \downarrow & & \\
\Omega^1_{\mathcal{X}/T} & \leftrightarrows & s^*\Omega^1_{\mathcal{Y}/T} & \leftrightarrows & s^*\Omega^1_{\mathcal{Y}/\mathcal{X}} \cong \mathcal{K}/\mathcal{K}^2 \\
 & & \downarrow & \swarrow^{d} & \\
 & & s^*\Omega^1_{\mathcal{Y}/\mathcal{Z}} & &
\end{array}$$

It is easy to see that d induces an injection of the geometric fibres at x, iff μ does. Hence the claim follows from the lemma.

Let us identify κ, $\kappa(x)$, $\kappa(y)$, and $\kappa(z)$. By duality it is enough to show that

$$\mathrm{Hom}_{\mathcal{O}_\mathcal{X}}(\Omega^1_{\mathcal{X}/T}, \kappa) \longrightarrow \mathrm{Hom}_{\mathcal{O}_\mathcal{Z}}(\Omega^1_{\mathcal{Z}/T}, \kappa)$$

is surjective.

We set $A = \mathcal{O}_{\mathcal{X},x}/m_x^2 + \pi\mathcal{O}_{\mathcal{X},x}$ and $B = \mathcal{O}_{\mathcal{Z},z}/m_z^2 + \pi\mathcal{O}_{\mathcal{Z},z}$. These are augmented artinian κ-algebras, such that the squares of the maximal ideals are zero. Our assertion is that the map

$$\mathrm{Der}_\kappa(A, \kappa) \longrightarrow \mathrm{Der}_\kappa(B, \kappa)$$

is surjective. This means that any commutative diagram

$$\begin{array}{ccc}
\mathcal{M} & \longrightarrow & \hat{\mathbf{M}}^{loc} \\
\uparrow & \nwarrow & \uparrow \\
Spec\ \kappa & \longrightarrow & Spec\ \kappa[\varepsilon]\,,
\end{array}$$

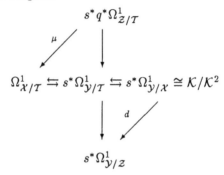

such that the vertical arrows are concentrated in x, respectively z, admits a diagonal arrow as indicated.
Indeed the left vertical arrow is given by a chain $\{\bar{X}_\Lambda\}$ of p-divisible groups over κ. The point y defines a rigidification of their crystals, i.e. an isomorphism of polarized chains

$$\bar{M}_\Lambda \simeq \Lambda \otimes_{\mathbf{Z}_p} \kappa.$$

Set $\bar{t}_\Lambda = \mathrm{Lie}\bar{X}_\Lambda$. Then the Hodge filtration defines the point z of the period space $\hat{\mathbf{M}}^{loc}$

$$\begin{array}{ccc} \bar{M}_\Lambda & \simeq & \Lambda \otimes_{\mathbf{Z}_p} \kappa \\ & \searrow \quad \swarrow & \\ & \bar{t}_\Lambda & \end{array}$$

The right vertical arrow in the diagram above gives a lifting of the last map

$$\Lambda \otimes_{\mathbf{Z}_p} \kappa[\varepsilon] \longrightarrow t_\Lambda.$$

By the horizontal isomorphism above $\bar{M}_\Lambda \otimes_\kappa \kappa[\varepsilon] \simeq \Lambda \otimes_{\mathbf{Z}_p} \kappa[\varepsilon]$ is identified with the value of the crystal of $\bar{X}_\Lambda$ at $\kappa[\varepsilon]$. Hence by Grothendieck and Messing we get a lifting X_Λ of $\bar{X}_\Lambda$, such that M_Λ, the Lie algebra of the universal extension of X_Λ, is identified with $\bar{M}_\Lambda \otimes_\kappa \kappa[\varepsilon]$. We get a rigidification

$$M_\Lambda \simeq \bar{M}_\Lambda \otimes_\kappa \kappa[\varepsilon] \xrightarrow{\sim} \Lambda \otimes_{\mathbf{Z}_p} \kappa[\varepsilon]$$

and hence an element of $\mathcal{N}(\kappa[\varepsilon])$. The point is, that this is the image of the point $\{X_\Lambda\} \in \breve{\mathcal{M}}(\kappa[\varepsilon])$ under $s : \breve{\mathcal{M}} \to \mathcal{N}$ because s is rigid of the first order.
It follows that the map on derivations is bijective. The formal étaleness follows immediately from corollary (3.35). □

We conjecture that $\breve{\mathcal{M}}$ is flat over $Spf\ O_{\breve{E}}$. By proposition (3.33) it is equivalent to ask whether the local model $\mathbf{M}^{loc}$ is flat over $O_{\breve{E}}$. At the end of this chapter we review some examples which support this conjecture.

3.36 In the definition (3.18) of the moduli data we had assumed that L is algebraically closed. We now want to consider an arbitrary sufficiently big perfect field L of characteristic p. We keep the notations of (3.18) except that we now impose an additional condition, namely that b is decent, cf. (1.8).

Lemma 3.37 *Assume that we are in case (PEL) and that b satisfies a decency equation with an integer $s > 0$, cf. (1.8). Assume that L contains the field $\mathbf{F}_{p^s}$. Let $K_0 = W(L)_{\mathbf{Q}_p}$. Then there is a unit $u \in W(L)$, such that the K_0-bilinear form*

$$\psi(v, w) = u(v, w), \qquad v, w \in N$$

is a polarization of the isocrystal N.

Proof: As in the case where L is algebraically closed (comp. (3.19)) our conditions imply that $\text{ord}_p\ c(b) = 1$. We define $\text{Nm}\, b \in G(\mathbf{Q}_{p^s})$ by the equation $(b\sigma)^s = (\text{Nm}\ b)\sigma^s$. For $v, w \in N$ we find the equation for the given symplectic form on V:

$$((b\sigma)^s v, (b\sigma)^s w) = c(\text{Nm}\ b)(\sigma^s v, \sigma^s w).$$

If we replace $(b\sigma)^s$ by $s\nu(p)\sigma^s$ in the equation, we obtain

$$c(\text{Nm}\ b) = p^s.$$

Let us denote by Nm_s the norm of the extension $W(\mathbf{F}_{p^s})/W(\mathbf{F}_p)$. Then we have $c(\text{Nm}\ b) = \text{Nm}_s c(b)$. The above equation takes the form:

$$\text{Nm}_s\ p^{-1}c(b) = 1.$$

By Hilbert Satz 90 this is equivalent to the existence of $u \in W(\mathbf{F}_{p^s})$, such that

$$c(b) = pu^{-1}\sigma(u)$$

Clearly this is a u as required by the lemma. □

From the corollary (1.10) we obtain:

Proposition 3.38 *The isomorphism class of the homogeneously polarized B-isocrystal*
$(N, \mathbf{Q}_p^{\times}\psi)$ depends only on the σ-conjugacy class $\bar{b}$.

Definition 3.39 *A set of data for moduli of p-divisible groups relative to a perfect field L in the case (EL) is a tuple $(F, B, O_B, V, b, \mu, \mathcal{L})$, that satisfies the conditions (i) (ii) and (iii) of the definition (3.18). Moreover we require that $b \in G(K_0)$ is decent, and that with respect to the inclusion of the*

Shimura field $E \subset K$, the residue class field of E is contained in L. This gives an O_E-algebra structure on L. In the case (PEL) we have in addition the nondegenerate antisymmetric pairing (,), that satisfies the conditions of definition 3.18. In this case we assume that L contains $\mathbf{F}_{p^s}$, where $s > 0$ appears in a decency equation for b.

Let $\breve{E}$ be the complete unramified extension of the Shimura field E with residue class field L. By proposition (3.38) there is a $\mathbf{Q}_p$–homogeneously polarized p-divisible group $\mathbf{X}$ over L whose isocrystal is $(N, \mathbf{Q}_p^\times \psi)$. With this $\mathbf{X}$ the moduli problem $\breve{\mathcal{M}}$ of definition (3.21) makes sense over $O_{\breve{E}}$ in this more general situation.

Corollary 3.40 *The functor $\breve{\mathcal{M}}$ of definition (3.21) associated to the data of definition (3.39) is representable by a formal scheme which is formally locally of finite type over $O_{\breve{E}}$.*

3.41 Our next aim is to define a completion of the formal scheme $\breve{\mathcal{M}}$ of definition (3.21) over the ring of integers O_E of the Shimura field for any data of definition (3.18) over an algebraically closed field L. For the following proposition we start with data of definition (3.18) in case L is algebraically closed or with data of definition (3.40) if L is an arbitrary sufficiently large perfect field (in the latter case b is decent).
Let s be any integer, such that the morphism $s\nu$ factors through $\mathbf{G}_m$. We set

$$\gamma_s = p^s (s\nu(p))^{-1}.$$

This is a quasi–isogeny of height $s \dim \mathbf{X}$ of the p-divisible group $\mathbf{X}$, which lies in the group $J(\mathbf{Q}_p)$. Let $\Gamma_s = \gamma_s^{\mathbf{Z}}$ be the cyclic group generated by γ_s. Hence Γ_s acts on the functor $\breve{\mathcal{M}}$.

Proposition 3.42 *Let $\mathcal{M}_s$ be the Zariski sheaf associated to the functor*

$$\breve{\mathcal{M}}(S)/\Gamma_s.$$

Then $\mathcal{M}_s$ is representable by a formal scheme, which is locally of finite type over $Spf\ O_{\breve{E}}$.

More explicitly let S be a connected scheme in $Nilp_{O_{\breve{E}}}$. Two data $(X_{\mathcal{L}}, \varrho)$ and $(X'_{\mathcal{L}}, \varrho')$ over S define the same point of $\mathcal{M}_s(S)$, if for some integer k the quasi–isogeny $\varrho\gamma_s^k\varrho'^{-1}$ lifts to an isomorphism $X'_\Lambda \to X_\Lambda, \Lambda \in \mathcal{L}$.
Since γ_s is in the center of the algebra *End* $\mathbf{X}$ the group $J(\mathbf{Q}_p)$ continues to act on $\mathcal{M}_s$.

Proof: We fix a member Λ_0 of the lattice chain $\mathcal{L}$. Let $\breve{\mathcal{M}}(n) \subset \breve{\mathcal{M}}$ be the subfunctor where ϱ_{Λ_0} is a quasi–isogeny of height n. This is an open and closed subfunctor and $\breve{\mathcal{M}}$ is a disjoint union:

$$\breve{\mathcal{M}} = \coprod_n \breve{\mathcal{M}}(n).$$

Let us first exclude the uninteresting case where $\mathbf{X}$ is etale. Since γ_s is a quasi–isogeny of height $s \dim \mathbf{X}$, the action of γ_s on $\breve{\mathcal{M}}$ is homogeneous of a nonzero degree. We get an isomorphism:

$$\coprod_{n=1}^{s\dim \mathbf{X}} \breve{\mathcal{M}}(n) \to \mathcal{M}_s.$$

In the case where $\mathbf{X}$ is etale the quasi–isogeny γ_s is the identity. Hence the result is trivial in this case.

Remark 3.43 : If the group $\mathbf{X}$ is etale, we are in the case (EL). Since our functor $\breve{\mathcal{M}}$ is empty unless dim $\mathbf{X}$ = dim V_1, we may assume that the morphism μ is trivial. Assume that L is a perfect field, and that $\bar{b} \in G(W(L)_{\mathbf{Q}})$ is a decent conjugacy class. Take a decent $b \in \bar{b}$. Then for a certain integer s, we have:

$$(b\sigma)^s = s\nu(p)\sigma^s = p^s\sigma^s, \qquad b \in G(\mathbf{Q}_{p^s}).$$

Since by Hilbert Satz 90 the cohomology group $H^1(\mathbf{Q}_{p^s}/\mathbf{Q}_p, G(\mathbf{Q}_{p^s}))$ is trivial, we conclude that $p \in \bar{b}$. By corollary 3.40 we find a formal scheme $\mathcal{M}_0$ over *Spf* $\mathbf{Z}_p$, such that $\mathcal{M}_0 \times$ *Spf* $W(L) = \breve{\mathcal{M}}$. The scheme $\mathcal{M}_0$ is a disjoint union of copies of *Spf* $\mathbf{Z}_p$. We have one copy for each chain of O_B-lattices in V conjugate to $\mathcal{L}$. This is the model of $\breve{\mathcal{M}}$ over the integers $\mathbf{Z}_p$of the Shimura field, which we would like to define in the general case.

3.44 We will now define some sort of descent data on the functors $\breve{\mathcal{M}}$ resp. $\mathcal{M}_s$. Let us start with a more general setting.

Let E be a finite extension of $\mathbf{Q}_p$. Let L be an algebraically closed extension of the residue class field κ of E. We denote by $\breve{E}$ the complete unramified extension of E with residue class field L. Let us denote by $\tau \in Gal(\breve{E}/E)$ the Frobenius automorphism. Let $S = (S, \phi)$ be an object of $Nilp_{O_{\breve{E}}}$, where $\phi : S \rightarrow Spf\ O_{\breve{E}}$ denotes the structure morphism. We will denote by $S_{[\tau]}$ the pair $(S, \tau\phi)$. We mean by τ the Frobenius automorphism of the scheme $Spec\ O_{\breve{E}}$. We consider a (contravariant) functor $\mathcal{G}$ on the category $Nilp_{O_{\breve{E}}}$. We denote by $\mathcal{G}^\tau$ the functor defined by:

$$\mathcal{G}^\tau(S) = \mathcal{G}(S_{[\tau]})$$

Definition 3.45 *A Weil descent datum for the functor $\mathcal{G}$ is an isomorphism of functors:*

$$\alpha : \mathcal{G} \longrightarrow \mathcal{G}^\tau$$

3.46 If we are given a functor $\mathcal{G}_0$ on the category $Nilp_{O_E}$, we have an obvious isomorphism for $S \in Nilp_{O_{\breve{E}}}$:

$$\alpha = \mathcal{G}_0(id) : \mathcal{G}_0(S) \rightarrow \mathcal{G}_0(S_{[\tau]}).$$

Hence we have a Weil descent datum on the restriction $\mathcal{G}$ of $\mathcal{G}_0$ to $Nilp_{O_{\breve{E}}}$. Let $S_0 \in Nilp_{O_E}$ be a scheme, and let $S = S_0 \times_{Spf\ O_E} Spf\ O_{\breve{E}}$. The Frobenius automorphism on the second factor defines an isomorphism:

$$\tau : S_{[\tau]} \longrightarrow S.$$

If $\mathcal{G}_0$ is a formal scheme we have an exact sequence:

$$\mathcal{G}_0(S_0) \longrightarrow \mathcal{G}(S) \overset{\mathcal{G}(\tau)}{\underset{\alpha}{\rightrightarrows}} \mathcal{G}(S_{[\tau]})$$

Hence the Weil descent datum determines $\mathcal{G}_0$ uniquely. We say that a Weil descent datum on a formal scheme $\mathcal{G}$ over $Spf\ O_{\breve{E}}$ is *effective*, if it is isomorphic to the descent datum defined by a formal scheme $\mathcal{G}_0$.

If $\mathcal{G}'$, α', $\mathcal{G}'_0$ is a second formal scheme over $Spf\ O_{\breve{E}}$ with an effective Weil descent datum, we obtain:

$$\mathrm{Hom}((\mathcal{G},\alpha),\ (\mathcal{G}',\alpha')) = \mathrm{Hom}(\mathcal{G}_0,\mathcal{G}'_0).$$

Let E_r be the unramified extension of degree r of E. Then α induces a Weil descent datum relative to the extension $\breve{E}/E_r$:

$$\alpha^{\tau^{r-1}} \ldots \alpha^{\tau}\alpha : \mathcal{G} \longrightarrow \mathcal{G}^{\tau^r}$$

We will call it a *power* of α, and denote it by α^r. If α is effective, then α^r is effective, and is the Weil descent datum defined by $\mathcal{G}_0 \times_{Spf\ O_E} Spf\ O_{E_r}$. The usual Galois-descent asserts that a converse is true:

Proposition 3.47 *Let $\mathcal{G}$ be a formal scheme, which is formally locally of finite type over $Spf\ O_{\breve{E}}$. Assume that the associated reduced scheme is quasiprojective. Let α be a Weil descent datum on $\mathcal{G}$. If some power α^r of α is effective, then α is effective. The same is true if $\mathcal{G}_{red}$ is a union of an ascending chain of quasiprojective subschemes.*

Proof: There is a unique formal scheme $\mathcal{G}_1$ over $Spf\ O_{E_r}$ and an isomorphism

$$\mathcal{G} = \mathcal{G}_1 \times_{Spf\ O_{E_r}} Spf\ O_{\breve{E}},$$

which induces on $\mathcal{G}$ the descent datum α^r. Since α commutes with α^r the morphism $\alpha : \mathcal{G} \to \mathcal{G}^{\tau}$ is obtained by base change from a morphism $\bar{\alpha} : \mathcal{G}_1 \to \mathcal{G}_1^{\tau}$. Clearly $\bar{\alpha}^r = 1$ holds. Hence $\bar{\alpha}$ is an ordinary Galois descent datum for the etale covering $Spf\ O_{E_r} \to Spf\ O_E$. We conclude by SGA 1, Exp.VIII Corollaire 7.6.

3.48 We will now define a Weil descent datum on the functor $\breve{\mathcal{M}}$. Let us denote by $\bar{\tau} : Spec\ L \to Spec\ L$ the Frobenius automorphism relative to the residue field κ of E. We have the Frobenius morphism of the p-divisible group $\mathbf{X}$ relative to κ,

$$Frob_E : \mathbf{X} \longrightarrow \bar{\tau}^*\mathbf{X}.$$

Let $S \in Nilp_{O_{\breve{E}}}$ be a scheme. We consider a point $(X_{\mathcal{L}}, \varrho)$ of $\breve{\mathcal{M}}$ over S. We can now define a point $(X^{\alpha}_{\mathcal{L}}, \varrho^{\alpha})$ in $\breve{\mathcal{M}}(S_{[\tau]})$ as follows:
We set $X^{\alpha}_{\mathcal{L}} = X_{\mathcal{L}}$, and we set ϱ^{α} to be the morphism:

$$\varrho\phi^*(Frob_E^{-1}) : \overline{\tau\phi}^*\mathbf{X} \to \bar{\phi}^*\mathbf{X} \to X_{\bar{S}}.$$

We obtain a Weil descent datum on the formal scheme $\breve{\mathcal{M}}$:

$$\begin{array}{rccc} \alpha: & \breve{\mathcal{M}}(S) & \longrightarrow & \breve{\mathcal{M}}(S_{[\tau]}) \\ & (X_{\mathcal{L}}, \varrho) & \mapsto & (X_{\mathcal{L}}^{\alpha}, \varrho^{\alpha}). \end{array}$$

It is obvious that the action of γ_s commutes with this Weil descent datum, hence we obtain an induced Weil descent datum on $\mathcal{M}_s$.

Theorem 3.49 *Let $\breve{\mathcal{M}}/Spf\ O_{\breve{E}}$ be the functor of definition (3.21), i.e. the field L is algebraically closed (but the conjugacy class $\bar{b}$ need not to be decent). Let $\mathcal{M}_s$ be the functor given by proposition 3.43. Then the above Weil descent datum on $\mathcal{M}_s$ is effective.*

Proof: One checks that the isomorphism class of the descent datum does not depend on the choice of $b \in \bar{b}$, and on the choice of $\mathbf{X}$. For any pair of natural numbers $s, t \in \mathbf{N}$ we have a canonical isomorphism:

$$\mathcal{M}_{st}/\Gamma_s = \mathcal{M}_s.$$

Since this isomorphism respects the descent data, it is enough to prove the theorem if s is sufficiently divisible.
Let us first consider the case, where the group G is connected. Then we may assume that there is a decent $b \in \bar{b}$, such that

$$(b\sigma)^s = s\nu(p)\sigma^s, \qquad b \in G(W(\mathbf{F}_{p^s})_{\mathbf{Q}})$$

We may assume that the residue class field of E is contained in $\mathbf{F}_{p^s}$. Let us denote by E_s the unramified extension of E with residue class field $\mathbf{F}_{p^s}$. To the element b and the representation V we have associated an isocrystal N_0 over $\mathbf{F}_{p^s}$. In the case (PEL) it carries a canonical $\mathbf{Q}_p$-homogeneous polarization. Let $\mathbf{X}_0$ be a p-divisible group with this isocrystal. By corollary (3.41) we get a formal scheme over $Spf\ O_{E_s}$, which we call $\breve{\mathcal{M}}^0$. We set:

$$\mathcal{M}_s^0 = \breve{\mathcal{M}}^0/\Gamma_s$$

This is a well-defined formal scheme over $Spf\ O_{E_s}$, if we exclude the case where the formal group $\mathbf{X}$ is étale. We have a canonical isomorphism:

$$\mathbf{X} = \mathbf{X}_0 \times_{Spec\ \mathbf{F}_{p^s}} Spec\ L.$$

The action of $\gamma_s \mathbf{V}^{-s}$ on the left hand side of this equation induces on the right hand side the action of σ^s via the second factor. Indeed, this is what the above equation says.
For the moduli schemes we get:

$$\mathcal{M}_s = \mathcal{M}_s^0 \times_{Spf\ O_{E_s}} Spf\ O_{\breve{E}}.$$

The canonical Weil descent datum on the right hand side of this equation, looks on the left hand side as follows:

$$(X, \varrho) \mapsto (X, \varrho\gamma_s Frob_{\mathbf{F}_{p^s}}^{-1}).$$

Since γ_s is the identity on $\mathcal{M}_s$ this is a power of our descent datum. Hence a power of our descent datum is representable. By proposition (3.48) we are done in the case, where G is connected.
In the case, where the group is not connected, we consider the embedding $G \to G^1 = GL_B(V)$. The additional data μ, $\bar{b} \in G(W(L)_{\mathbf{Q}})$, and $\mathcal{L}$ in the definition for the functor $\breve{\mathcal{M}}$, may also serve as additional data for the group G^1. Hence we get a formal scheme $\breve{\mathcal{M}}^1$. Since the group G^1 is connected, the Weil descent datum is effective for the formal scheme $\mathcal{M}_s^1$. The reader convinces himself that $\mathcal{M}_s \subset \mathcal{M}_s^1$ is a closed subscheme. Our theorem follows from the following lemma:

Lemma 3.50 *Let E be local field of residue characteristic p. Let $\breve{E}$ be an unramified extension of E with algebraically closed residue field. Let us denote by τ the Frobenius automorphism of $\breve{E}$ over E. Let V be a scheme over O_E and assume that p is locally nilpotent on V. Let $\tilde{Z} \subset V_{O_{\breve{E}}}$ be a closed τ-invariant subscheme. Then $\tilde{Z}$ is obtained by base change from a closed subscheme $Z \subset V$.*

Proof: We may assume that $V = Spec\ A$ is affine. Let $\tilde{I}$ be the ideal of $\tilde{Z}$ in $A \otimes_{O_E} O_{\breve{E}}$. We have to show that $\tilde{I}$ is generated by elements of A. Let M be any submodule of A and set $\tilde{M} = M \otimes_{O_E} O_{\breve{E}}$. It is enough to show that $\tilde{N} = \tilde{I} \cap \tilde{M}$ is generated as an $O_{\breve{E}}$-module by elements of M. If M is annihilated by a prime element π of E this is a well-known fact from linear algebra (Bourbaki, Algebre II, §8,7). Since the residue field of $\breve{E}$ is algebraically closed, τ—id is surjective on $\tilde{M}$. The general case now follows by induction on the power of π annihilating M. Indeed, let $\pi^n\ M = (0)$. We find generators of $\tilde{N} \cap \pi^{n-1}\tilde{M}$ lying in M. Considering

$\tilde{N}/\tilde{N} \cap \pi^{n-1}\tilde{M} \subset \tilde{M}/\pi^{n-1}\tilde{M}$ we find by induction elements of $\tilde{N}$ which are τ–invariant modulo $\tilde{N} \cap \pi^{n-1}\tilde{M}$, which generate $\tilde{N}/\tilde{N} \cap \pi^{n-1}\tilde{M}$. Since τ—id is surjective on $\tilde{N} \cap \pi^{n-1}\tilde{M}$ we can take these elements τ–invariant. Hence we have found generators of $\tilde{N}$ in M. □

Definition 3.51 *We will denote by $\mathcal{M}$ the pro-formal scheme over O_E that is the projective limit of the formal schemes $\mathcal{M}_s$. It is equipped with a left action of the group $J(\mathbf{Q}_p)$.*

The last statement says that the action of $J(\mathbf{Q}_p)$ on $\mathcal{M}_s$ commutes with the descent datum in (3.49). This is true because any endomorphism commutes with the Frobenius morphism.

3.52 We will use the height of ϱ to split $\breve{\mathcal{M}}$ into a disjoint union of open and closed formal subschemes. We have a natural map of groups of $\mathbf{Q}_p$-rational characters

$$X^*_{\mathbf{Q}_p}(G) \longrightarrow X^*_{\mathbf{Q}_p}(J).$$

Indeed let χ be a $\mathbf{Q}_p$–rational character of G. For a $\mathbf{Q}_p$-algebra R we get a map

$$J(R) \hookrightarrow G(R \otimes K_0) \xrightarrow{\chi} (R \otimes K_0)^\times .$$

An element $x \in J(R)$ satisfies $b\sigma(x)b^{-1} = x$. Since χ commutes with σ, we obtain $\sigma\chi(x) = \chi(x)$ and hence a morphism of functors in R

$$\tilde{\chi} : J \longrightarrow \mathbf{G}_{m,\mathbf{Q}_p} .$$

Let $\Delta = \mathrm{Hom}_{\mathbf{Z}}(X^*_{\mathbf{Q}_p}(G), \mathbf{Z})$ be the dual. We define a map

$$\omega_J : J(\mathbf{Q}_p) \longrightarrow \Delta$$

by the equation

$$< \omega_J(x), \chi >= \mathrm{ord}_p\, \tilde{\chi}(x) .$$

Our aim is to define a map

$$\varkappa : \breve{\mathcal{M}} \longrightarrow \Delta ,$$

which is equivariant with respect to ω_J.

Let us give an explicit basis for the group Δ. Clearly it is enough to do this in the cases where F is either a field or $*$ is an involution of the second kind and $F = F_0 \times F_0$, where F_0 is a field, and $*$ interchanges the two factors.

Consider first the case where F is a field. We set $i = i(B) = \sqrt{[B : F]}$. Let us denote by $\mathbf{n}$ the composition of the maps

$$\mathrm{End}_B(V) \xrightarrow{Nm^0} F \xrightarrow{Nm_{F/\mathbf{Q}_p}} \mathbf{Q}_p \,, \tag{3.2}$$

where Nm^0 denotes the reduced norm of the central simple F-algebra $\mathrm{End}_B(V)$. Then $\mathbf{n}$ defines a character of the group G, which we denote by the same letter.

We have the relation

$$\mathbf{n}(g)^i = \det_{\mathbf{Q}_p}(g; V), \quad g \in G(\mathbf{Q}_p)\,.$$

In the case (EL) the character $\mathbf{n}$ is a generator of $X^*_{\mathbf{Q}_p}(G)$. In the case (PEL) we have the character (1.38)

$$c : G \longrightarrow \mathbf{G}_{m,\mathbf{Q}_p},$$

which is related to $\mathbf{n}$ by

$$\mathbf{n}(g)^2 = c(g)^{\dim_{\mathbf{Q}_p} V/i}.$$

Then $X^*_{\mathbf{Q}_p}(G)$ is the $\mathbf{Z}$–module generated by $\mathbf{n}$ and c and the above relation between them.

Finally we consider the case of an involution of the second kind, where $F = F_0 \times F_0$. Then there is an isomorphism $B \cong D \times D^{opp}$, such that $*$ induces on the right hand side the involution $(d_1, d_2) \mapsto (d_2, d_1)$. Moreover there is a left D–module W and an isomorphism of $D \times D^{opp}$–left modules $V \cong W \oplus W^*$, such that the given form on V induces the natural pairing $W \times W^* \to \mathbf{Q}_p$ between W and its $\mathbf{Q}_p$–dual W^*. Let $\mathbf{n}$ and $\mathbf{n}^*$ be the maps $\mathrm{End}_D W \to \mathbf{Q}_p$ resp. $\mathrm{End}_{D^*} W^* \to \mathbf{Q}_p$ defined by (3.2). Then $X^*_{\mathbf{Q}_p}(G)$ is generated by the characters $\mathbf{n}$, $\mathbf{n}^*, c$ subject to the single relation

$$\mathbf{n} \cdot \mathbf{n}^* = c^{\dim_{\mathbf{Q}_p} W/i(D)}.$$

The definition of the map $\varkappa$ in the case (EL) and F a field is as follows. For a point $(X_{\mathcal{L}}, \varrho), \varrho = \{\varrho_\Lambda\}$ of $\breve{\mathcal{M}}$ over a connected scheme S it suffices to define the integer

$$\breve{\mathbf{n}}(X_{\mathcal{L}}, \varrho) = < \varkappa(X_{\mathcal{L}}, \varrho), \mathbf{n} >, \tag{3.3}$$

such that for $g \in J(\mathbf{Q}_p)$

$$\breve{\mathbf{n}}(g(X_{\mathcal{L}}, \varrho)) = < \omega_J(g), \mathbf{n} > + \breve{\mathbf{n}}(X_{\mathcal{L}}, \varrho). \tag{3.4}$$

Lemma 3.53 *Let F be a field. Let $\varrho : X \to Y$ a quasi-isogeny of p-divisible O_B-modules over an algebraically closed field P, which both satisfy the condition 3.21 (iv). Then the height of ϱ is an integral multiple of $i(B) \cdot f(F/\mathbf{Q}_p)$, where $f(F/\mathbf{Q}_p)$ is the index of inertia.*

Proof: Let M resp. N denote the Cartier modules of X resp. Y. Using the decompositions $M = \oplus M^\phi, N = \oplus N^\phi$ from the proof of 3.24, we get maps induced by ϱ

$$M^\phi \otimes \mathbf{Q}_p \longrightarrow N^\phi \otimes \mathbf{Q}_p.$$

The orders of the determinants of these maps are well defined. Since the maps commute with $\mathbf{V}$ we conclude by 3.21 (iv), that these orders are independent of ϕ. □

For the construction of $\varkappa$ we may assume that $\breve{\mathcal{M}}$ is not empty. Then we may take for $\mathbf{X}$ a p-divisible O_B-module, which satisfies 3.21 (iv). We fix a particular Λ and set

$$\breve{\mathbf{n}}(X_{\mathcal{L}}, \varrho) = -\frac{1}{i(B)} \text{height } \varrho_\Lambda \in \mathbf{Z}.$$

We have to verify the identity (3.4),

$$-\frac{1}{i(B)} \text{height } \varrho_\Lambda g^{-1} = < \omega_J(g), \mathbf{n} > -\frac{1}{i(B)} \text{height } \varrho_\Lambda.$$

This follows from

$$i(B) < \omega_J(g), \mathbf{n} > = \text{ord}_p \det(g; V \otimes K_0) = \text{height } g.$$

Next we consider the case (PEL), where F is a field. We may assume $\breve{\mathcal{M}}$ is not empty and choose a point $(X_{\mathcal{L}}, \varrho)$ over L. We may take $\mathbf{X} = X_\Lambda$ for a fixed lattice Λ. We choose the polarization $\lambda : \mathbf{X} \to \hat{\mathbf{X}}$ in such a way that it induces an isomorphism $X_\Lambda \to \widehat{X_{\hat{\Lambda}}}$. Then we have height $\lambda = \log_p |\hat{\Lambda}/\Lambda|$.

We have to define equivariant maps in the sense of (3.4)

$$\breve{c}, \breve{\mathbf{n}} : \breve{\mathcal{M}} \longrightarrow \mathbf{Z},$$

such that $2\breve{\mathbf{n}} = ((\dim_{\mathbf{Q}_p} V)/i(B))\breve{c}$. This amounts to the assertion that 2 height ϱ_Λ is divisible by $\dim_{\mathbf{Q}_p} V$. But we have

$$2 \,\text{height}\ \varrho_\Lambda = \text{height}\ \varrho_{\hat{\Lambda}} + \text{height}\ \varrho_\Lambda - \log_p |\hat{\Lambda}/\Lambda|.$$

Hence the divisibility follows from 3.21 (v).
Finally we consider the case $F = F_0 \times F_0$, $B = D \times D^{opp}$. We may assume $\mathbf{X} = \mathbf{Y} \times \hat{\mathbf{Y}}$, where $\mathbf{Y}$ is an O_D–module and $\hat{\mathbf{Y}}$ the dual $O_{D^{opp}}$–module. We fix a lattice of the form $\Lambda_0 \oplus \hat{\Lambda}_0 \subset W \oplus W^*$ of the multichain $\mathcal{L}$. For a point $(X_\mathcal{L}, \varrho)$ of $\breve{\mathcal{M}}$ we have quasi–isogenies

$$\alpha : \mathbf{Y} \longrightarrow X_{\Lambda_0}, \quad \beta : \hat{\mathbf{Y}} \longrightarrow X_{\hat{\Lambda}_0},$$

such that $\alpha \times \beta = \varrho_{\Lambda_0 \oplus \hat{\Lambda}_0}$. The quasi–isogeny $\beta\hat{\alpha}$ is up to a constant in $\mathbf{Q}_p$ an isomorphism. Then we define maps

$$\breve{\mathbf{n}}\, \breve{\mathbf{n}}^*, \breve{c} : \breve{\mathcal{M}} \longrightarrow \mathbf{Z}$$

by $\breve{\mathbf{n}} = -1/i(D) \cdot \text{height}\ \alpha$, $\breve{\mathbf{n}}^* = -1/i(D) \cdot \text{height}\ \beta$, and

$$\breve{c} = -\frac{1}{\dim_{\mathbf{Q}_p} W} \text{height}\ \beta\hat{\alpha}.$$

This gives the desired $J(\mathbf{Q}_p)$–equivariant morphism

$$\varkappa : \breve{\mathcal{M}} \longrightarrow \Delta.$$

3.54 In the end of this chapter we discuss some examples. Drinfeld[Dr2] first considered a functor $\breve{\mathcal{M}}$ in the case (EL) in the following situation (compare 1.44). Let $B = D$ be a central division algebra over F with invariant $1/d$. Let $V = D$ considered as a D-module. In the following we will also use the right D-module structure on V. It gives us an identification $G(\mathbf{Q}_p) \cong (D^{opp})^\times$. We keep the notation $\tilde{F}, \pi, \Pi, \tau, \varepsilon : F \to \bar{\mathbf{Q}}_p$ from (1.44). Since the invariant is $1/d$ we have $\Pi^d = \pi$ and the ring of integers in D is

$$O_D = O_{\tilde{F}}[\Pi], \quad \Pi x = \tau(x)\Pi, \quad x \in O_{\tilde{F}}.$$

In this example L will be the field $\bar{\mathbf{F}}_p$, which we identify with the residue class field of $\bar{\mathbf{Q}}_p$. We write $W = W(L)$ for the Witt vectors. Then $K_0 \subset \mathbf{C}_p$, the completion of $\bar{\mathbf{Q}}_p$. Let K/K_0 be a finite Galois extension contained in $\mathbf{C}_p$, such that $\varepsilon(F) \subset K$. Then we have the decompositions

$$\begin{aligned} D \otimes_{\mathbf{Q}_p} K &= \prod_{\eta: F \rightarrow K} D \otimes_{F,\eta} K \\ V \otimes_{\mathbf{Q}_p} K &= \bigoplus_{\eta: F \rightarrow K} V_\eta \,. \end{aligned}$$

We take for $V_0 \subset V \otimes_{\mathbf{Q}_p} K$ a K-vectorspace of dimension d, which is invariant by the action of D (from the left), and such that $V_0 \subset V_\varepsilon$. For $V_1 \subset V \otimes_{\mathbf{Q}_p} K$ we take any complementary space invariant by the action of D. Let $\mu : \mathbf{G}_m \rightarrow G$ be the cocharacter with weight decomposition

$$V \otimes_{\mathbf{Q}_p} K = V_0 \oplus V_1 \,.$$

It is the cocharacter μ defined under (1.44).
Finally we will define the structure of a crystal on $O_D \otimes_{\mathbf{Z}_p} W$, such that the induced isocrystal is (V, Φ) in the notation of (1.44). This will give us the σ–conjugacy class of $b \in G(K_0)$.
Let F^t be the maximal unramified extension of $\mathbf{Q}_p$ contained in F. We use the decomposition

$$O_D \otimes_{\mathbf{Z}_p} W \cong \prod_{\alpha: F^t \rightarrow K_0} O_D \otimes_{O_{F^t},\alpha} W \,.$$

Let $u \in O_D \otimes_{\mathbf{Z}_p} W$ be the element, whose components u_α with respect to this decomposition are defined as follows

$$u_\alpha = \begin{cases} \Pi & \text{if } \alpha = \varepsilon | F^t \\ 1 & \text{if } \alpha \neq \varepsilon | F^t \end{cases}$$

Let $\tilde{\mathbf{V}}$ be the σ^{-1}-linear operator on $O_D \otimes_{\mathbf{Z}_p} W$ defined by $\tilde{\mathbf{V}}x = \sigma^{-1}(x) \cdot u, \;\; x \in O_D \otimes_{\mathbf{Z}_p} W$.
Let us denote by

$$\tilde{\mathbf{M}} = (O_D \otimes_{\mathbf{Z}_p} W, \tilde{\mathbf{V}}, \tilde{\mathbf{F}}) \tag{3.5}$$

the crystal. The action of O_D from the left gives a homomorphism

$$\iota : O_D \longrightarrow \operatorname{End} \tilde{\mathbf{M}}.$$

The action of O_F via ι on the $\bar{\mathbf{F}}_p$–vectorspace $\tilde{\mathbf{M}}/\mathbf{V}\tilde{\mathbf{M}}$ is given by the homomorphism $\bar{\varepsilon} : O_F \to \bar{\mathbf{F}}_p$ induced by ε.

3.55 In this context it is more convenient to replace $\tilde{\mathbf{M}}$ by a $\tau - W_F(L)$-crystal $\mathbf{M}$ (1.40). The procedure is as follows.
Let L be any perfect field with an O_F-algebra structure $\bar{\varepsilon} : O_F \to L$. From $\bar{\varepsilon}$ we get a map $\varepsilon : O_{F^t} \to W(L)$. We set $W_F(L) = O_F \otimes_{O_{F^t},\varepsilon} W(L)$. It is a complete discrete valuation ring with residue class field L, that is unramified over O_F. The quotient field is $K_F(L)$ (1.40). Let τ be the Frobenius automorphism of $W_F(L)$ relative to O_F.
A $\tau - W_F(L)$–crystal is a free $W_F(L)$-module of finite rank M with a τ^{-1}-linear operator $V : M \to M$, such that $VM \supset \pi M$.

Proposition 3.56 *The category of $\tau - W_F(L)$–crystals is equivalent to the category of crystals $\tilde{M}$ over L with an action $\iota : O_F \to \operatorname{End} \tilde{M}$, such that the action of O_F on $\tilde{M}/\tilde{\mathbf{V}}\tilde{M}$ induced by ι coincides with the action given by $\bar{\varepsilon}$. Objects of the latter category are called O_F-crystals.*

This is well-known (Drinfeld [Dr2]). We only indicate how to obtain a $\tau - W_F(L)$-crystal from the crystal $\tilde{M}$ over L. Now $\tilde{M}$ is an $O_F \otimes_{\mathbf{Z}_p} W(L)$-module. From the decomposition

$$O_F \otimes_{\mathbf{Z}_p} W(L) = \prod_{i \in \mathbf{Z}/f} O_F \otimes_{O_{F^t},\sigma^{-i}\varepsilon} W(L)\,,$$

where $f = [F^t : \mathbf{Q}_p]$ we deduce a decomposition

$$\tilde{M} = \bigoplus_{i \in \mathbf{Z}/f} M_i\,.$$

One checks that $\tilde{\mathbf{V}} M_i \subset M_{i+1}$. The condition that the action ι on the tangent space $\tilde{M}/\tilde{\mathbf{V}}\tilde{M}$ be via ε is equivalent to the condition that the operators

$$\tilde{\mathbf{V}} : M_{i-1} \longrightarrow M_i, \quad i \neq 0$$

are bijective. We set $\mathbf{V} = \tilde{\mathbf{V}}^f$ and $M = (M_0, \mathbf{V})$. Then we have an identification $\tilde{M}/\tilde{\mathbf{V}}\tilde{M} = M/\mathbf{V}M$. Since this O_F-module is annihilated by π, we obtain in fact a $\tau - W_F(L)$–crystal:

$$\pi M \subset \mathbf{V}M \subset M\,.$$

□

We have

$$[F:\mathbf{Q}_p]\ \mathrm{rank}_{W_F(L)}M = \ \mathrm{rank}_{W(L)}\tilde{M}\,.$$

We call $\mathrm{rank}_{W_F(L)}M$ the *F-height.* Let $\tilde{\varphi}:(\tilde{M},\iota)\longrightarrow(\tilde{M}',\iota')$ be an isogeny of O_F-crystals. Then $\tilde{\varphi}$ induces a morphism φ of $\tau - W_F(L)$–crystals. We have

$$\frac{1}{f}\cdot\mathrm{height}\ \tilde{\varphi} = \frac{1}{f}\cdot\ \mathrm{length}_{W(L)}\tilde{M}'/\tilde{\varphi}(\tilde{M}) = \ \mathrm{length}_{W_F(L)}M'/\varphi(M)\,.$$

We denote this number by F-height φ, comp. (3.53).
O_F-crystals arise from p-divisible groups of the following type.

Definition 3.57 *Let S be a scheme over O_F. A p-divisible O_F-module over S is a p-divisible group X over S with an action $\iota: O_F\longrightarrow End\ X$, such that the action of O_F on the tangent space Lie X induced by ι is given by the structure morphism $O_F\longrightarrow\mathcal{O}_S$.*

3.58 The $\tau - W_F(L)$ crystal associated to the crystal $\tilde{\mathbf{M}}$ defined by (3.5) is

$$\mathbf{M} = O_D\otimes_{O_{F^t},\varepsilon}W(\bar{\mathbf{F}}_p)\,,$$

with the operator $\mathbf{V}$ defined by $\mathbf{V}m = \tau^{-1}(m)\cdot\Pi$. We will denote by $\mathbf{X}$ the associated p-divisible group over $\bar{\mathbf{F}}_p$ with the action $\iota: O_D\longrightarrow \mathrm{End}\ \ \mathbf{X}$. Let $\mathcal{L}$ be the chain of lattices in $V = D$, which consists of the lattices $\{\Pi^k O_D\}, k\in\mathbf{Z}$.
We have thus a complete set of data $(F, D, O_D, V = D, b, \mu, \mathcal{L})$ of type (3.18) (EL) which we call the *Drinfeld example.* The associated Shimura field E is $\varepsilon(F)\subset\mathbf{C}_p$. We note that ε induces an isomorphism of the field $K_F(\bar{\mathbf{F}}_p)$ with $\breve{E}$.
Let $\breve{\mathcal{M}}$ be the associated formal scheme over $Spf\ O_{\breve{E}}$. Let us denote by S a scheme over $Spf\ O_{\breve{E}}$. In the example at hand the condition (3.21) (iv) on the determinants may be expressed in a different way. The decomposition

$$O_{\tilde{F}^t} \otimes_{\mathbf{Z}_p} \mathcal{O}_S = \prod_{\alpha:\tilde{F}^t \to \breve{E}} \mathcal{O}_S$$

induces a decomposition of the tangent spaces of the p-divisible groups X involved

$$\text{Lie } X = \bigoplus_\alpha \text{Lie}^\alpha X \,.$$

Here again α ranges over all embeddings $\tilde{F}^t \to \breve{E}$. If $\alpha|F^t = \varepsilon|F^t$, there is a unique extension $\varepsilon_\alpha : \tilde{F} \to \breve{E}$ of α, such that $\varepsilon_\alpha|F = \varepsilon$.
We claim that the condition (3.21) (iv) is equivalent to the following:
If $\alpha|F^t \neq \varrho|F^t$ the $\mathcal{O}_S$-module $\text{Lie}^\alpha X$ *is zero. If $\alpha|F^t = \varepsilon|F^t$ the $\mathcal{O}_S$-module* $\text{Lie}^\alpha X$ *is locally free of rank* 1 *and $O_{\tilde{F}}$ acts on* $\text{Lie}^\alpha X$ *via ε_α.*
A formal p-divisible group X over an $O_{\breve{E}}$-scheme S with an O_D-action $\iota : O_D \to \text{End } X$ that satisfies the condition above will be called (following Drinfeld) a *special formal O_D-module.* We show that X is a special formal O_D-module if 3.21 (iv) is satisfied and leave the converse to the reader.
To do this we restrict our attention to an element $a \in O_{\tilde{F}}$ in 3.21 (iv). We denote by $\mathbf{W}$ the scheme over $\mathbf{Z}_p$, whose S-valued points are $O_{\tilde{F}} \otimes_{\mathbf{Z}_p} \Gamma(S, \mathcal{O}_S)$. The decomposition above induces a decomposition

$$\mathbf{W}_{O_{\breve{E}}} = \prod_\alpha \mathbf{W}_\alpha \,.$$

If $\alpha|F^t = \varepsilon|F^t$ we define a function $p_\alpha : \mathbf{W}_\alpha \to \mathbf{A}^1$ by the map ε_α : $O_{\tilde{F}} \otimes_{\tilde{F}^t,\alpha} \mathcal{O}_S \to \mathcal{O}_S$.
For the remaining $\alpha : F^t \to \breve{E}$ we set $p_\alpha = 1$. The right hand side of 3.21 (iv) restricted to $\mathbf{W}_K$ is Πp_α.
On the other hand consider the decomposition:

$$\text{Lie } X = \bigoplus_\alpha \text{Lie}^\alpha X \,.$$

We set

$$q_\alpha(a) = \det_{\mathcal{O}_S}(a\,; \text{Lie}^\alpha X) \text{ for } a \in O_{\tilde{F}} \otimes_{O_{\tilde{F}^t},\alpha} \mathcal{O}_S \,.$$

This is a function on $\mathbf{W}_\alpha$ which is by definition 1 if $\text{Lie}^\alpha X = 0$. Then the left hand side of 3.21 (iv) is the product of the q_α. Hence the condition is

$$\prod q_\alpha = \prod p_\alpha .$$

Since the functions p_α do not vanish everywhere there are constants $k_\alpha \in (\breve{E})^\times$ such that $q_\alpha = k_\alpha p_\alpha$. The degree of q_α is the dimension of $\mathrm{Lie}^\alpha X$. Hence we get

$$\mathrm{rank}_{\mathcal{O}_S} \mathrm{Lie}^\alpha X = \begin{cases} 0 & \alpha|F^t \neq \varepsilon|F^t \\ 1 & \alpha|F^t = \varepsilon|F^t . \end{cases}$$

Hence for $\alpha|F^t = \varepsilon|F^t$ the function q_α is a homomorphism

$$O_{\tilde{F}} \otimes_{O_{\tilde{F}^t},\alpha} \mathcal{O}_S \longrightarrow \mathcal{O}_S .$$

Then the equation $q_\alpha = k_\alpha p_\alpha$ implies that this homomorphism is ε_α. This completes the verification that X is a special formal O_D-module.

3.59 We may thus replace the condition 3.21 (iv) in the definition of the functor $\breve{\mathcal{M}}$ by the condition that X is a special formal O_D-module. In [Dr2] Drinfeld considered the subfunctor given by the condition that ϱ is of height zero.

It is obvious that the notion of a special formal O_D-module makes sense over any O_E-scheme S. If $\iota : O_D \to \mathrm{End}\, X$ is a special formal O_D-module then the action of $\iota(f)$ on Lie X for $f \in O_F$ coincides with the action of $\varepsilon(f) \in O_E \to \mathcal{O}_S$.

Let S be the spectrum of an algebraically closed field L. We denote by M the $\tau - W_F(L)$-crystal associated to X. Since O_D acts on M, we have on M the structure of an $O_D \otimes_{O_F,\varepsilon} W_F(L)$-module. We choose an embedding

$$\tilde{\varepsilon} : O_{\tilde{F}} \longrightarrow O_{\breve{E}} \longrightarrow W_F(L) \tag{3.6}$$

which extends ε. Let

$$M_i = \left\{ m \in M \, ; \, \iota(\tilde{f})m = \tau^{-i}\tilde{\varepsilon}(\tilde{f})m \right\} .$$

Then we have

$$M = \oplus M_i .$$

The operator $\iota(\Pi)$, which we denote also by Π, is $W_F(L)$-linear and homogeneous of degree 1,

$$\Pi : M_i \longrightarrow M_{i+1}\,, \quad \Pi^d = \pi\,;$$

while $\mathbf{V}$ is homogeneous of degree 1 and τ^{-1}-linear

$$\mathbf{V} : M_i \longrightarrow M_{i+1}\,, \quad \mathbf{V}(wm) = \tau^{-1}(w)\mathbf{V}m,\ w \in W_F(L).$$

The $W_F(L)$-modules M_i are free, and $M_i/\mathbf{V}M_{i-1}$ are 1-dimensional L-vector spaces. The length of the $W_F(L)$-module $M_i/\Pi M_{i-1}$ is independent of i. Since we are interested in special formal O_D-modules that are isogenous to $\mathbf{X}$, we will assume that this length is 1 or equivalently that $\mathrm{rank}_{W_F(L)} M_i = d$, i.e. F-height $X = d^2$. The isogeny class of X is uniquely determined by the $\tau - K_F(L)$–isocrystal $(M_0 \otimes \mathbf{Q}_p, \mathbf{V}^{-1}\Pi)$.

Lemma 3.60 *Over an algebraically closed field L any two special formal O_D-modules of F-height d^2 are isogenous. The group $J(\mathbf{Q}_p)$ defined by (3.22) is isomorphic to $GL_d(F)$.*

Proof: By a theorem of Dieudonné there is a unique isotypic isocrystal of slope zero and height d. Therefore it suffices to see that $(M_0 \otimes \mathbf{Q}_p, \mathbf{V}^{-1}\Pi)$ is isotypic of slope zero. Consider the maps induced by multiplication with $\iota(\Pi)$,

$$\Pi : M_i/\mathbf{V}M_{i-1} \longrightarrow M_{i+1}/\mathbf{V}M_i.$$

Since $\Pi^d = \pi = 0$ in L, we obtain that there is an index i, such that the above map is zero. We give a definition before finishing the proof.

Definition 3.61 *Let X be a special formal O_D-module over an $O_{\breve{E}}$-scheme S. We have a decomposition of the tangent space of X*

$$\mathrm{Lie}X = \bigoplus_{i \in \mathbf{Z}/d\mathbf{Z}} \mathrm{Lie}^i X.$$

Here $\iota(\tilde{f})$ for $\tilde{f} \in \tilde{F}$ acts on $\mathrm{Lie}^i(X)$ via the homomorphism $\tau^{-i}\tilde{\varepsilon} : O_{\tilde{F}} \rightarrow O_{\breve{E}} \rightarrow \mathcal{O}_S$.
We call the index i critical for X if the map

$$\Pi : \mathrm{Lie}^i X \longrightarrow \mathrm{Lie}^{i+1} X$$

is zero.

For $S = Spec\ L$ we have $\mathrm{Lie}^i X = M_i / V M_{i-1}$. We have seen that there is a critical index i for X. In this case $\Pi M_i \subset V M_i$, and since both modules have the same index in M_i we obtain $\Pi M_i = V M_i$. Hence $M_i \subset M_0 \otimes \mathbf{Q}_p$ is a lattice stable by $V^{-1}\Pi$. Therefore $M_0 \otimes \mathbf{Q}_p$ is isotypical of slope zero. □

The fact that there is only one isogeny class is the reason the formal scheme $\breve{\mathcal{M}}$ is p-adic, as we are going to prove now.
Let R be a complete noetherian local ring of characteristic p, which is equipped with a $O_{\breve{E}}$-algebra structure. Then we have an injection $\bar{\mathbf{F}}_p \to O_{\breve{E}}/pO_{\breve{E}} \to R$. Let us denote by L the residue class field of R.

Proposition 3.62 *Let X be a special formal O_D-module over R. Then any quasi-isogeny $\mathbf{X}_L \to X_L$ extends uniquely to a quasi-isogeny $\mathbf{X}_R \to X$.*

Before proving this we note:

Corollary 3.63 *In the Drinfeld example $\breve{\mathcal{M}}$ is a p-adic formal scheme locally of finite type over $Spf\ O_{\breve{E}}$.*

Proof: It is enough to show by (2.2) that $Z = \breve{\mathcal{M}} \times_{Spf\ O_{\breve{E}}} Spec\ O_{\breve{E}}/pO_{\breve{E}}$ is a scheme. Since Z is a formal scheme, it is enough to verify that a sheaf of ideals of definition is locally nilpotent.
Consider a point z of Z and let R be the completion of the local ring $\mathcal{O}_{Z,z}$. The special formal O_D-module X given over $Spf\ \mathcal{O}_{Z,z}$ extends to a special formal O_D-module on $Spec\, \mathcal{O}_{Z,z}$ and hence on $Spec\, R$, which we also denote by X. By the proposition there is a unique quasi-isogeny $\mathbf{X}_R \to X$, which extends the given quasi-isogeny over the closed point. By definition of $\breve{\mathcal{M}}$ we get a morphism $Spec\, R \to Spf\ \mathcal{O}_{Z,z}$, such that the following diagram is commutative

$$\begin{array}{ccc} Spec\, R & \longrightarrow & Spf\ \mathcal{O}_{Z,z} \\ & \searrow & \downarrow \\ & & Spec\, \mathcal{O}_{Z,z} \end{array}$$

But this means that an ideal of definition $\mathcal{I} \subset \mathcal{O}_{Z,z}$ is nilpotent in R. Therefore $\mathcal{I}$ is nilpotent itself, and the corollary is proved. □

3.64 For the proof of proposition (3.62) we shall use Cartier theory (Drinfeld [Dr2], Zink [Z2]). Let R be an O_F-algebra. We denote by $\mathbf{E}_R = \mathbf{E}$ the Cartier ring of R relative to O_F and a prime element $\pi \in O_F$. Then $\mathbf{E}_R$ is the set of all formal sums

$$\sum_{i,j \geq 0} \mathbf{V}^i [a_{ij}] \mathbf{F}^j ,$$

where i, j are integers and $a_{ij} \in R$. One requires that for fixed i only finitely many a_{ij} are non–zero.

One has the relations $\mathbf{F}[a] = [a^q]\mathbf{F}, [a]\mathbf{V} = \mathbf{V}[a^q]$, where q is the number of elements in the residue class field κ of O_F. Furthermore $\mathbf{E}_R$ is a O_F-algebra. The structure morphism maps a $(q-1)^{\text{th}}$ root of unity $\xi \in O_F$ to $[\xi] \in \mathbf{E}_R$. Moreover we have $\mathbf{FV} = \pi$ and

$$[a] + [b] = [a+b] + \sum_{i=1}^{\infty} \mathbf{V}^i [P_i(a,b)] \mathbf{F}^i ,$$

where the P_i are universal polynomials.

The category of formal O_F-modules (compare Definition (3.57)) over R is equivalent to the category of reduced Cartier modules. Let us assume that we are given a homomorphism $O_{\tilde{F}} \to R$ that extends the structure morphism $O_F \to R$. Then $\mathbf{E}_R$ is a $O_{\tilde{F}}$-algebra.

The category of special formal O_D-modules over R is equivalent to the category of triples (M, M_i, Π), where M is a reduced $\mathbf{E}_R$-module, $M = \bigoplus_{i \in \mathbf{Z}/d} M_i$ is a grading of the abelian group M and $\Pi : M \to M$ is a $\mathbf{E}_R$-module homomorphism. One requires that the following conditions are satisfied. The operators $[a], \mathbf{V}, \mathbf{F}, \Pi$ act homogeneously on M and have the degrees $\deg[a] = 0, \deg \mathbf{V} = 1, \deg \mathbf{F} = -1, \deg \Pi = 1$. We have $\Pi^d = \pi$. The R-modules $M_i/\mathbf{V}M_{i-1}$ are locally free of rank 1.

We note that the decomposition

$$M/\mathbf{V}M = \bigoplus_{i \in \mathbf{Z}/d} M_i/\mathbf{V}M_{i-1}$$

is exactly the decomposition in definition (3.61) for the corresponding special formal O_D-module X.

Assume that the R-modules $M_i/\mathbf{V}M_{i-1}$ are free for $i \in \mathbf{Z}/d$. Let $m_i \in M$, $i \in \mathbf{Z}/d$, be a $\mathbf{V}$-basis of M such that $m_i \in M_i$. Then the elements Πm_i may be uniquely expressed as follows

$$\Pi m_i = \sum_{n\geq 0} \mathbf{V}^n[a_{i,n}]m_{i-n+1}, \quad i \in \mathbf{Z}/d. \tag{3.7}$$

Conversely for any set of elements $a_{i,n} \in R, n \geq 0, i \in \mathbf{Z}/d$ such that $\prod_{i\in \mathbf{Z}/d} a_{i,0} = \pi$ there is a unique (M, M_i, Π) with a $\mathbf{V}$-basis m_i, such that the equations above are satisfied.

Let X be a special formal O_D-module over R and assume that the index $i \in \mathbf{Z}/d$ is critical for X (3.61). Then we have $\Pi M_i \subset \mathbf{V}M_i$ and because $\mathbf{V}$ is injective we get an operator

$$U = \mathbf{V}^{-1}\Pi : M_i \longrightarrow M_i.$$

Let R' be a R-algebra. We denote by $X_{R'}$ the special formal O_D-module obtained by base change, and by $M_{R'}$ its Cartier module.

With this assumption we have the following

Lemma 3.65 (Drinfeld): *For any $n \geq 0$ the functor which associates to a R-algebra R' the set of invariants of the operator U*

$$R' \longmapsto \left((M_{R'})_i/\mathbf{V}^{nd}(M_{R'})_i\right)^U$$

is representable by a scheme étale over $Spec\, R$.

Proof: Since the question is local on $Spec\, R$ we may assume that a $\mathbf{V}$-basis exists. Any element of $(M_{R'})_i/\mathbf{V}^{nd}(M_{R'})_i$ has a unique representation

$$\sum_{s=0}^{nd-1} \mathbf{V}^s[x_s]m_{i-s}, \quad x_s \in R'. \tag{3.8}$$

This identifies the functor $R' \mapsto (M_{R'})_i/\mathbf{V}^{nd}(M_{R'})_i$ with the affine space $\mathbf{A}_R^{nd}$ and U with an endomorphism of $\mathbf{A}_R^{nd}$. Hence the functor in (3.65) is representable by a scheme of finite presentation over $Spec\, R$. We show by the infinitesimal criterion that this scheme is étale.

Let $R' \to R''$ be a surjective homomorphism of R-algebras with nilpotent kernel $\mathbf{a}$. We have to show that the map

$$(M_{R'})_i/\mathbf{V}^{nd}(M_{R'})_i \twoheadrightarrow (M_{R''})_i/\mathbf{V}^{nd}(M_{R''})_i \tag{3.9}$$

induces a bijection of the U-invariants. This follows if we prove that U is nilpotent on the kernel of (3.9). An element of the kernel may be expressed in the form (3.8), where $x_s \in \mathbf{a}$. It is enough to see, that for sufficiently large N

$$\Pi^N \left(\sum_{s=0}^{nd-1} \mathbf{V}^s [x_s] m_{i-s} \right) = 0,$$

where the equality is meant in $M_{R'}$.
But for given $t \in \mathbf{N}$ we have $\Pi^N m_{i-s} \in \mathbf{V}^t M$. The desired equation is therefore clear for large t. □

Let us denote by $\eta_i^X[n]$ the étale scheme given by Lemma (3.65) on $Spec\, R$. It commutes with base change $R \to R'$:

$$\eta_i^{X_{R'}}[n] = \eta_i^X[n] \times_{Spec\, R} Spec\, R' .$$

Furthermore $\eta_i^X[n]$ has an O_F-module structure given by that of M. If X is of the minimal possible F-height d^2, we see by the case of an algebraically closed field that $\eta_i^X[n]$ is locally for the étale topology isomorphic to the constant scheme associated to $(O_F/\pi^n O_F)^d$.

3.66 We are now ready to prove proposition (3.62). Let us start with the case that there is an index i critical for X. By assumption F-height X = F-height $\mathbf{X} = d^2$. Hence $\eta_i^X[n]$ is a finite scheme locally isomorphic to $(O_F/\pi^n O_F)^d$ as a scheme with O_F-action. Since we have a quasi–isogeny $\mathbf{X}_L \to X_L$ this scheme is constant over $Spec\, L$ and hence over $Spec\, R$. Let $\gamma_1, \ldots, \gamma_d \in \varprojlim_n \eta_i^X[n]$ be a O_F-module basis. Then $\gamma_1, \ldots, \gamma_d \in \varprojlim_n M/\mathbf{V}^{nd} M = M$, and we have $\Pi\gamma_i = \mathbf{V}\gamma_i$.
The Cartier module $\mathbf{M}_R$ of the special formal O_D-module $\mathbf{X}_R$ is given by the equations $\Pi m_k = \mathbf{V} m_k, k \in \mathbf{Z}/d$ in the sense of (3.7). We get a map of Cartier modules

$$\begin{array}{ccll} \mathbf{M}_R & \longrightarrow & M & \\ m_k & \longmapsto & \mathbf{V}^{k-i}\gamma_k, & k = i, i+1, \ldots, i+d-1 \end{array}$$

Using the fibre criterion for isogenies (Zink [Z3]) we see that the morphism of formal groups $\mathbf{X}_R \to X$ induced by this map is an isogeny of height $d(d-1)$.

If we multiply this isogeny $\mathbf{X}_R \to X$ by a suitable quasi–isogeny of $\mathbf{X}$, we get the desired lifting of $\mathbf{X}_L \to X_L$.
Let us now consider the general case. Then $S = Spec\, R$ is the union of the closed subsets $S_i \subset S$, where the index i is critical. Let us more generally consider two p–divisible groups X and Y over a noetherian scheme S, where p is locally nilpotent. By the rigidity property (see after definition (2.8)) it makes sense to speak of a quasi–isogeny from X to Y over a closed subset $T \subset S$.

Lemma 3.67 *Let S be a union of finitely many closed subsets $S = S_1 \cup \ldots \cup S_r$. Assume we are given quasi–isogenies $\varphi_i : X_{S_i} \to Y_{S_i}$ for $1, \ldots, r$, such that φ_i and φ_j agree on $S_i \cap S_j$ for any i and j. Then there is a unique quasi–isogeny $\varphi : X \to Y$ over S, which induces the φ_i.*

Proof: It is enough to prove this for a covering with two closed subsets $S = S_1 \cup S_2$. We may suppose that S is affine, $S = Spec\, A$, $S_1 = Spec\, A_1$, $S_2 = Spec\, A_2$, $S_1 \cap S_2 = Spec\, B$. Since nilpotent elements play no role, we may assume that A is the difference kernel $A \to A_1 \times A_2 \rightrightarrows B$. We see that for two flat affine group schemes G and H over A there is an exact sequence $0 \to \mathrm{Hom}\,(G, H) \to \mathrm{Hom}\,(G_{A_1}, H_{A_1}) \oplus \mathrm{Hom}\,(G_{A_2}, H_{A_2}) \rightrightarrows \mathrm{Hom}\,(G_B, H_B)$. The lemma follows easily.

Since we have shown that the quasi–isogeny $\mathbf{X}_L \to X$ uniquely lifts to any S_i the proposition (3.62) follows from this lemma. □

3.68 We give now the description of the p-adic formal scheme $\mathcal{M}$, which is due to Drinfeld.
We recall the definition of the Bruhat-Tits-building $\mathcal{B}$ of $PGL_d(F)$. It is a simplicial complex, whose 0-simplices are equivalence classes $\bar{\eta}$ of O_F-lattices modulo homothety. For two lattices η and η' we define the logarithmic index

$$\log[\eta : \eta'] = \mathrm{length}_{O_F}\eta/(\eta \cap \eta') - \mathrm{length}_{O_F}\eta'/(\eta \cap \eta').$$

A simplex is a set of 0-simplices $\Delta = \{\bar{\eta}_{i_0}, \ldots, \bar{\eta}_{i_r}\}, 0 \leq i_0 < i_1 < \cdots < i_r < d$ such that there are representatives $\eta_{i_k} \in \bar{\eta}_{i_k}$, such that

$$\pi\eta_{i_r} \subset \eta_{i_0} \subset \cdots \subset \eta_{i_r}. \qquad (3.10)$$

Since we have different homothety classes of lattices the inclusions are proper. We assume that the indices are chosen in such a way that $\log[\eta_{i_k} :$

$\eta_{i_{k-1}}] = i_k - i_{k-1}, k = 1, \ldots r$. Of course there are many choices of indices $i_0, \ldots, i_r$ for the same simplex Δ.
Let us fix an integer h. Then for any simplex Δ there is a unique choice of indices and a unique choice of representatives $\eta_{i_k} \in \bar{\eta}_{i_k}$ satisfying (3.10) and such that

$$\log\left[\eta_{i_k} : O_F^d\right] = i_k - h\,. \tag{3.11}$$

We view $\mathcal{B} \times \mathbf{Z}$ as a disjoint union of $\mathbf{Z}$ copies of $\mathcal{B}$. Then a simplex of $\mathcal{B} \times \{h\}$ is a chain of lattices (3.10) such that (3.11) holds.
The group $GL_d(F)$ acts naturally on $\mathcal{B} \times \mathbf{Z}$, such that the action on the second factor is translation by $\mathrm{ord}_F \det g, g \in GL_d(F)$ and the action on the first factor is via the projection $GL_d \to PGL_d$. More explicitly, if $\Delta \times \{h\}$ is a simplex of $\mathcal{B} \times \{h\}$, which looks in the canonical representation (3.10) as follows

$$\pi\eta_{i_r} \subset \eta_{i_0} \subset \cdots \subset \eta_{i_r}\,,$$

then the canonical representation of $g\Delta$ is of the form

$$\pi\lambda_{i_r} \subset \lambda_{i_0} \subset \cdots \subset \lambda_{i_r}\,,$$

where $\lambda_{i_k} = g\eta_{i_k}$.

3.69 Let us consider the category $Nilp_{O_F}$ of O_F-schemes S, such that $p \in \mathcal{O}_S$ is locally nilpotent. For any simplex Δ of $\mathcal{B} \times \mathbf{Z}$ we are going to define a functor $\hat{U}_\Delta$ on $Nilp_{O_F}$. We always assume that Δ is given in its canonical representation (3.11).
A point of $\hat{U}_\Delta(S)$ is given by an isomorphism class of diagrams

$$\begin{array}{ccccccccc} \eta_{i_0} & \subset & \eta_{i_1} & \subset & \cdots & \subset & \eta_{i_r} & \xrightarrow{\pi} & \eta_{i_0} \\ \varphi_{i_0}\downarrow & & \varphi_{i_1}\downarrow & & & & \downarrow & & \downarrow \\ \mathcal{L}_{i_0} & \to & \mathcal{L}_{i_1} & \to & & \to & \mathcal{L}_{i_r} & \to & \mathcal{L}_{i_0} \end{array} \tag{3.12}$$

Here we denote by $\mathcal{L}_{i_k}$ invertible $\mathcal{O}_S$-modules. The lower horizontal arrows are $\mathcal{O}_S$-module homomorphisms. The φ_{i_k} are O_F-module homomorphisms. We require that for any $n \in \eta_{i_k} \backslash \eta_{i_{k-1}}$ the section $\varphi_{i_k}(n)$ of $\mathcal{L}_{i_k}$ vanishes nowhere on S.
It is not difficult to check that $\hat{U}_\Delta$ is a formal π-adic scheme over $Spf\, O_F$ isomorphic to

$$Spf\ \left(O_F\left[T_0,\ldots,T_r,U_{r+1},\ldots,U_d,U_{r+1}^{-1},\ldots,U_d^{-1}\right]/T_0\cdot\ldots\cdot T_r-\pi\right)^{\wedge}. \tag{3.13}$$

The symbol $^\wedge$ denotes the π-adic completion.
Fix an index k and let $\Delta' = \Delta\backslash\{\eta_{i_k}\}$. Consider the morphism $\hat{U}_{\Delta'} \to \hat{U}_\Delta$, which associates to a point $\eta_{i_l} \to \mathcal{L}_{i_l}$, $l = 0,\ldots,r$ and $l \neq k$, the point

$$\begin{array}{ccccccccc} \eta_{i_0} & \to\cdots\to & \eta_{i_{k-1}} & \to & \eta_{i_k} & \to & \eta_{i_{k+1}} & \to\cdots \\ \downarrow & & \downarrow & & \downarrow & & \downarrow & \\ \mathcal{L}_{i_0} & \to & \mathcal{L}_{i_{k-1}} & \to & \mathcal{L}_{i_{k+1}} & = & \mathcal{L}_{i_{k+1}} & \to\cdots \end{array}$$

Then $\hat{U}_{\Delta'} \to \hat{U}_\Delta$ is an open immersion.
Let us denote by $\hat{\Xi}_F^d$ the union of the formal schemes $\hat{U}_\Delta$ with respect to the open immersions defined above.

Proposition 3.70 *The formal scheme $\hat{\Xi}_F^d$ is separated over O_F.*

Proof: A scheme X over O_F is separated if there exists an open covering $\{U_i\}_{i\in I}$, such that for all $i,j \in I$ the canonical immersions $U_i \cap U_j \to U_i \times_{O_F} U_j$ are closed. Since we have $\hat{U}_\Delta \cap \hat{U}_\Gamma = \hat{U}_{\Delta\cap\Gamma}$ by definition of $\hat{\Xi}$, we need to verify that the canonical morphism $\hat{U}_{\Delta\cap\Gamma} \to \hat{U}_\Delta \times_{O_F} \hat{U}_\Gamma$ is a closed immersion. We leave this to the reader. □

3.71 There is a left action of $GL_d(F)$ on $\hat{\Xi}_F^d$. An element $g \in GL_d(F)$ transforms a S-valued point $\varphi_\bullet : \eta_\bullet \to \mathcal{L}_\bullet$ to $\varphi_\bullet g^{-1} : g\eta_\bullet \to \mathcal{L}$. The units in the center of $GL_d(F)$ act trivially.
The projection $\mathcal{B} \times \mathbf{Z} \to \mathbf{Z}$ induces a natural morphism

$$\hat{\Xi}_F^d \longrightarrow \underline{\mathbf{Z}},$$

where $\underline{\mathbf{Z}}$ denotes the constant formal scheme over $Spf\, O_F$ associated to $\mathbf{Z}$. Following Drinfeld we denote the fibre over 0 of this morphism by $\hat{\Omega}_F^d$. We note that the fibres over different connected components $n : Spf\, O_F \to \underline{\mathbf{Z}}$ are all canonically isomorphic. Indeed for a simplex $\Delta \in \mathcal{B} \times \mathbf{Z}$ the functor $\hat{U}_\Delta$, depends only on the projection of Δ to $\mathcal{B}$. Therefore we obtain a canonical isomorphism

$$\hat{\Xi}_F^d \longrightarrow \hat{\Omega}_F^d \times \mathbf{Z}.$$

The additive group $\mathbf{Z}$ acts via the second factor on $\hat{\Xi}^d_F$. The action of $m \in \mathbf{Z}$ takes a point $\eta_{i_k} \to \mathcal{L}_{i_k}$ to the point $\{\eta'_{i_k+m} \to \mathcal{L}'_{i_k+m}\}$, where $\eta'_{i_k+m} = \eta_{i_k}$, $\mathcal{L}'_{i_k+m} = \mathcal{L}_{i_k}$. We call this the *translation by m.*

Theorem 3.72 (Drinfeld): *There is an isomorphism of formal schemes*

$$\breve{\mathcal{M}} \longrightarrow \hat{\Xi}^d_F \times_{Spf\, O_F} Spf\, O_{\breve{E}},$$

where the morphism $Spf\, O_{\breve{E}} \to Spf\, O_F$ is given by ϱ. For a suitable isomorphism $GL_d(F) \simeq J(\mathbf{Q}_p)$ this map is equivariant. The Weil descent datum on $\breve{\mathcal{M}}$ gives on the right hand side the composite of the canonical Weil descent datum and translation by 1. The translation by m on $\hat{\Xi}^d_F \times_{Spf\, O_F} Spf\, O_{\breve{E}}$ induces on $\breve{\mathcal{M}}$ the morphism which associates to a point (X, ϱ) the point $(X^{\Pi^m}, \Pi^{-m}\varrho)$, where Π^{-m} here denotes the morphism $X \to X^{\Pi^m}$ defined by (3.20).

For the proof we refer to Drinfeld (loc. cit.).

3.73 We will give here a few comments on the proof which we will use later. Let us denote by $\mathbf{M}$ the $\tau - W_F(\bar{\mathbf{F}}_p)$ crystal of $\mathbf{X}$,

$$\mathbf{M} = O_D \otimes_{O_F} W_F(\bar{\mathbf{F}}_p).$$

Note that there is a unique isomorphism $W_F(\bar{\mathbf{F}}_p) \simeq O_{\breve{E}}$ that induces the given $O_{\breve{E}}$-algebra structure on $\bar{\mathbf{F}}_p$, and such that

$$O_F \longrightarrow W_F(\bar{\mathbf{F}}_p) \longrightarrow O_{\breve{E}}$$

is the embedding ε. We use this isomorphism to identify $W_F(\bar{\mathbf{F}}_p)$ and $O_{\breve{E}}$. With respect to the choice of the extension $\tilde{\varepsilon}$ of ε to $O_{\tilde{F}}$ (3.6) we have the decomposition

$$\mathbf{M} = \bigoplus \mathbf{M}_i.$$

We may write

$$\mathbf{M}_i = W_F(\bar{\mathbf{F}}_p) \otimes_{\tau^{-i}\tilde{\varepsilon}, O_{\tilde{F}}} O_D.$$

Any index $i \in \mathbf{Z}/d$ is critical with respect to $\mathbf{M}$, i.e. $\mathbf{V}\mathbf{M}_i = \Pi\mathbf{M}_i$. The operator $\mathbf{V}^{-1}\Pi : \mathbf{M}_i \to \mathbf{M}_i$ is given by the formula

$$\mathbf{V}^{-1}\Pi(w\otimes x)=\tau(w)\otimes \Pi x\Pi^{-1}, \quad w\in W_F(\bar{\mathbf{F}}_p), x\in O_D\,.$$

The invariants $\Lambda_i \subset \mathbf{M}_i$ of this operator consist exactly of the elements

$$w\otimes \Pi^j, \quad w\in O_F, \quad 0\leq j<d\,.$$

Let us denote by $e_1,\ldots,e_d$ the standard basis of F^d.
If n is an integer we denote by $\tilde{n}$ the integer given by the conditions $0\leq \tilde{n}<d$ and $n\equiv \tilde{n} \bmod d$. We use the same notation for $n\in \mathbf{Z}/d$. We define O_F–linear embeddings

$$\begin{array}{ccc}\Lambda_i & \longrightarrow & F^d\\ \Pi^j & \longmapsto & \pi^{\varepsilon(i,j)}e_{\widetilde{i+j+1}}\,,\end{array}$$

where $\varepsilon(i,j)=0$ if $\tilde{i}+\tilde{j}<d$, and $\varepsilon(i,j)=1$ if $\tilde{i}+\tilde{j}\geq d$. Let us denote the image of this embedding by λ_i. Then we have a commutative diagram

$$\begin{array}{ccccccccccc}\Lambda_{d-1} & \stackrel{\Pi}{\rightarrow} & \Lambda_0 & \stackrel{\Pi}{\rightarrow} & \Lambda_1 & \cdots\rightarrow & \Lambda_{d-1} & \stackrel{\Pi}{\rightarrow} & \Lambda_0 & & \\ \pi\downarrow\simeq & & \downarrow\simeq & & \downarrow\simeq & & \downarrow\simeq & & \pi^{-1}\downarrow\simeq & & \cdots\\ \pi\lambda_{d-1} & \subset & \lambda_0 & \subset & \lambda_1 & \subset & \lambda_{d-1} & \subset & \pi^{-1}\lambda_0 & & \end{array}$$

3.74 Let X be a special formal O_D-module over a $O_{\breve{E}}$-scheme S of characteristic p. Assume $\varrho:\mathbf{X}_S\rightarrow X$ is a quasi–isogeny that gives a point of $\breve{\mathcal{M}}$. Let $\eta_i^X[n]$ be the étale sheaf of lemma (3.65), and

$$\eta_i^X=\varprojlim_n \eta_i^X[n]\,.$$

The quasi–isogeny ϱ gives a map $\lambda_i\rightarrow \eta_i^X$. We use it to identify the constant sheaf η_i^X with a lattice in F^d. We have the canonical map

$$\eta_i^X\longrightarrow M_i\longrightarrow M_i/\mathbf{V}M_i=\mathrm{Lie}^i X\,, \tag{3.14}$$

where $\oplus M_i$ denotes the Cartier module of X.
Assume that the maps $\Pi:\mathrm{Lie}^j X\rightarrow \mathrm{Lie}^{j+1}X$ are isomorphisms for $j\neq i$. Then the morphism in (3.72) associates to ϱ the point of $\hat{U}_{\{\eta_i\}}$ given by (3.14). The canonical index of the 0-simplex $\{\eta_i\}$ in the sense of (3.11) is $\tilde{i}$.

If S is the spectrum of a perfect field L, the morphism in (3.72) looks on the L-valued points as follows. Let $0 \leq i_0 < \cdots < i_r < d$ be representatives for the critical indices of X. Again the quasi–isogeny ϱ allows us to identify $\eta^X_{i_k}$ with a lattice in F^d. The canonical maps (3.14) fit into a commutative diagram

$$\begin{array}{ccccccc}
\eta_{i_0} & \subset & \eta_{i_1} & \subset\cdots\subset & \eta_{i_r} & \xrightarrow{\Pi} & \eta_{i_0} \\
\downarrow & & \downarrow & & \downarrow & & \downarrow \\
\mathrm{Lie}^{i_0} X & \underset{0}{\rightarrow} & \mathrm{Lie}^{i_1} X & \underset{0}{\rightarrow}\cdots\underset{0}{\rightarrow} & \mathrm{Lie}^{i_r} X & \underset{0}{\rightarrow} & \mathrm{Lie}^{i_0} X\,.
\end{array}$$

It is easy to check that this is a point of $\hat{U}_\Delta$, where $\Delta \subset \mathcal{B} \times \mathbf{Z}$ is the simplex $\{\eta_{i_k}\}$ with the canonical indices i_k.

3.75 To make this definition work over any base S, Drinfeld proposes the following construction.

Let again $S = Spec\, L$. Let $i \in \mathbf{Z}/d$ and j be the first critical index that follows i. If i is critical, j is by definition equal to i. Let $s \geq 0$ be the smallest integer, such that $i + s \equiv j \bmod d$.

Consider in $M_i^{s+1} = M_i \oplus \cdots \oplus M_i$ the abelian subgroup U_i^{s+1}, which is generated by elements of the form

$$(0, \ldots, 0, Vx, -\Pi x, 0, \ldots, 0)\,.$$

Then there is an isomorphism

$$\begin{array}{ccc}
M_i^{s+1}/U_i^{s+1} & \xrightarrow{\sim} & M_j \\
(x_0, \ldots, x_s) & \longmapsto & \Pi^s x_0 + \Pi^{s-1} V x_1 + \cdots + V^s x_s.
\end{array}$$

The operator $\mathbf{V}^{-1}\Pi$ on M_j induces on the left hand side of this isomorphism an operator of the form

$$(x_0 \ldots x_s) \longmapsto \mathbf{L} x_0 + (x_1, \ldots, x_s, 0) \tag{3.15}$$

where $\mathbf{L} : M_i \rightarrow M_i^{s+1}/U_i^{s+1}$ is a homomorphism of abelian groups. One defines an operator

$$\varphi : M_i^d/U_i^d \longrightarrow M_i^d/U_i^d$$

by the formula $\varphi(x_0, \ldots, x_{d-1}) = \mathbf{L}x_0 + (x_1, \ldots, x_{d-1}, 0)$, where $\mathbf{L}$ is the composite of the $\mathbf{L}$ above with $M_i^{s+1}/U_i^{s+1} \to M_i^d/U_i^d$.
The invariants $(M_i^d/U_i^d)^{\varphi}$ are the same as those of the operator (3.15) on M_i^{s+1}/U_i^{s+1}, i.e. via the isomorphism with M_j, equal to η_j^X.
The construction of φ makes sense for the Cartier module of a special formal O_D-module X over any base S and gives the desired morphism (3.72). By the case of a perfect field it is radical and surjective. The étaleness is verified in a standard way from the Grothendieck-Messing criterion.

3.76 Drinfeld's theorem shows that the local equations of the formal scheme $\breve{\mathcal{M}}$ are given by (3.13). We may also obtain the local equations by computing $\mathbf{M}^{loc}$, and then applying proposition (3.33).
Let us consider the definition (3.27) in our case. The chain of lattices $\mathcal{L}$ consists of a single homothety class $O_D \subset V$ of O_D-lattices. Assume we are given a point $O_D \to t$ of $\mathbf{M}^{loc}$ over a $O_{\breve{E}}$-scheme S. We fix an embedding $\tilde{\varepsilon} : \tilde{F} \to \breve{E}$. Again by condition (3.27, (i)) t is a direct sum of line bundles

$$t = \bigoplus_{i \in \mathbf{Z}/d} t^i \, .$$

Here $O_{\tilde{F}} \subset O_D$ acts on t^i via the embedding $O_{\tilde{F}} \xrightarrow{\tau^{-i}\tilde{\varepsilon}} O_{\breve{E}} \to \mathcal{O}_S$, and $\Pi \in O_D$ acts as an operator of degree 1. Let $\Gamma \subset O_D$ be the free O_F-submodule with basis $1, \Pi, \ldots, \Pi^{d-1}$. To give a $O_{\tilde{F}}$-linear map $O_D \to t^i$ is equivalent to giving a O_F-linear map $\varphi_i : \Gamma \to t^i$. The condition that the φ_i define a O_D-linear map amounts to the requirement that the following diagram is commutative

$$\begin{array}{ccccc} \cdots \longrightarrow \Gamma & \xrightarrow{\Pi} & \Gamma \longrightarrow \cdots \\ \varphi_i \downarrow & & \downarrow \varphi_{i+1} \\ \cdots \longrightarrow t_i & \longrightarrow & t_{i+1} \longrightarrow \cdots \end{array} \tag{3.16}$$

Hence $\breve{\mathbf{M}}^{loc}$ is the functor of those diagrams (3.16) which satisfy the condition that $\varphi_i \otimes \mathcal{O}_S$ is surjective.
The φ_i define a closed embedding

$$\breve{\mathbf{M}}^{loc} \longrightarrow \mathbf{P}^{d-1}_{O_{\breve{E}}} \times \ldots \times \mathbf{P}^{d-1}_{O_{\breve{E}}} \, ,$$

where we have one copy of the projective space for each $i \in \mathbf{Z}/d$. Let us denote by $T_0^{(i)}, \ldots, T_{d-1}^{(i)}$ the homogeneous coordinates on the ith copy. Then $\breve{\mathbf{M}}^{loc}$ is the closed subscheme given by the equation

$$\begin{aligned} T_{j-1}^{(i)} T_k^{(i+1)} &= T_j^{(i+1)} T_{k-1}^{(i)} \qquad j < k < 0 \\ T_{j-1}^{(i)} T_0^{(i+1)} &= \pi T_j^{(i+1)} T_{d-1}^{(i)} \qquad j < 0. \end{aligned}$$

The scheme $\breve{\mathbf{M}}^{loc}$ has indeed the same local equations as those given by (3.13).

Proposition 3.77 *Let $\mathcal{M}$ be the pro-formal scheme over $Spf\, O_E$ (3.52) associated to the Drinfeld example. Then there is an isomorphism of functors over $Spf\, O_E$*

$$\mathcal{M} \longrightarrow \hat{\Omega}_F^d \times_{Spf\, O_F} Spf\, O_{\breve{E}} \,.$$

Proof: By Drinfeld's theorem (3.72) the right hand side represents the following functor G on the category $Nilp_{O_E}$. A point of $G(S)$ is a triple (φ, X, ϱ) up to isomorphism. Here $\varphi : S \to Spf\, O_{\breve{E}}$ is a morphism over $Spf\, O_E$, X is a special formal O_D-module over S. Let us denote as before the reduction modulo p by a bar. Then ϱ is a quasi-isogeny of height zero of special formal O_D-modules

$$\varrho : \bar{\varphi}^* \mathbf{X} \longrightarrow X_{\bar{S}} \,.$$

For $\mathbf{X}$ we take the special formal O_D-module given by the $\tau - W_F(L)$-crystal associated to (3.5).
Let $S \in Nilp_{O_{\breve{E}}}$ be an object and $\psi : S \to Spf\, O_{\breve{E}}$ the structure morphism. We define a morphism of functors

$$\begin{aligned} \breve{\mathcal{M}}(S) &\longrightarrow G \times_{Spf\, O_E} Spf\, O_{\breve{E}} . \\ (X, \varrho) &\longmapsto (\tau^m \psi, X, \varrho \bar{\psi}^* (\mathrm{Frob}_{\bar{E}}^{-m})) \times \psi \end{aligned}$$

Here m denotes the F-height of ϱ which is locally constant. Recall that on $\breve{\mathcal{M}}$ we have a Weil descent datum given by

$$\begin{aligned} \breve{\mathcal{M}}(S) &\longrightarrow \breve{\mathcal{M}}(S_{[\tau]}) \\ (X, \varrho) &\longmapsto (X, \varrho \bar{\psi}^* (\mathrm{Frob}_{\bar{E}}^{-1})) \,. \end{aligned}$$

We give $G \times_{Spf\, O_E} Spf\, O_{\breve{E}}$ its canonical Weil-descent datum. One checks immediately using the expression for the descent datum given by Drinfeld's theorem (3.72) that the functor morphism above respects these descent data. Since the p-divisible group $\mathbf{X}$ is isoclinic, we get that the action of γ_{nd^2s} on $\breve{\mathcal{M}}(S)$ (see 3.41) is multiplication by p^{ds}. On the right hand side of (3.77) we get an action of γ_{nd^2s} on G. The natural morphism $G \to Spf\, O_{\breve{E}}$ becomes equivariant, if γ_{nd^2s} acts on $Spf\, O_{\breve{E}}$ by τ^{nd^2s}.
If we denote by G_s respectively $Spf\, O_{E_s}$ the quotients with respect to the action of γ_{nd^2s} on G respectively $Spf\, O_{\breve{E}}$, we get a cartesian diagram

$$\begin{array}{ccc} G & \longrightarrow & Spf\, O_{\breve{E}} \\ \downarrow & & \downarrow \\ G_s & \longrightarrow & Spf\, O_{E_s} \end{array} .$$

Taking this into account, we obtain

$$\mathcal{M} \cong \varprojlim G_s \cong G .$$

In terms of the functor G the action of $J(\mathbf{Q}_p)$ on $\mathcal{M}$ takes the following form

$$g(\varphi, X, \varrho) \longmapsto \left(\tau^{-\mathrm{ord}_F \det g}\varphi, X, \varrho g^{-1}\mathrm{Frob}_E^{\mathrm{ord}_F \det g}\right) ,$$

where $g \in J(\mathbf{Q}_p) \cong GL_d(F)$.

3.78 We now discuss an example for the rational data ($F = B, V = F^d$) introduced in (1.47). In this case we take as a lattice chain $\mathcal{L}$ the multiples of the standard lattice O_F^d. An argument similar to the one used in the analysis of the Drinfeld example (3.58) shows that our moduli problem $\breve{\mathcal{M}}$ (for $L = \bar{\mathbf{F}}_p$) is given as follows. Let $\mathbf{X}$ be a p–divisible group over $\bar{\mathbf{F}}_p$ with isocrystal equal to $(V \otimes K_0, b\sigma)$. A point of $\breve{\mathcal{M}}$ with values in $S \in Nilp_{O_{\breve{E}}}$ is given by a pair (X, ϱ) consisting of a p–divisible group over S with an action of O_F and a quasi–isogeny $\varrho : \mathbf{X} \times_{Spec\, \bar{\mathbf{F}}_p} \bar{S} \to X \times_S \bar{S}$ which commutes with the action of O_F. The determinant condition is equivalent to the condition that the induced action of O_F on $Lie\, X$ is the natural one, after identifying F with $E = \varepsilon(F)$. In other words, the fibres of X are formal O_F–modules of dimension 1 and F-height d, comp. [HG2]. The corresponding infinitesimal deformation functor was studied by Lubin and Tate [LT], and Drinfeld [Dr1], comp. [HG2].

Proposition 3.79 *The formal scheme $\breve{\mathcal{M}}$ is (non–canonically) isomorphic to a disjoint sum of copies of Spf $W(\bar{\mathbf{F}}_p)\,[[T_1,\ldots,T_{d-1}]]$,*

$$\breve{\mathcal{M}} \simeq \coprod_{h\in\mathbf{Z}} Spf\ W(\bar{\mathbf{F}}_p)\,[[T_1,\ldots,T_{d-1}]]$$

Proof: Let $(X,\varrho) \in \breve{\mathcal{M}}(\bar{\mathbf{F}}_p)$ with height $\varrho = h$. Let $\tilde{\varrho} : \mathbf{X} \to \mathbf{X}$ be a quasi–isogeny of height $-h$. Then

$$\varrho\tilde{\varrho} : \mathbf{X} \longrightarrow X$$

is a quasi–isogeny of height 0 between formal O_F–modules of dimension 1 over an algebraically closed field and hence an isomorphism (comp. [HG2]). We obtain a bijection given by the height,

$$\breve{\mathcal{M}}(\bar{\mathbf{F}}_p) = \mathbf{Z}.$$

Let $\breve{\mathcal{M}}^{(h)}$ be the open and closed subfunctor of $\breve{\mathcal{M}}$ where the height of ϱ is equal to h. Then $\breve{\mathcal{M}}^{(h)}(\bar{\mathbf{F}}_p)$ has only one point. Since $\breve{\mathcal{M}}^{(h)}$ is formally locally of finite type over $Spf\ O_{\breve{E}}$ it follows that $\breve{\mathcal{M}}$ is of the form

$$\breve{\mathcal{M}}^{(h)} = Spf\ A,$$

where A is a complete local ring with residue field $\bar{\mathbf{F}}_p$. Let (X_0,ϱ_0) be the special fibre of the universal object over $Spf\ A$ and let $(\tilde{X},\tilde{\varrho})$ be its universal formal deformation over $Spf\ W(\bar{\mathbf{F}}_p)\,[[T_1,\ldots,T_{d-1}]]$, comp. [HG2]. Then $(\tilde{X},\tilde{\varrho})$ defines an object of $\breve{\mathcal{M}}^{(h)}$, i.e. a local homomorphism

$$A \longrightarrow W(\bar{\mathbf{F}}_p)\,[[T_1,\ldots,T_{d-1}]]\,.$$

Conversely the universal object over $\breve{\mathcal{M}}^{(h)}$ is an infinitesimal deformation, i.e. defines a local homomorphism,

$$W(\bar{\mathbf{F}}_p)[[T_1,\ldots,T_{d-1}]] \longrightarrow A.$$

Universality shows that these are mutually inverse isomorphisms. □

3.80 We now consider the example (1.50). It is of type (EL) with $F = B = \mathbf{Q}_p$ and with $V = \mathbf{Q}_p^{2n}$. Let the chain $\mathcal{L}$ be given by the multiples of the standard lattice $\Lambda_0 = \mathbf{Z}_p^{2n}$. Let $\mathbf{X}$ be the p–divisible group over $\bar{\mathbf{F}}_p$,

$$\mathbf{X} = \hat{\mathbf{G}}_m^n \times (\mathbf{Q}_p/\mathbf{Z}_p)^n.$$

Its isocrystal is equal to $(V \otimes K_0, b\sigma)$.

Proposition 3.81 *The formal scheme $\breve{\mathcal{M}}$ is isomorphic to a disjoint sum of copies of the formal spectrum of the ring of formal power series in n^2 variables with coefficients in $W(\bar{\mathbf{F}}_p)$,*

$$\breve{\mathcal{M}} = \coprod_{GL_n(\mathbf{Q}_p)/GL_n(\mathbf{Z}_p)\times GL_n(\mathbf{Q}_p)/GL_n(\mathbf{Z}_p)} Spf\ W(\bar{\mathbf{F}}_p)\,[[T_{11},\ldots,T_{nn}]]\,.$$

Proof: Let $S \in Nilp_{W(\bar{\mathbf{F}}_p)}$ and let $(X,\varrho) \in \breve{\mathcal{M}}(S)$. The existence of the quasi-isogeny ϱ shows that the function

$$s \longmapsto \text{sep. rank } X[1]_s, \quad s \in S$$

which associates to a point s the separable rank of the finite group scheme of p-division in X_s is constant. It follows [Gr1], II. 4.9. that X is an extension of an ind-étale p-divisible group by a connected p-divisible group,

$$0 \longrightarrow X^0 \longrightarrow X \longrightarrow X^{et} \longrightarrow 0.$$

The quasi-isogeny respects this extension structure and induces therefore quasi-isogenies

$$\varrho^0 : \hat{\mathbf{G}}^n_{m\bar{S}} \longrightarrow X^0_{\bar{S}}, \quad \varrho^{et} : (\mathbf{Q}_p/\mathbf{Z}_p)^n_{\bar{S}} \longrightarrow X^{et}_{\bar{S}}.$$

The fibre of ϱ^{et} resp. ϱ^0 in a point $s \in S$ defines an element $g_1(s) \in GL_n(\mathbf{Q}_p)/GL_n(\mathbf{Z}_p)$ resp. $g_2(s) \in GL_n(\mathbf{Q}_p)/GL_n(\mathbf{Z}_p)$ and the functions

$$s \longmapsto g_i(s), \quad i = 1,2$$

are locally constant on S. Correspondingly we obtain a decomposition of $\breve{\mathcal{M}}$ into a disjoint sum

$$\breve{\mathcal{M}} = \coprod_{(g_1,g_2)} \breve{\mathcal{M}}^{(g_1,g_2)},$$

where $\breve{\mathcal{M}}^{(g_1,g_2)}$ is the open and closed subfunctor where the functions $s \mapsto g_i(s)$ are constant of value g_i.
Let $(X,\varrho) \in \breve{\mathcal{M}}^{(g_1,g_2)}(\bar{\mathbf{F}}_p)$. Then (X,ϱ) is isomorphic to

$$(\mathbf{X}, \tilde{g}_1 \times \tilde{g}_2 : \mathbf{X} \longrightarrow \mathbf{X}),$$

where $\tilde{g}_i \in GL_n(\mathbf{Q}_p)$ are representatives of $g_i \in GL_n(\mathbf{Q}_p)/GL_n(\mathbf{Z}_p)$. Therefore $\breve{\mathcal{M}}^{(g_1,g_2)}(\bar{\mathbf{F}}_p)$ consists of a single point. The rest of the argument is similar to that used in proposition (3.55), with the result of Lubin–Tate being replaced by the fact that the universal formal deformation of $(\mathbf{X}, \varrho)$ is represented by the formal torus $\hat{T}$ with character group $\mathbf{Z}^{n^2}$, cf. [DI]. More precisely ([DI], p.131), let

$$e_1, \ldots, e_n \text{ be the standard basis of } \Lambda_{0-} = \Lambda_0 \cap V_-,$$
$$e_{n+1}, \ldots, e_{2n} \text{ be the standard basis of } \Lambda_{0+} = \Lambda_0 \cap V_+.$$

Let

$$q_{ij} \in \mathrm{Hom}\,(\Lambda_{0-}, \Lambda_{0+}), \quad i, j = 1, \ldots, n$$

be the element which sends e_i into e_{n+j} and all other basis elements to zero. Put $T_{ij} = q_{ij} - 1$. Then the universal deformation space is canonically isomorphic to

$$Spf\, W(\bar{\mathbf{F}}_p)[[T_{11}, \ldots, T_{nn}]].$$

3.82 We call a set of data $(B, F, O_B, V, b, \mathcal{L})$ of type (EL) *unramified* if B is a product of matrix algebras over unramified extensions of $\mathbf{Q}_p$ and if the multichain $\mathcal{L}$ is a product of chains of lattices consisting of multiples (by powers of p) of a single lattice. In the case (PEL) we require in addition that in each of the factor chains there is one member which is selfdual with respect to the given alternating form. In other words, in the unramified case the data $\mathcal{L}$ is completely determined by giving a single O_B–lattice Λ in V which in case (PEL) is supposed to be selfdual. In the unramified case the Shimura field E associated to a set of data of our moduli problem, $(F, B, O_B, V, b, \mu, \mathcal{L})$, is an unramified extension of $\mathbf{Q}_p$ and hence $\breve{E} = K_0(L)$. An object of our moduli problem over $S \in Nilp_{O_{K_0}}$ is a pair (X, ϱ) consisting of a p–divisible group with O_B–action over S and a quasi–isogeny

$$\varrho : \mathbf{X} \times_{Spec\, L} \bar{S} \longrightarrow X \times_S \bar{S}.$$

The conditions of (3.21) reduce in this case to the determinant condition and, in case (PEL), to the condition that the given polarization on $\mathbf{X}$ induce on X a multiple by a power of p of a principal polarization. The unramified case is considered by Kottwitz in [Ko3]. He shows by an application of the deformation theory of Grothendieck–Messing that the representing formal scheme $\breve{\mathcal{M}}$ is formally smooth over $Spf\ O_{K_0}$ ([Ko3],§5). In particular, in this case the flatness conjecture before (3.36) is obviously true.

Appendix: Normal forms of lattice chains

We will give the proofs of the theorems (3.11) and (3.16).

We start with the proof of theorem (3.11). Clearly we may assume that $B \cong M_n(D)$, where D is a central division algebra over a local field F. Moreover, we assume that $O_B = M_n(O_D)$. We consider $O_D \subset O_B$ as a subalgebra by the diagonal embedding. We will fix a prime element Π of D. Let $\mathcal{L} = \{\Lambda_i\}_{i \in \mathbf{Z}}$ be the given chain of O_B-lattices in V. Then *a chain of $O_B \otimes \mathcal{O}_T$-modules of type $(\mathcal{L})$* on an affine scheme $T = Spec\, R$ is given by the following data (corollary 3.7):

A sequence of $O_B \otimes_{\mathbf{Z}_p} R$-modules

$$\cdots \xrightarrow{\varrho} M_i \xrightarrow{\varrho} M_{i+1} \xrightarrow{\varrho} \cdots \qquad i \in \mathbf{Z},$$

and for any $i \in \mathbf{Z}$ a periodicity isomorphism

$$\theta : M_{i-r} \xrightarrow{\sim} M_i^{\Pi}$$

such that the following conditions are satisfied:

1. *locally on $Spec\, R$ there exist isomorphisms of $O_B \otimes_{\mathbf{Z}_p} R$-modules*

$$M_i \simeq \Lambda_i \otimes R, \qquad M_i/\varrho(M_{i-1}) \simeq \Lambda_i/\Lambda_{i-1} \otimes R$$

2. $\theta\varrho = \theta\varrho, \quad \varrho^r = \Pi\theta.$

Remarks A.1 The first condition says that M_i is locally on $Spec\, R$ a free $O_D \otimes R$-module of the same rank as the D-module V.

Let $\kappa(D) = O_D/\Pi O_D$ be the residue class field of D. Then the second condition says that locally on $Spec\, R$ the $\kappa(D) \otimes R$-module $M_i/\varrho(M_{i-1})$ is free of the same rank as the $\kappa(D)$-vector space Λ_i/Λ_{i-1}.

Lemma A.2 *For $k < r$ the following sequence of $O_B \otimes_{\mathbf{Z}_p} R$-modules is exact and splits locally on Spec R for any $i \in \mathbf{Z}$*

$$0 \longrightarrow M_i/\varrho^k(M_{i-k}) \longrightarrow M_{i+1}/\varrho^{k+1}(M_{i-k}) \longrightarrow M_{i+1}/\varrho(M_i) \longrightarrow 0\,.$$

Especially there exists locally on Spec R isomorphisms $M_i/\varrho^k(M_{i-k}) \simeq \Lambda_i/\Lambda_{i-k} \otimes R$ of $O_B \otimes_{\mathbf{Z}_p} R$-modules.

Proof: We may assume that $B = D$. Moreover we may assume that the $\kappa(D) \otimes R$-modules $M_{i+1}/\varrho(M_i)$ are free. Clearly the sequence of $\kappa(D) \otimes R$-modules is exact on the right and the surjection

$$M_{i+1}/\varrho^{k+1}(M_{i-k}) \longrightarrow M_{i+1}/\varrho(M_i)$$

splits. Hence we get surjections

$$M_i/\varrho^k(M_{i-k}) \oplus M_{i+1}/\varrho(M_i) \longrightarrow M_{i+1}/\varrho^{k+1}(M_{i-k})\,.$$

Hence by induction $M_i/\varrho^k(M_{i-k})$ is the quotient of an $\kappa(D) \otimes_{\mathbf{Z}_p} R$-module $F_{i,k}$ which is locally on $Spec\, R$ free, and has the same rank as the $\kappa(D)$-vector space Λ_i/Λ_{i-k}.

To see that $M_i/\varrho^k(M_{i-k})$ is locally on $Spec\, R$ a free $\kappa(D) \otimes_{\mathbf{Z}_p} R$-module of the same rank as Λ_i/Λ_{i-k} we apply descending induction on k. For $k = r$ this follows from our assumptions. Assuming by induction that $M_{i+1}/\varrho^{k+1}(M_{i-k})$ is locally on $Spec\, R$ free of the given rank, we obtain a surjection of projective modules of the same rank

$$F_{i,k} \oplus M_{i+1}/\varrho(M_i) \longrightarrow M_{i+1}/\varrho^{k+1}(M_{i-k})\,.$$

This is then also injective. Hence $F_{i,k} \to M_i/\varrho^k(M_{i-k})$ is an isomorphism. The exactness on the left of the sequence asserted in the lemma is immediate.

□

Let us denote by $\bar{M}_i$ the $\kappa(D) \otimes R$-module

$$\bar{M}_i = \kappa(D) \otimes M_i = M_i/\Pi M_i = M_i/M_{i-r}$$

Corollary A.3 *For any integer i and any k such that $0 \leq k \leq r$ there is an exact sequence*

$$\bar{M}_{i+k-r} \xrightarrow{\varrho^{r-k}} \bar{M}_i \xrightarrow{\varrho^k} \bar{M}_{i+k}$$

Proof: Indeed by the lemma we have an injection:

$$Coker\ \varrho^{r-k} = M_i/M_{i+k-r} \xrightarrow{\varrho^k} M_{i+k}/M_{i+k-r} = \bar{M}_{i+k} \quad .$$

□

We obtain a trivial example of a chain of $O_B \otimes R$-modules if we tensor the chain $\mathcal{L}$ by R.

Proposition A.4 *Let T be a scheme over $\mathbf{Z}_p$, such that p is locally nilpotent on T. Let $\{M_i\}$ be a chain of $O_B \otimes_{\mathbf{Z}_p} \mathcal{O}_T$-modules of type $(\mathcal{L})$.*

Then locally on T the chain $\{M_i\}$ is isomorphic to $\mathcal{L} \otimes R$. Moreover, the functor on the category of T-schemes

$$T' \longrightarrow Aut\ (\{M_i \otimes_R \mathcal{O}_{T'}\})$$

is representable by a smooth group scheme over T.

Proof: Let $T = Spec\, R$ be affine. We may assume that $B = D$ is a division algebra. If N is an O_D-module we denote by $\bar{N}$ the O_D-module $N/\Pi N = \kappa(D) \otimes_{O_D} N$.

We choose a $\kappa(D) \otimes R$-linear section s of the surjection $\bar{M}_i \to M_i/\varrho(M_{i-1})$. Let $\bar{U}_i \subset \bar{M}_i$ be the image of s. We may lift $\bar{U}_i$ to a direct summand U_i of the $O_D \otimes R$-module M_i. Clearly the U_i may be chosen to be periodic, i.e. for each i the morphism θ induces an isomorphism

$$\theta : U_{i-r} \longrightarrow U_i^{\Pi} .$$

The maps ϱ induce an obvious map

$$\bigoplus_{k=r-1}^{0} U_{i-k} \longrightarrow M_i \,. \tag{A.1}$$

We claim that this map is surjective. By Nakayama's lemma we need to verify this modulo Π.
Let us denote by M'_{i-k} the image of M_{i-k} in $\bar{M}_i$ for the given i. We obtain a flag by direct summands

$$0 \subset M'_{i-r+1} \subset \cdots \subset M'_{i-1} \subset \bar{M}_i \,. \tag{A.2}$$

By the lemma we have isomorphisms

$$M_{i-k}/\varrho(M_{i-k-1}) \simeq M'_{i-k}/M'_{i-k-1} \,.$$

Hence the images of $\bar{U}_{i-k}$ in $\bar{M}_i$ define a splitting of the flag (A.2). This shows that (A.1) is a surjection mod Π. Since (A.1) is a surjection of projective modules of the same rank it is an isomorphism.
In terms of the U_i the map $M_i \to M_{i+1}$ looks as follows

$$\bigoplus_{k=0}^{r-1} U_{i-k} \longrightarrow \bigoplus_{k=-1}^{r-2} U_{i-k} \,.$$

On the summand U_{i-k} for $k \neq r-1$ this map induces the identity to the corresponding summand on the right hand side. On U_{i-r+1} it induces the map

$$\Pi\, \theta : \; U_{i-r+1} \longrightarrow U_{i+1} \,.$$

From this we see that any two chains of type $(\mathcal{L})$ are locally isomorphic, since locally on T the $O_D \otimes \mathcal{O}_T$-module U_i is free of the same rank as the $\kappa(D)$-vector space Λ_i/Λ_{i-1}.

The representability of the functor of automorphisms by a scheme of finite type over T is obvious.

Definition A.5 *We call the modules $\{U_i\}$ a splitting of the chain $\{M_i\}$. They are characterized by the property that U_i is a direct summand of M_i such that $\bar{U}_i$ maps isomorphically to $M_i/\varrho(M_{i-1})$, and the U_i are periodic with respect to θ.*

Let $R \to S$ be a surjection with nilpotent kernel. Let $\{V_i\}$ be a splitting of $\{M_i\} \otimes_R S$. Then any set of liftings of the V_i to direct summands U_i of M_i which is periodic is a splitting for $\{M_i\}$. Therefore any given splitting $\{V_i\}$ lifts. The formal smoothness of the functor Aut is a consequence. Indeed, let α be an automorphism of $\{M_i\} \otimes_R S$. We find liftings U_i and U_i' of the splittings V_i respectively $\alpha(V_i)$. Then the isomorphism $V_i \to \alpha(V_i)$ lifts to an isomorphism $U_i \to U_i'$ which is periodic. It gives the lifting of α to an automorphism of $\{M_i\}$. This completes the proof of the proposition. □

A.6 Let us now turn to the case of polarized chains (theorem 3.16). Without loss of generality we assume that the invariants F_0 of the involution $*$ on F form a field. We do a case by case verification according to the following list.

(I) $F = F_0 \times F_0$ and the involution on F induces the obvious transposition. There is a central division algebra D over F_0, such that

$$B = M_n(D) \times M_n(D^{opp})$$

and the involution on B is given by

$$(d_1, d_2)^* = (d_2, d_1), \quad \text{where} \quad d_1, d_2 \in M_n(D) = M_n(D)^{opp}.$$

(II) $B = M_n(F), \quad F = F_0$.

(III) $B = M_n(F)$ and F/F_0 is a quadratic extension.

(IV) $B = M_n(D)$ where D is a quaternion algebra over F and $F = F_0$.

(We remark that on $M_n(D)$ there are no involutions of the second kind).

We note that the case where $B = \mathbf{Q}$ and $\mathcal{L}$ is a maximal selfdual chain was treated by de Jong [dJ1].

A.7 Let us consider the case (I). We have the decomposition

$$B \simeq M_n(D) \times M_n(D^{opp}). \tag{A.3}$$

We may choose the isomorphism (A.3) in such a way that it takes the maximal order O_B to $M_n(O_D) \times M_n(O_D^{opp})$. From (A.3) we get a decomposition of the representation V:

$$V = W \oplus \widetilde{W}.$$

The spaces W and $\widetilde{W}$ are isotropic subspaces of V with respect to $(\ ,\ )$. Hence $(\ ,\ )$ puts these two spaces in duality

$$W \times \widetilde{W} \longrightarrow \mathbf{Q}_p.$$

More precisely $(\ ,\ )$ identifies $\widetilde{W}$ with the dual space $W^* = \mathrm{Hom}_{\mathbf{Q}_p}(W, \mathbf{Q}_p)$ with its natural $M_n(D^{opp})$-action from the left.
A multichain of O_B-lattices in V is a pair $\mathcal{L}, \widetilde{\mathcal{L}}$, where $\mathcal{L}$ is a chain of $M_n(O_D)$-lattices in W and $\tilde{\mathcal{L}}$ is a chain of $M_n(O_D^{opp})$-lattices in $\widetilde{W}$. The multichain is selfdual if and only if the two chains $\mathcal{L}$ and $\widetilde{\mathcal{L}}$ are dual to each other.
A polarized multichain of $O_B \otimes \mathcal{O}_T$-modules on a scheme T of type $(\mathcal{L}, \widetilde{\mathcal{L}})$ is a pair of chains, a chain of $M_n(O_D) \otimes \mathcal{O}_T$-modules of type $(\mathcal{L})$ and a chain of $M_n(O_D^{opp}) \otimes \mathcal{O}_T$-modules of type $(\widetilde{\mathcal{L}})$, which are dual to each other.

Lemma A.8 *The functor which associates to a polarized multichain $\{M_\Lambda\}_{\Lambda \in \mathcal{L}}, \{M_{\widetilde{\Lambda}}\}_{\widetilde{\Lambda} \in \widetilde{\mathcal{L}}}$ of type $(\mathcal{L}, \widetilde{\mathcal{L}})$ the unpolarized chain $\{M_\Lambda\}_{\Lambda \in \mathcal{L}}$ of type $(\mathcal{L})$ is an equivalence of categories.*

Proof: We have an obvious quasi-inverse functor: If $\Lambda \in \mathcal{L}$ and hence $\Lambda^* \in \widetilde{\mathcal{L}}$, we put

$$M_{\Lambda^*} = \mathrm{Hom}_{\mathcal{O}_T}(M_\Lambda, \mathcal{O}_T)$$

with the natural $M_n(O_D^{opp}) \otimes_{\mathbf{Z}_p} \mathcal{O}_T$-module structure. We have to check that M_{Λ^*} is of type $\widetilde{\mathcal{L}}$. But this is obvious, since we know that locally on T the chain $\{M_\Lambda\}$ is isomorphic to $\mathcal{L} \otimes R$. □

This lemma shows that the case (I) of theorem (3.16) reduces to the unpolarized theorem (3.11).

A.9 For the other cases we recall some basic facts on quadratic forms in the generality needed for the proof.
Let R be a commutative unitary ring and S a unitary R-algebra, which need not be commutative. Assume that we are given an involution $s \longmapsto \check{s}$

of S, i.e. a R-algebra anti–isomorphism of order 2 on S. Let us assume that 2 is invertible in R. Let M and N be left S-modules and L a S-bimodule. *A sesquilinear form* is a biadditive map

$$\Phi : M \times N \longrightarrow L$$

that satisfies the relations

$$\Phi(sm, n) = \Phi(m, n)\check{s}$$
$$\Phi(m, sn) = s\Phi(m, n).$$

We will say that a S-module is R-locally free, if it is locally free with respect to the Zariski-topology on $Spec\, R$. We note that we can view any right S-module as a left S-module by restriction of scalars with respect to the involution. For example the right S-module $\mathrm{Hom}_{S-}(M, L)$ becomes a left module by the rule

$$(s\varphi)(m) = \varphi(m)\check{s}, \quad \varphi \in \mathrm{Hom}_{S-}(M, L)\,.$$

With this convention a sesquilinear form is a homomorphism of left S–modules

$$\begin{array}{ccc} M & \longrightarrow & \mathrm{Hom}_{S-}(N, L) \\ m & \longmapsto & \Phi(m, -)\,. \end{array} \tag{A.4}$$

If we view M as a right module by our convention, this is also equivalent to giving a homomorphism of left S-modules

$$\begin{array}{ccc} N & \longrightarrow & \mathrm{Hom}_{-S}(M, L) \\ n & \longmapsto & \Phi(-, n). \end{array} \tag{A.5}$$

Φ is said to be *perfect*, if (A.4) and (A.5) are isomorphisms.
In the case $M = N = S$ a hermitian form is given by an element $s \in S$:

$$\Phi(m, n) = ns\check{m}. \tag{A.6}$$

Then the condition (A.4) (respectively (A.5)) says that s has a right inverse (respectively a left inverse). Hence perfect means that s is a unit.
A hermitian respectively antihermitian form on a S-module M is a sesquilinear form

$$\Phi : M \times M \longrightarrow S\,,$$

that satisfies $\Phi(m_1, m_2) = \pm\Phi(m_2, m_1)\check{}$. We will also say that Φ is an ε-hermitian form, where $\varepsilon = +1$ if Φ is hermitian and $\varepsilon = -1$ if Φ is antihermitian. We note that for such a form the conditions that (A.4) and (A.5) are isomorphisms are clearly equivalent.
If $\mathcal{I}$ is a two-sided ideal in S invariant under the involution, we can consider the reduction of the ε–hermitian form Φ modulo $\mathcal{I}$:

$$\bar{\Phi} : M/\mathcal{I}M \times M/\mathcal{I}M \longrightarrow S/\mathcal{I}\,.$$

If M is a projective S-module and Φ is perfect, so is $\bar{\Phi}$.

Definition A.10 *Let M be a finite projective S-module with a ε-hermitian form Φ which is perfect. Let N be a isotropic direct summand of M. An isotropic complement to N is a direct summand C of M, which is isotropic and such that the pairing*

$$\Phi : N \times C \longrightarrow S$$

is perfect.

We note that the restriction of Φ to the orthogonal complement $(N \oplus C)^\perp$ of $N \oplus C$ is perfect, and that

$$M = N \oplus C \oplus (N \oplus C)^\perp\,.$$

Lemma A.11 *An isotropic complement exists.*

Proof: Since Φ is perfect, we get a split surjection

$$M \twoheadrightarrow \mathrm{Hom}_{S_-}(N, S) \longrightarrow 0$$

Let C the image of a section. Then

$$\Phi : N \times C \longrightarrow S$$

is perfect. Any sesquilinear form on C is of the form $\Phi(\alpha(c_1), c_2)$, where α ranges over the homomorphisms $\mathrm{Hom}_{S_-}(C, N)$. Hence there is α such that $-\frac{1}{2}\Phi(c_1, c_2) = \Phi(\alpha(c_1), c_2)$. But then

$$C' = \{c + \alpha(c); c \in C\}$$

is the isotropic complement we are looking for. □

Proposition A.12 *Let M be a finite R-locally free S-module and Φ be a perfect ε-hermitian form on M. Let $\mathcal{I} \subset S$ be a two–sided nilpotent ideal, which is invariant under the involution. Then $\bar{\Phi} = \Phi$ modulo $\mathcal{I}$ is a perfect ε-hermitian form on the module $\bar{M} = M/\mathcal{I}M$ over the R-algebra $\bar{S} = S/\mathcal{I}$. Assume that $\bar{N}$ is a R-locally free isotropic direct summand of $\bar{M}$. Then there is an isotropic R-locally free direct summand N of M such that $\bar{N} = N/\mathcal{I}N$.*

Proof: As usual we may assume that $\mathcal{I}^2 = 0$. Let us start by assuming that there is a lifting N of $\bar{N}$ to a direct summand of M, which is not necessarily isotropic. We write

$$M = N \oplus C.$$

If N' is any other lifting of $\bar{N}$, the projection to C defines a map $N' \longrightarrow M = N \oplus C \longrightarrow C$.
The image of this map lies in $\mathcal{I}C$. Hence $\mathcal{I}N'$ is mapped to zero and we get a map

$$\alpha : \bar{N} \longrightarrow \mathcal{I} \cdot C = \mathcal{I} \otimes_{\bar{S}} \bar{C}\,.$$

We have

$$N' = \{n + \alpha(n) \mid n \in N\}\,.$$

Therefore we get a bijection between the liftings and the set

$$\mathrm{Hom}_{\bar{S}}(\bar{N}, \mathcal{I} \otimes_{\bar{S}} \bar{C}). \tag{A.7}$$

If we replace C by C_1, such that $\bar{C}_1 = \bar{C}$ the map α that belongs to N' does not change. Hence the obstruction to lifting $\bar{N}$ lies in the cohomology H^1 of the sheaf on *Spec* R defined by (A.7) and is therefore zero.
The sesquilinear form $\bar{\Phi}(\alpha(n_1), n_2)$ on $\bar{N} \times \bar{N}$ is given by the homomorphism

$$\begin{array}{ccc} \bar{N} \longrightarrow \mathcal{I} \otimes_{\bar{S}} \bar{C} \longrightarrow & \mathcal{I} \otimes_{\bar{S}} \mathrm{Hom}_{\bar{S}_-}(\bar{N}, \bar{S}) \\ & | \simeq \\ & \mathrm{Hom}_{S_-}(\bar{N}, \mathcal{I})\,, \end{array}$$

where the last isomorphism of left S-modules maps $i \otimes \varphi$ to the homomorphism $n \longmapsto \varphi(n)\check{i}$. Since the second map of the last diagram is surjective, any $\mathcal{I}$-valued sesquilinear form of $\bar{N}$ may be represented in the form $\Phi(\alpha(n_1), n_2)$. We get the isotropic lifting if we choose α such that $\bar{\Phi}(\alpha(n_1), n_2) = -\frac{1}{2}\Phi(n_1, n_2) \in \mathcal{I}$. □

Proposition A.13 *Let M be a finite left S-module and Φ a ε-hermitian perfect form on M. Assume that $\mathcal{I} \subset S$ is a nilpotent two-sided ideal of S, which is invariant under the involution. Assume that $\bar{e}_1, \ldots, \bar{e}_n$ is an orthogonal basis of the $\bar{S}$-module $\bar{M}$. Let $a_i \in S$ be liftings of the elements $\bar{\Phi}(e_i, e_i) \in \bar{S}$, such that $a_i = \varepsilon \check{a_i}$. (Liftings with this property always exist.) Then there is an orthogonal basis $e_1, \ldots, e_n$ of the S-module M with respect to Φ, such that e_i modulo $\mathcal{I} = \bar{e}_i$ and $\Phi(e_i, e_i) = a_i$.*

Proof: Let $e'_1, \ldots, e'_n \in M$ be any liftings of the elements $\bar{e}_1, \ldots, \bar{e}_n$. Then $\Phi(e'_i, e'_j) = a_i\delta_{ij} + x_{ij}$, where $x_{ij} \in \mathcal{I}$ and $x_{ij} = \check{x}_{ji}$ respectively $x_{ij} = -\check{x}_{ji}$, depending on whether $\varepsilon = 1$ or $\varepsilon = -1$.
For arbitrary elements $m_1, \ldots, m_n \in M$ we have the formula $\Phi(e'_i + m_i, e'_j + m_j) = a_i\delta_{ij} + x_{ij} + \Phi(m_i, e'_j) \pm \Phi(m_j, e'_i)\check{}$. By perfectness any linear form on M with values in $\mathcal{I}$ is of the type $\Phi(m, -)$ for some $m \in \mathcal{I}M$. Hence we may choose $m_i \in \mathcal{I}M$ in such a way that $\Phi(m_i, e'_j)$ becomes an arbitrary given matrix with coefficients in $\mathcal{I}$. To finish the proof it is therefore enough to find $y_{ij} \in \mathcal{I}$, such that $x_{ij} + y_{ij} \pm \check{y}_{ji} = 0$. Clearly $y_{ij} = -\frac{x_{ij}}{2}$ does the job. □

The usual orthogonalization procedure takes in our setting the following form. Assume that S is a finite R-algebra. Then by Nakayama's lemma the condition that a fixed element $s \in S$ is a unit is represented by an open subscheme of $Spec\, R$. For our purposes it is enough to make the strong assumption that for any point $x \in Spec\, R$ the ring $S \otimes_R \kappa(x)$ modulo its radical is a product of fields. In particular this ensures that the kernel of a surjection of R-locally free S-modules is again R-locally free.

Proposition A.14 *Suppose S satisfies the assumptions just made. Let M be a finite R-locally free S-module and Φ be a perfect ε-hermitian form on S. In the case $\varepsilon = -1$, we assume that there is a unit $u \in S$, such that $u = -\check{u}$.*

Then locally for the Zariski topology on Spec R the S-module M admits an orthogonal base, i.e., (M, Φ) is an orthogonal sum of forms of the type (A.6).

Proof: It is enough to find locally an element $m \in M$, such that $\Phi(m, m)$ is a unit in S. Indeed our assumptions ensure that the orthogonal complement of Sm is again R-locally free. Hence we may restrict to the case, where R is a field. Let J be the radical of S. Then by proposition (A.13) it is enough to show our assertion for S/J. Hence we may assume that S is a product of fields. Then one reduces easily to the case, where either S is a field or $S = S_0 \times S_0$, where S_0 is a field and the involution interchanges the factors. In these cases the proof is standard linear algebra. □

A.15 Let us consider the case (II) where $B = M_n(F)$ and $*$ is an involution of the first kind. We will assume that $O_B = M_n(O_F)$. Let us fix once for all a prime element $\pi \in O_F$.
We may index our chain of lattices $\mathcal{L}$ by $\mathbf{Z}$:

$$\cdots \subset \Lambda_i \subset \Lambda_{i+1} \subset \cdots$$

The periodicity condition says that there is an integer r, such that

$$\pi \Lambda_i = \Lambda_{i-r} \tag{A.8}$$

for all $i \in \mathbf{Z}$. We call r the period of the lattice chain $\mathcal{L}$.
The selfduality condition means that there is an integer a such that

$$\Lambda_i^* = \Lambda_{-i+a} \,.$$

We may index our chain in such a way that $a = 0$ or $a = 1$.

Let $T = Spec\ R$ be an affine $\mathbf{Z}_p$–scheme, where p is locally nilpotent. We consider polarized chains $M_{\mathcal{L}}$ of $M_n(O_F) \otimes \mathcal{O}_T$-modules on T. We set $M_i = M_{\Lambda_i}$ and get a sequence of $M_n(O_F) \otimes \mathcal{O}_T$-modules:

$$\cdots \xrightarrow{\varrho} M_i \xrightarrow{\varrho} M_{i+1} \xrightarrow{\varrho} \cdots, \qquad i \in \mathbf{Z}.$$

We denote the periodicity isomorphism θ_π simply by θ :

$$\theta : M_i \longrightarrow M_{i-r}.$$

We have the relations

$$\varrho^r \theta = \pi \;, \quad \theta \varrho = \varrho \theta.$$

The polarization on the chain $\{M_i\}$ is given by a set of perfect bilinear forms

$$\mathcal{E}_i : M_i \times M_{-i+a} \longrightarrow O_T.$$

By Morita equivalence we write

$$V = F^n \otimes_F W\,.$$

Then there is a chain of O_F-lattices $\mathcal{G} = \{\Gamma_i\}$ of the F-vectorspace W, such that

$$\Lambda_i = O_F^n \otimes_{O_F} \Gamma_i.$$

We will see that the chain $\mathcal{G}$ is selfdual for a suitable inner product on W.

Proposition A.16 *Up to multiplication by a unit in O_F there is a unique perfect bilinear form*

$$\phi_F : O_F^n \times O_F^n \longrightarrow O_F$$

that satisfies the equation

$$\phi_F(Ax, y) = \phi_F(x, A^* y), \quad x, y \in O_F^n,\; A \in M_n(O_F)\,. \tag{A.9}$$

Moreover ϕ_F is either alternating or symmetric.

Proof: One knows (e.g. [M1]) that any involution on $M_n(F)$ may be written

$$A^* = C^{-1}\, {}^tA\, C \quad \text{where} \quad {}^tC = \pm C\,.$$

The bilinear form

$$\tilde{\phi}_F : F^n \times F^n \longrightarrow F$$

defined by $\tilde{\phi}_F(x, y) = {}^t x C y$ satisfies the equality (A.9). We set $\Gamma = O_F^n \subset F^n$. The dual lattice Γ^* with respect to $\tilde{\phi}_F$ is an $M_n(O_F)$-module. Hence by Morita equivalence there is an $f \in F$, such that

$$\Gamma^* = f\Gamma\,.$$

The bilinear form $\phi_F = f\tilde{\phi}_F$ is the perfect pairing we are looking for.

Corollary A.17 *There is a perfect $\mathbf{Z}_p$-bilinear pairing*

$$\phi : O_F^n \times O_F^n \longrightarrow \mathbf{Z}_p$$

that satisfies the equation

$$\phi(Ax, y) = \phi(x, A^* y), \quad x, y \in O_F^n,\ A \in M_n(O_F)\,.$$

Any other pairing ϕ' with this property is of the form

$$\phi'(x, y) = \phi(fx, y)\,,$$

for a suitable $f \in O_F^\times$.

Proof: Let ϑ_F be a generator of the different ideal of F over $\mathbf{Q}_p$. Then the pairings ϕ and ϕ_F determine each other by the equation

$$\phi(fx, y) = tr_{F/\mathbf{Q}_p} \vartheta_F^{-1} f \phi_F(x, y)\,.$$

□

Proposition A.18 *Let $M_1 = O_F^n \otimes_{O_F} N_1$ and $M_2 = O_F^n \otimes_{O_F} N_2$ be left $O_B \otimes R$-modules which are projective and finite.*
Then there is a bijection between perfect R-bilinear forms

$$\mathcal{E} : M_1 \times M_2 \longrightarrow R\,,$$

that satisfy the equation

$$\mathcal{E}(b^* m_1, m_2) = \mathcal{E}(m_1, b m_2), \quad b \in O_B$$

and perfect $O_F \otimes R$-bilinear forms

$$\mathcal{B} : N_1 \times N_2 \longrightarrow O_F \otimes R.$$

The forms $\mathcal{B}$ and $\mathcal{E}$ determine each other uniquely by the equation

$$\mathcal{E}(u_1 \otimes n_1, u_2 \otimes n_2) = \phi(u_1, u_2 \mathcal{B}(n_1, n_2)), \quad n_i \in N_i,\ u_i \in O_F^n.$$

We omit the easy proof. A slight modification of this proposition may be applied to our form $(\ ,\ )$ on $V = F^n \otimes_F W$ and provides us with a F-bilinear form ζ on W, which satisfies the equation:

$$(u \otimes w, u' \otimes w') = \phi(u, \zeta(w, w') \cdot u') \quad u, u' \in F^n,\ w, w' \in W. \qquad \text{(A.10)}$$

The form ζ is symmetric if ϕ is antisymmetric and vice versa. The chain $\mathcal{G}$ is selfdual with respect to the form ζ,

$$\Gamma_i^* = \Gamma_{-i+a}.$$

Consider a polarized chain of $M_n(O_F) \otimes R$-modules of type $(\mathcal{L})$. Then the Morita equivalence provides us with a chain of $O_F \otimes R$-modules $\{N_i\}_{i \in \mathbf{Z}}$,

$$M_i = O_F^n \otimes_{O_F} N_i.$$

Clearly the chain $\{N_i\}$ is of type $(\mathcal{G})$, i.e. there are locally on $Spec\, R$ isomorphisms of $O_F \otimes R$-modules:

$$N_i = \Gamma_i \otimes R, \qquad N_i/\varrho(N_{i-1}) \cong \Gamma_i/\Gamma_{i-1} \otimes R. \qquad \text{(A.11)}$$

Moreover by proposition (A.18) we get perfect $O_F \otimes \mathcal{O}_T$-bilinear pairings

$$\mathcal{B}_i : N_i \times N_{-i+a} \longrightarrow O_F \otimes \mathcal{O}_T$$

such that the following equations are satisfied:

1. $\mathcal{B}_i(\varrho(n), n') = \mathcal{B}_{i-1}(n, \varrho(n')), \quad n \in N_{i-1},\ n' \in N_{-i+a+1}.$

2. $\mathcal{B}_i(n, n') = \pm \mathcal{B}_{-i+a}(n', n), \quad n \in N_i,\ n' \in N_{-i+a}.$

   Here the sign depends only on the type of the involution with which we started.

3. If $\theta : N_i \to N_{i-r}$ is the period isomorphism, then

$$\mathcal{B}_i(n, \theta(n')) = \mathcal{B}_{i-r}(\theta(n), n'), \quad n \in N_i,\ n' \in N_{-i+r+a}.$$

We call a set of perfect pairings $\mathcal{B}_i$, which satisfy these relations a *polarization of the chain* $\{N_i\}$ *of type* $(\mathcal{G})$.

Proposition A.19 *The above functor which associates to a polarized chain $\{M_i\}, \mathcal{E}$ of $M_n(O_F) \otimes R$-modules of type $(\mathcal{L})$ the polarized chain $\{N_i\}, \mathcal{B}$ of type $(\mathcal{G})$ is an equivalence of categories.* □

Remarks A.20 We may take indices in $\mathbf{Z}/(r)$, if we define

$$N_{\bar{k}} = \varprojlim_{k \in \bar{k}} (N_k, \theta) \qquad \bar{k} \in \mathbf{Z}/(r).$$

The condition 3) simply says that $\mathcal{B}_{\bar{k}}$ is well-defined, and we may forget about it. The condition $\varrho^r = \pi\theta$ on the period morphism becomes $\varrho^r = \pi$. We call $\{N_{\bar{k}}\}$ a *polarized circle of $O_F \otimes \mathcal{O}_T$-modules.*

By proposition (A.19) we may reformulate the theorem (3.16) in the case (II) at hand:

Proposition A.21 *Let $\{N_i\}_{i\in\mathbf{Z}}, \mathcal{B}$ and $\{N'_i\}_{i\in\mathbf{Z}}, \mathcal{B}'$ be two polarized chains of $O_F \otimes \mathcal{O}_T$-modules of type $(\mathcal{G})$.*
Then locally for the etale topology on T these two polarized chains are isomorphic and moreover the following functor on the category of T-schemes

$$T' \mapsto Isom((\{N_i\}, \mathcal{B}), (\{N'_i\}, \mathcal{B}')) \tag{A.12}$$

is representable by a smooth affine scheme over T.
If the form ζ (A.10) is antisymmetric or if the period r of $\mathcal{G}$ and the number a both are even, the two polarized chains are even isomorphic locally for the Zariski topology.

Let us first treat the case $r = 1$. We may moreover assume that $a = 0$. Then the chains in proposition (A.21) are already determined by the modules N_0 and N'_0. To ease the notation we set $N = N_0$, $\mathcal{B} = \mathcal{B}_0$, $N' = N'_0$, and $\mathcal{B}' = \mathcal{B}'_0$. The $O_F \otimes R$-module N, N' are R-locally free of the same rank, and the pairings $\mathcal{B}$ and $\mathcal{B}'$ are perfect. We need to verify that the pairs $(N, \mathcal{B})$ and $(N', \mathcal{B}')$ are locally isomorphic for the etale topology.
Let us first consider the case, where the forms $\mathcal{B}$ and $\mathcal{B}'$ are symmetric. By the existence of an orthogonal base (proposition A.14) we are reduced to the case, where N, N' are free of rank 1. Clearly it is enough to see that there is locally for the etale topology an orthonormal vector in N. The pair $(N, \mathcal{B})$ is given by a unit $s \in O_F \otimes R$. We have to check that after an etale covering of Spec R the element s becomes a square. This is obvious since the residue characteristic is different from 2.

If the form $\mathcal{B}$ is antisymmetric, the usual proof in linear algebra works to show:

Lemma A.22 *Locally for the Zariski topology on Spec R the symplectic $O_F \otimes R$-module has a standard symplectic base $e_1, \ldots, e_{2r}$, i.e.*

$$\mathcal{B}(e_i, e_{j+r}) = \delta_{ij}, \qquad i, j = 1, \ldots, r.$$

Hence the pairs $(N, \mathcal{B})$ and $(N', \mathcal{B}')$ are locally isomorphic for the Zariski topology in the antisymmetric case.
The representability of the functor (A.12) is obvious. The smoothness is a consequence of the propositions (A.13) in the symmetric case and (A.12) in the antisymmetric case.

Before we go on with the proof of (A.21) let us add a remark about some nonperfect forms over a ring R respectively an O_F-algebra R.

Definition A.23 *Let N be a finite projective R-module and $\mathcal{Q}$ be a symmetric or antisymmetric bilinear form on N.*
We call $\mathcal{Q}$ semiperfect, iff the cokernel of the corresponding map $N \longrightarrow N^* = Hom_R(N, R)$ is a projective R-module.

Lemma A.24 *$(N, \mathcal{Q})$ is semiperfect, iff $(N, \mathcal{Q})$ may be written as an orthogonal sum*

$$(N, \mathcal{Q}) \simeq (M, \mathcal{B}) \oplus (K, 0)$$

where $(M, \mathcal{B})$ is perfect. □

Lemma A.25 *Let R be an O_F-algebra, where π is locally nilpotent in R. Let $(N, \mathcal{Q})$ be a symmetric or antisymmetric form on a finite projective R-module. Suppose that the map $N \longrightarrow N^*$ has a cokernel that is annilated by π and is a projective R/π-module. Then $(N, \mathcal{Q})$ may be written locally as a direct orthogonal sum*

$$(N, \mathcal{Q}) = (M, \mathcal{B}) \oplus (C, \pi\mathcal{P}),$$

where $\mathcal{B}$ is a perfect pairing on M and $\mathcal{P}$ is a perfect pairing on C.

Proof: Let $\bar{N} = N \otimes_{O_F} R/\pi$. Let $\bar{M} \subset \bar{N}$ the subspace of $\bar{N}$ given by the first lemma. We lift $\bar{M}$ to a direct summand M of N. Then $\mathcal{Q}$ induces a perfect pairing $\mathcal{B}$ on M. Let C be the orthogonal complement of M. We get a decomposition $N = M \oplus C$.

Then C^*/C is isomorphic to N^*/N and hence has the same rank as C. Therefore the image of C in C^* is πC^*. Let $\alpha : C \to C^*$ be the map induced by $\mathcal{Q}$. We find a lifting

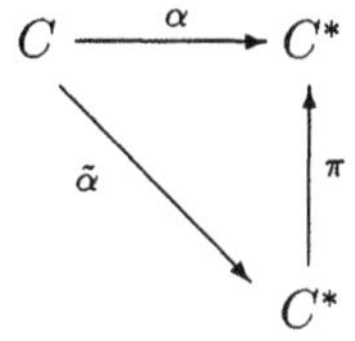

Then $\frac{\tilde{\alpha}+\tilde{\alpha}^*}{2}$ or $\frac{\tilde{\alpha}-\tilde{\alpha}^*}{2}$ defines the pairing $\mathcal{P}$ we are looking for. In order to show that $\mathcal{P}$ is perfect, we have to ensure that $\tilde{\alpha}$ may be chosen to be surjective. Then $\frac{\tilde{\alpha}\pm\tilde{\alpha}^*}{2} = \tilde{\alpha} + \frac{\tilde{\alpha}^*\mp\tilde{\alpha}}{2}$ is surjective because $\tilde{\alpha}^* \mp \tilde{\alpha} = 0 \bmod \pi$. Locally the map α may be written by choosing a basis in each module: $R^n \xrightarrow{\pi X} R^n$ where $X \in M_n(R)$. Since the image is $\pi \cdot R^n$ there is a matrix Y such that $\pi XY = \pi E$, where E is the unit matrix in $M_n(R)$. We have to show that in each point of $Spec\, R$ we may find a matrix Z, such that $\pi Z = 0$ and $X + Z$ is invertible. Since $\pi(E - XY) = 0$ it is enough to find locally matrices U_1 and U_2 which are invertible, and such that $X + U_1(E - XY)U_2$ is invertible. But over a field, for a suitable choice of basis, we have $X = diag\,(1, \ldots, 1, 0, \ldots, 0)$ and for suitable U_1 and U_2 we have $U_1(E - XY)U_2 = diag\,(0, \ldots, 0, 1, \ldots, 1)$ Hence we get what we want since obviously $rkX + rk(E - XY) \geq n$. □

We now continue with the proof of proposition (A.21). Let us consider the case where the period r is arbitrary and $a = 0$. Then $\mathcal{B}_0$ is a perfect symmetric or antisymmetric bilinear form on N_0,

$$\mathcal{B}_0 : N_0 \times N_0 \longrightarrow O_F \otimes \mathcal{O}_T\,.$$

For $r - 1 \geq t \geq 0$ let us denote by F_{-t} the image of N_{-t} in $N_0/\pi N_0$. We obtain a flag of $O_F/\pi O_F \otimes \mathcal{O}_T$-modules, whose quotients are locally free on T,

$$0 \subsetneqq F_{-r+1} \subsetneqq \cdots \subsetneqq F_0 = N_0/\pi N_0. \tag{A.13}$$

Let $\bar{\mathcal{B}}_0$ be the perfect pairing $\mathcal{B}_0$ modulo π on $N_0/\pi N_0$.

Lemma A.26 *F_{-t} is the orthogonal complement of F_{-r+t} with respect to $\bar{\mathcal{B}}_0$.*

Proof: The orthogonal complement of F_{-r+t} is the image of the following submodule of N_0 in F_0:

$$\{n_0 \in N_0\,;\, \mathcal{B}_0(\varrho^t(N_{-t}), n_0) \in \pi O_F \otimes \mathcal{O}_T\}$$

$$= \{n_0 \in N_0\,;\, \mathcal{B}_{-t}(N_{-t}, \varrho^t(n_0)) \in \pi O_F \otimes \mathcal{O}_T\}.$$

Since $\mathcal{B}_{-t}$ is perfect the last condition is equivalent to $\varrho^t(n_0) \in \pi N_t = \varrho^r(N_{t-r})$. By the corollary (A.3) we get $n_0 \in \varrho^{r-t}(N_{t-r})$. □

Hence F_{-t} for $t \geq \frac{r}{2}$ is an isotropic subspace of F_0. Moreover $\bar{\mathcal{B}}_0$ induces perfect pairings

$$N_{-t}/N_{-t-1} \times N_{-r+t+1}/N_{-r+t} \longrightarrow O_F/\pi O_F \otimes \mathcal{O}_T\,. \tag{A.14}$$

We are assuming that $a = 0$. Let us now take indices $i \in \mathbf{Z}/(r)$. Because of the perfect pairings

$$\mathcal{B}_{-i} : N_{-i} \times N_{i-r} \longrightarrow O_F \otimes \mathcal{O}_T$$

we may forget about N_{-i} and $\mathcal{B}_i$ for i in the open intervall $i \in (0, \frac{r}{2})$ without loosing information. We get a chain

$$N_0 \longrightarrow N_{-r+1} \longrightarrow \cdots \longrightarrow N_{-t} \longrightarrow N_0 \longrightarrow N_{-r+1} \longrightarrow \tag{A.15}$$

such that moving once around in the corresponding circle is multiplication by π. Here $t = \left[\frac{r+1}{2}\right]$. Let us denote the map $\varrho^{r-t} : N_0 \to N_{-t}$ by α and the map $\varrho^t : N_{-t} \to N_0$ by β.

For $0 < k < t$ we will identify N_{-k} with N^*_{k-r} by the formula $n_{-k} \longmapsto n^*_{-k}$, where

$$n^*_{-k}(n_{k-r}) = \mathcal{B}_{-k}(n_{-k}, n_{k-r}).$$

Equivalently we have the dual identification $N_{k-r} \to N^*_{-k}$ which is given by the formula $n_{k-r} \to n^*_{k-r}$ where

$$n^*_{k-r}(n_{-k}) = \mathcal{B}_{k-r}(n_{k-r}, n_{-k}) = \mathcal{B}_{-k}(n_{-k}, n_{k-r}) \,.$$

The maps $N_{-t} \to N^*_{-t}$ and $N_0 \to N^*_0 = N_0$ which come from the identifications above and ϱ define bilinear forms $\mathcal{Q}$ on N_{-t} and $\mathcal{B}$ on N_0. We have

$$\begin{array}{lcl} \mathcal{Q}(n_{-t}, n'_{-t}) & = & \left\{ \begin{array}{ll} \mathcal{B}_{-t}(n_{-t}, \varrho(n'_{-t})) & r \text{ odd} \\ \mathcal{B}_{-t}(n_{-t}, n'_{-t}) & r \text{ even} \end{array} \right. \\ \mathcal{B} & = & \mathcal{B}_0 \,. \end{array}$$

Lemma A.27 *To give a polarized chain of* $O_F \otimes \mathcal{O}_T$*-modules on* T *of type* $(\mathcal{G})$ *with period* r *and* $a = 0$ *is equivalent to giving the following data:*

1. *A circular diagram of locally on* T *free* $O_F \otimes \mathcal{O}_T$*-modules*

$$\begin{array}{ccc} N_{-r+1} \xrightarrow{\ \varrho\ } \cdots \xrightarrow{\ \varrho\ } N_{-t} & & t = \left[\frac{r+1}{2}\right] \\ {\scriptstyle\varrho}\nwarrow \qquad \swarrow{\scriptstyle\beta} & & \\ N_0 & & \end{array} \tag{A.16}$$

2. $O_F \otimes \mathcal{O}_T$*-bilinear pairings*

$$\begin{array}{llclcl} \mathcal{B}: & N_0 & \times & N_0 & \longrightarrow & O_F \otimes \mathcal{O}_T \\ \mathcal{Q}: & N_{-t} & \times & N_{-t} & \longrightarrow & O_F \otimes \mathcal{O}_T \end{array}$$

such that the following conditions are satisfied.

(i) There exist isomorphisms locally on T

$$\begin{array}{lcll} N_i & \simeq & \Gamma_i \otimes \mathcal{O}_T & i = -t, \ldots -r+1, 0 \\ N_i/N_{i-1} & \simeq & \Gamma_i/\Gamma_{i-1} \otimes \mathcal{O}_T & i = -t, \ldots, -r+1 \\ N^*_{-t}/N_{-t} & \simeq & \left\{ \begin{array}{c} \Gamma_{-t+1}/\Gamma_{-t} \otimes \mathcal{O}_T \\ 0 \end{array} \right. & \begin{array}{l} r \textit{ odd} \\ r \textit{ even} \end{array} \end{array}$$

The bilinear form $\mathcal{B}$ *is perfect.*

(ii) We have the identity

$$\mathcal{B}(\beta(x), y) = \mathcal{Q}(x, \alpha(y))$$

where $\alpha : N_0 \to N_{-t}, \beta : N_{-t} \to N_0$ *are the obvious maps in the diagram.*

(iii) Going once around the circle above starting from any point is multiplication by π.

Proof: We have to recover from the data of the lemma the modules N_k, $k \in \mathbf{Z}/r\mathbf{Z}$ and the bilinear forms

$$\mathcal{B}_k : N_k \times N_{-k} \longrightarrow O_F \otimes O_T .$$

This is simply accomplished by taking $N_k = N^*_{-k}$ for $k = 1, \ldots, \left[\frac{r}{2}\right] - 1$. We get the chain

$$N_0 \to N_{-r+1} \to \cdots \to N_{-t} \xrightarrow{q} N^*_{-t} \to N^*_{-t+1} \to \cdots \to N^*_{-1} \to N_0. \tag{A.17}$$

Here q is the map $q(n_{-t})(n'_{-t}) = \mathcal{Q}(n_t, n'_{-t})$. The maps which are symmetric around the arrow q are by definition dual to each other. Hence we have that $N^*_{-t} \to N_0 = N^*_0$ is the map α^*. We have $\alpha^* q = \beta$:

$$\mathcal{B}(\alpha^* q(n_t), n_0) = q(n_t)(\alpha(n_0)) = \mathcal{Q}(n_t, \alpha(n_0)) = \mathcal{B}(\beta(n_t), n_0) .$$

The existence of the isomorphisms (A.11) is obvious by duality. □

A.28 To prove the proposition (A.21) let us work with the circles above. First we consider the case $r = 2$. In this case $\mathcal{B}_0 \bmod \pi$ on $N_0/\pi N_0$ induces a perfect pairing

$$N_{-1}/\pi N_0 \times N_0/N_{-1} \xrightarrow{\mathcal{B}} O_F/\pi O_F \otimes R$$

and hence these modules have the same rank. Our circle of modules has the form:

$$N_0 \xrightarrow{\alpha} N_{-1} \xrightarrow{\beta} N_0 \xrightarrow{\alpha} N_{-1} \quad \alpha\beta = \beta\alpha = \pi .$$

Here the maps α and β are dual with respect to the perfect pairings $\mathcal{Q}$ and $\mathcal{B}$. The sequence is exact mod π by lemma (A.2). We construct an isotropic direct summand U_0 of N_0 as follows.
The image of N_{-1} in $\bar{N}_0 = N/\pi N_0$ is an isotropic direct summand. Let $\bar{U}_0$ be an isotropic complement. By the rank conditions we have $\mathrm{Im} N_{-1} \oplus \bar{U}_0 = \bar{N}_0$. Next we lift $\bar{U}_0$ to an isotropic subspace U_0 of N_0 (proposition A.12). By the same process we define an isotropic direct summand U_{-1} of N_{-1}.
The modules U_0 and U_{-1} are a splitting of our chain in the sense of definition (A.5) by isotropic subspaces. Hence we know

$$\begin{aligned} N_0 &= U_0 \oplus \beta(U_{-1}) \\ N_{-1} &= U_{-1} \oplus \alpha(U_0). \end{aligned}$$

Moreover we see that $\alpha(U_0)$ and $\beta(U_{-1})$ are isotropic subspaces. Hence our chain may be written as

$$U_0 \oplus U_{-1} \xrightarrow{id \oplus \pi} U_0 \oplus U_{-1} \xrightarrow{\pi \oplus id} U_0 \oplus U_{-1}, \tag{A.18}$$

where U_0 and U_{-1} are isotropic subspaces with respect to the bilinear forms and the pairing between them is given by the formula

$$\mathcal{B}(x, \beta(y)) = \mathcal{Q}(\alpha(x), y), \quad x \in U_0,\ y \in U_{-1}\,.$$

Hence in the case $r = 2$ any two chains of modules of type $(\mathcal{G})$ are locally isomorphic for the Zariski topology.

A.29 Let us now consider the case, where r is even. Then the modules $N_0 \to N_{-\frac{r}{2}} \to N_0$ form with the given bilinear forms a circle of period $r = 2$. Hence the last sequence may be written in the form (A.18). Hence our chain looks as follows

$$N_0 = U_0 \oplus U_0^* \xrightarrow{\alpha = id \oplus \pi} N_{-\frac{r}{2}} = U_0 \oplus U_0^* \xrightarrow{\beta = \pi \oplus id} N_0 = U_0 \oplus U_0^* \tag{A.19}$$

$$N_0 \subset N_{-r+1} \subset \cdots \subset N_{-\frac{r}{2}} \subset N_0 \tag{A.20}$$

Since α is the identity on U_0 we may write

$$N_{-k} = U_0 \oplus \hat{N}_{-k}, \qquad k = \frac{r}{2} + 1, \ldots, r - 1\,.$$

Hence we get a chain

$$U_0^* \longrightarrow \hat{N}_{-r+1} \longrightarrow \cdots \longrightarrow \hat{N}_{-\frac{r}{2}-1} \longrightarrow U_0^*. \tag{A.21}$$

We see that going once around the circle here is multiplication by π. Hence this is an unpolarized chain of a given type, which clearly determines the circle (A.19) which we started with up to isomorphism. We conclude that for r even and $a = 0$ (or a even) two polarized chains are locally isomorphic for the Zariski topology.

A.30 Next we need to show that in the case r even and $a = 0$ the functor of isomorphisms is formally smooth. First we look at the case $r = 2$. In order to write our circle $N_0 \to N_{-1} \to N_0$ in the form (A.18) it is enough to take for U_0 and U_{-1} any splitting of the underlying unpolarized chain, with the additional property that U_0 and U_{-1} are totally isotropic in N_0 respectively N_{-1}. Let us call U_0 and U_{-1} an *isotropic splitting*.

Let $R \to S$ be a surjection with nilpotent kernel. Since we can lift isotropic direct summands in the affine case (proposition A.12), we can do the same with isotropic splittings. We may conclude as in the proof of proposition (A.4) . In the case r even we need to lift isotropic splittings of the chain $N_0 \to N_{-\frac{r}{2}} \to N_0$ over S and then splittings of the chain (A.21) over S, which is possible by what we have shown.

A.31 Next let us consider the case, where r is odd. Again we look first at $r = 3$. In this case we start with a circle

$$N_0 \xrightarrow{\alpha} N_{-2} \xrightarrow{\beta} N_0 . \tag{A.22}$$

We are given a perfect pairing $\mathcal{B}$ on N_0 and a pairing $\mathcal{Q}$ on N_{-2} which are both symmetric or antisymmetric, such that the cokernel of the induced map $q : N_{-2} \to N_{-2}^*$ is locally on $Spec\ R$ a free $O_F/\pi \otimes R$-module. We have $\mathcal{B}(\beta(x), y) = \mathcal{Q}(x, \alpha(y))$. Moreover we require that $\alpha\beta = \beta\alpha = \pi$ and that

$$rk_{O_F/\pi \otimes R}\ N_{-2}/N_0 + rk_{O_F/\pi \otimes R}\ N_0/N_{-2} = rk_{O_F \otimes R}\ N_0 = rk_{O_F \otimes R}\ N_2. \tag{A.23}$$

These last conditions mean that we have an unpolarized $O_F \otimes R$-chain if we forget about $\mathcal{B}$ and $\mathcal{Q}$.

As we have remarked, we may insert N_{-2}^* in the diagram (A.22),

$$N_0 \xrightarrow{\alpha} N_{-2} \xrightarrow{q} N^*_{-2} \xrightarrow{\varrho} N_0. \qquad \text{(A.24)}$$

Explicitly q and ϱ are given by the equations

$$\begin{aligned} q(x)(z) &= \mathcal{Q}(x,z) && x,z \in N_{-2} \\ \mathcal{B}(\varrho(x^*),y) &= x^*(\alpha(y)); && x \in N_{-2}, y \in N_0. \end{aligned}$$

Then we have $\varrho q = \beta$ and (A.24) becomes an unpolarized chain if we disregard $\mathcal{B}$ and $\mathcal{Q}$. Especially, turning once around the circle is multiplication by π.

We will now construct a special isotropic splitting of the diagram (A.22). Let us denote by $\mathring{N}_{-2}$ and $\mathring{N}^*_{-2}$ the images of β and ϱ in $\bar{N}_0 = N_0/\pi N_0$,

$$0 \subset \mathring{N}_{-2} \subset \mathring{N}^*_{-2} \subset \bar{N}_0.$$

Then for the pairing $\mathcal{B}$ mod π, we have that $\mathring{N}_{-2}$ is isotropic and $\mathring{N}^*_{-2} = (\mathring{N}_{-2})^{\perp}$ is the orthogonal complement by lemma (A.26). Take any isotropic complement $\bar{U}_0$ of $\mathring{N}_{-2}$. Then the induced pairing

$$\mathring{N}_{-2} \times \bar{U}_0 \longrightarrow O_F/\pi O_F \otimes R$$

is perfect and moreover $\mathring{N}^*_{-2} \oplus \bar{U}_0 = \bar{N}_0$. Hence $\bar{U}_0 \rightarrow N_0/N^*_{-2}$ is an isomorphism. We lift $\bar{U}_0$ to an isotropic subspace U_0 of N_0.
Next we construct an isotropic subspace $U_{-2} \subset N_{-2}$. We look at the flag induced on $\bar{N}_{-2} = N_{-2}/\pi N_{-2}$,

$$0 \subset \breve{N}^*_{-2} \subset \breve{N}_0 \subset \bar{N}_{-2}.$$

Then $\breve{N}^*_{-2}$ is the kernel for the pairing $\mathcal{Q}$ mod π on $\bar{N}_2$ by corollary (A.3) and $\breve{N}_0$ is a totally isotropic subspace. Consider any complement $\bar{C} \subset \bar{N}_2$ of $\breve{N}^*_{-2}$. Then we lift $\bar{C}$ to a direct summand C of N_2. By construction the restriction of $\mathcal{Q}$ to C is perfect. The image of $\breve{N}_0$ by the projection to $\bar{C}$ is isotropic. We take an isotropic complement and lift it to a direct summand U_2 of C. Then the map $\bar{U}_{-2} \rightarrow \bar{N}_{-2}/\breve{N}_0$ is by construction an isomorphism.

$\mathcal{B}$ defines a perfect duality between U_{-2} and U_0, since mod π it reduces to the perfect duality by (A.14) :

$$N_{-2}/N_0 \times N_0/N^*_{-2} \xrightarrow{\mathcal{B}(\beta x,y)} O_F/\pi O_F \otimes R.$$

Let us denote by $V \subset N_0$ the orthogonal complement of $\beta(U_{-2}) \oplus U_0$ in N_0. Hence $N_0 = \beta(U_{-2}) \oplus U_0 \oplus V$. Then $\bar{U}_0 \oplus \bar{V}$ is mapped isomorphically to N_0/N_{-2}. Hence the spaces U_{-2} and $\bar{U}_0 \oplus \bar{V}$ are a splitting of our original chain (A.22). Therefore the map α restricted to $U_0 \oplus V$ is an isomorphism onto a direct summand of N_{-2}, and β restricted to U_{-2} is an isomorphism onto a direct summand of N_0. We see that the diagram (A.22) may be identified with a diagram of the form:

$$U_{-2} \oplus U_0 \oplus V \xrightarrow[\alpha]{} U_{-2} \oplus U_0 \oplus V \xrightarrow[\beta]{} U_{-2} \oplus U_0 \oplus V.$$

Here α is the identity on $U_0 \oplus V$ and β is the identity on U_{-2}. The restriction of α to U_{-2} is multiplication by π and so is the restriction of β to $U_0 \oplus V$. We see easily that with respect to $\mathcal{Q}$ the spaces $U_{-2} \oplus U_0$ and V are orthogonal:

$$\mathcal{Q}(u_{-2}, \alpha(v)) = \mathcal{B}(\beta(u_{-2}), v) = \mathcal{B}(u_{-2}, v) = 0$$
$$\mathcal{Q}(u_0, \alpha(v)) = \mathcal{B}(\beta(u_0), v) = \mathcal{B}(\pi u_0, v) = 0\,.$$

Moreover U_{-2} and U_0 are totally isotropic for $\mathcal{B}$ and $\mathcal{Q}$ and the pairings $\mathcal{B}, \mathcal{Q} : U_{-2} \times U_0 \to O_F \otimes R$ are the same by

$$\mathcal{B}(\beta(u_{-2}), u_0) = \mathcal{Q}(u_{-2}, \alpha(u_0)).$$

Finally we have for $v_1, v_2 \in V$

$$\mathcal{Q}(v_1, v_2) = \mathcal{Q}(\alpha(v_1), \alpha(v_2)) = \mathcal{B}(\beta\alpha(v_1), v_2) = \pi\mathcal{B}(v_1, v_2) \qquad \text{(A.25)}$$

From this we conclude that the whole situation is up to isomorphism determined by the perfect symmetric or antisymmetric form $(V, \mathcal{B})$ and the number $rkU_0 = rkU_{-2}$. But locally for the étale topology the isomorphism class of $(V, \mathcal{B})$ is determined by rkV. Since these ranks are fixed by the type $(\mathcal{G})$ we conclude that two chains of type $(\mathcal{G})$ are locally isomorphic for the étale topology.

A.32 The proof of the formal smoothness goes as usual. Let $R \to S$ be a surjection with nilpotent kernel.Let α be an automorphism of the polarized chain $\{N_i\} \otimes S$. We choose a splitting $U_{-2,S}$, $U_{0,S}$ by isotropic direct summands. Let V_S the orthogonal complement of $U_{-2,S} \oplus U_{0,S}$. The image under α gives another splitting of $\{N_i\} \otimes S$,

$$U'_{-2,S} = \alpha(U_{-2,S}), \quad U'_{0,S} = \alpha(U_{0,S}), \quad V'_S = \alpha(V_S).$$

We take liftings of the isotropic subspaces $U_{0,S}, U'_{0,S}$ and $U_{-2,S}, U'_{-2,S}$ to isotropic subspaces U_0, U'_0 of N_0 and U_{-2}, U'_{-2} of N_{-2}. Clearly the isomorphism α lifts

$$\tilde{\alpha} : U_0 \longrightarrow U'_0, \qquad \tilde{\alpha} : U_{-2} \longrightarrow U'_{-2}.$$

We denote by V respectively V' the orthogonal complements of $U_0 \oplus U_{-2}$ respectively $U'_0 \oplus U'_{-2}$ with respect to $\mathcal{B}$. These are liftings of the spaces V_S respectively V'_S. Hence all we need to show is that the isomorphism of perfect bilinear forms

$$\alpha : (V, \mathcal{B})_S \longrightarrow (V', \mathcal{B})_S$$

lifts to an isomorphism

$$\tilde{\alpha} : (V, \mathcal{B}) \longrightarrow (V', \mathcal{B})$$

But this follows by the propositions (A.13) and (A.14).

A.33 Next we must treat the case $r > 3$, where r is odd and $a = 0$. We set $t = \left[\frac{r+1}{2}\right]$. We start with a circle of lemma (A.27),

$$N_0 \longrightarrow N_{-r+1} \longrightarrow \cdots \longrightarrow N_{-t} \longrightarrow N_0, \tag{A.26}$$

and with symmetric or antisymmetric bilinear forms $\mathcal{B}$ on N_0 and $\mathcal{Q}$ on N_{-t}.

Then the chain

$$N_0 \longrightarrow N_{-t} \longrightarrow N_0 \tag{A.27}$$

with the pairing $\mathcal{B}$ and $\mathcal{Q}$ is exactly of the type (A.22). Therefore (A.27) may be written as follows:

$$U_{-t} \oplus U_0 \oplus V \overset{\pi \oplus id_{U_0 \oplus V}}{\longrightarrow} U_{-t} \oplus U_0 \oplus V \overset{id_{U_{-t}} \oplus \pi}{\longrightarrow} U_{-t} \oplus U_0 \oplus V.$$

Because of the $id_{U_0 \oplus V}$ in this diagram, we may write

$$N_{-k} = N'_{-k} \oplus U_0 \oplus V, \qquad k = \left[\frac{r}{2}\right] + 1, \ldots, r-1.$$

We get an unpolarized chain of a certain type,

$$U_{-t} \longrightarrow N'_{-r+1} \longrightarrow \cdots \longrightarrow N'_{-[\frac{r}{2}]-2} \longrightarrow U_{-t}.$$

We conclude by the same arguments as in the case where the period is even.

A.34 Let us now assume that $a = 1$. Let us start with a polarized chain of $O_F \otimes \mathcal{O}_T$-modules

$$\cdots N_{-i} \xrightarrow{\varrho} N_{-i+1} \xrightarrow{\varrho} \cdots \xrightarrow{\varrho} N_0 \xrightarrow{\varrho} N_1 \longrightarrow \cdots \tag{A.28}$$

and perfect pairings $\mathcal{B}_{-i} : N_{-i} \times N_{i+1} \to O_F \otimes \mathcal{O}_T$. Assume that we have period r, so that we may take the indices in $\mathbf{Z}/r\mathbf{Z}$. We define the bilinear form $\mathcal{B}$ on N_0 by

$$\mathcal{B}(n_0, n_0') = \mathcal{B}_0(n_0, \varrho(n_0')).$$

The chain induces a filtration on $\bar{N}_0 = N_0/\pi N_0$ (compare (A.13)),

$$0 \subset F_{-r+1} \subset \cdots \subset F_{-1} \subset \bar{N}_0 .$$

By assumption this chain of $O_F/\pi O_F \otimes R$-modules splits. Moreover one easily verifies, following the proof of lemma (A.26):

Lemma A.35 *With respect to the bilinear form $\bar{\mathcal{B}} = \mathcal{B} \bmod \pi$ on $\bar{N}_0$, the orthogonal complement of F_{-k} is F_{-r+k+1}, i.e.*

$$F_{-k}^{\perp} = F_{-r+k+1} \qquad k = 0, \ldots, -r+1 .$$

In particular F_{-r+1} is the kernel of the pairing $\bar{\mathcal{B}}$ and

$$\bar{\mathcal{B}} : N_0/N_{-r+1} \times N_0/N_{-r+1} \longrightarrow O_F/\pi \otimes R$$

is perfect. □

Let us start with the case $r = 4$, which is the essential one. Hence we have a chain

$$N_0 = N_{-4} \longrightarrow N_{-3} \longrightarrow N_{-2} \longrightarrow N_{-1} \longrightarrow N_0 . \tag{A.29}$$

By $\mathcal{B}_0$ respectively $\mathcal{B}_{-2}$ the modules N_0 and N_{-3} respectively N_{-2} and N_{-1} are in perfect duality. In addition to $\mathcal{B}$ on N_0 we define a bilinear form $\mathcal{Q}$ on N_{-2},

$$\mathcal{Q}(n_{-2}, n'_{-2}) = \mathcal{B}_{-2}(n_{-2}, \varrho(n'_{-2})).$$

First we construct isotropic splittings $U_0 \subset N_0$ for $\mathcal{B}$ respectively $U_{-2} \subset N_{-2}$ for $\mathcal{Q}$.
Indeed $N_{-2}/N_{-3} = F_{-2}/F_{-3} \subset \bar{N}_0/F_{-3}$ is an isotropic subspace. We write its inverse image in $\bar{N}_0$ as $\bar{V}_0 \oplus F_{-3}$. Let us choose an isotropic complement $\tilde{U}_0 \subset \bar{N}_0/F_{-3}$. We write the inverse image of $\tilde{U}_0$ in $\bar{N}_0$ as $\bar{U}_2 \oplus F_{-3}$. Next we lift $\bar{V}_0 \oplus \bar{U}_0$ to a direct summand M of N_0. Then $\mathcal{B}$ is perfect on M and hence there is an isotropic lifting $U_0 \subset M$ of $\bar{U}_0$. The canonical map $\bar{U}_0 \to N_0/N_1$ is an isomorphism, because the orthogonal complement of N_{-2}/N_{-3} in N_0/N_{-3} is N_{-1}/N_{-3}.
The construction of U_{-2} is the same. From the unpolarized case, we know that $N_{-2} \to N_0$ induces an injection on U_{-2}. We will identify U_{-2} with its image. Moreover we identify U_0 with its image under the map $N_0 \to N_{-2}$. By the lemma (A.35) we see that $\mathcal{B}$ and $\mathcal{Q}$ define perfect pairings

$$U_0 \times U_{-2} \longrightarrow O_F \otimes \mathcal{O}_T \,. \tag{A.30}$$

It is easily checked that these pairings coincide. Let us denote by V the orthogonal complement of $U_0 \oplus U_{-2} \subset N_0$ with respect to $\mathcal{B}$. Because $\mathcal{B}$ is perfect on $U_0 \oplus U_{-2}$ we have a decomposition

$$N_0 = U_0 \oplus U_{-2} \oplus V.$$

For N_{-2} we get a similar decomposition

$$N_{-2} = U_0 \oplus U_{-2} \oplus W \,.$$

Let us identify $N_{-3} = N_0^*$ with $U_0 \oplus U_{-2} \oplus V^*$ via the given perfect pairing on $U_0 \oplus U_{-2}$ induced by $\mathcal{B}$ or $\mathcal{Q}$. Doing the same to N_{-1} we may write $N_{-1} = U_0 \oplus U_{-2} \oplus W^*$.
Since the map $N_0 \to N_{-3} = N_0^*$ is induced by the form $\mathcal{B}$ it induces a map

$$U_0 \oplus U_{-2} \oplus V \xrightarrow{id \oplus \alpha} U_0 \oplus U_{-2} \oplus V^*,$$

where $\alpha : V \to V^*$ is some symmetric or antisymmetric map.

In the same way we may write the map $N_{-2} \longrightarrow N_{-1}$ as

$$U_0 \oplus U_{-2} \oplus W \stackrel{id\oplus\beta}{\longrightarrow} U_0 \oplus U_{-2} \oplus W^*$$

where $\beta : W \longrightarrow W^*$ is again a symmetric or antisymmetric map.

Next we will check that under the map $N_{-2}^* = N_{-1} \longrightarrow N_0$ the subspace W^* is mapped to V. Let $l \in W^*$. We have to check

$$\mathcal{B}(u_0, \varrho(l)) = 0, \quad \mathcal{B}(u_{-2}, \varrho(l)) = 0 .$$

We interpret l as a linear form on N_{-2}, such that $l(U_0) = 0,\ l(U_{-2}) = 0$. Note that by definition of the identification $N_{-2}^* = N_{-1}$, we have

$$\mathcal{B}_2(n_{-2}, l) = l(n_{-2}) .$$

Hence we get

$$\mathcal{B}(u_0, \varrho(l)) = \mathcal{B}_0(u_0, \varrho^2(l)) = \mathcal{B}_{-2}(\varrho^2(u_0), l) = l(u_0) = 0.$$

We used that by definition $\varrho^2 : N_0 \longrightarrow N_{-2}$ is on U_0 the identity.

$$\begin{aligned} &\mathcal{B}(u_{-2}, \varrho(l)) = \mathcal{B}_0(\varrho^2(u_{-2}), \varrho^2(l)) = \mathcal{B}_{-2}(\varrho^4(u_{-2}), l) \\ &= \pi\mathcal{B}_{-2}(u_{-2}, l) = \pi l(u_{-2}) = 0 . \end{aligned}$$

We conclude that the map $N_{-1} \longrightarrow N_0$ is of the form

$$U_0 \oplus U_{-2} \oplus W^* \stackrel{\pi\oplus id_{U_{-2}}\oplus\gamma}{\longrightarrow} U_0 \oplus U_{-2} \oplus V,$$

for some map $\gamma : W^* \longrightarrow V$. Since U_0 is a splitting, the cokernel of this map is $U_0/\pi U_0$. Therefore $\gamma : W^* \longrightarrow V$ is an isomorphism. Because $N_{-3} \longrightarrow N_{-2}$ is the dual map to $N_{-1} \longrightarrow N_0$, we have identified all maps in the diagram (A.29). Identifying U_{-2} with U_0^* by (A.30) our original chain looks as follows:

$$U_0 \oplus U_0^* \oplus V \stackrel{id\oplus\alpha}{\longrightarrow} U_0 \oplus U_0^* \oplus V^* \stackrel{id_{U_0}\oplus\pi\oplus id_{V^*}}{\longrightarrow} U_0 \oplus U_0^* \oplus V^* \stackrel{id\oplus\beta}{\longrightarrow} \tag{A.31}$$

$$U_0 \oplus U_0^* \oplus V \stackrel{\pi\oplus id_{U_0^*}\oplus V}{\longrightarrow} U_0 \oplus U_0^* \oplus V .$$

The perfect pairings $\mathcal{B}_i$ are the obvious ones. Hence the whole polarized chain is up to isomorphism determined by the diagram

$$V \xrightarrow{\alpha} V^* \xrightarrow{\beta} V, \text{ where } \alpha\beta = \beta\alpha = \pi, \tag{A.32}$$

and where the maps α and β are both symmetric or antisymmetric. By construction of V it is easily checked, that V^*/V and V/V^* are locally on T free $O_F/\pi O_F \otimes \mathcal{O}_T$-modules and that

$$rk\ V^*/\alpha(V) + rk\ V^*/\beta(V) = rk\ V/\pi V.$$

Therefore we are exactly in the situation $r = 2$ and $a = 1$. Hence by lemma (A.35) the sequence (A.32) becomes exact if we reduce it mod π,

$$\bar{V} \xrightarrow{\bar{\alpha}} \bar{V}^* \xrightarrow{\bar{\beta}} \bar{V} \xrightarrow{\bar{\alpha}} \bar{V}^*. \tag{A.33}$$

Let us take a direct summand $\bar{M}$ of $\bar{V}$ which is complementary to Ker $\bar{\alpha}$. Then lift $\bar{M}$ to a direct summand M of V. Let us denote by K the kernel of the canonical map

$$V \xrightarrow{\alpha} V^* \longrightarrow M^*.$$

Hence we have an orthogonal decomposition with respect to the bilinear form $\mathcal{Q}_\alpha$ defined by α,

$$V = K \oplus M.$$

We rewrite the diagram (A.33) as follows

$$K \oplus M \xrightarrow{\alpha_K \oplus \alpha_M} K^* \oplus M^* \xrightarrow{\beta} K \oplus M \xrightarrow{\alpha_K \oplus \alpha_M} K^* \oplus M^*.$$

We claim that β respects the direct sum decomposition. Firstly

$$\beta(M^*) = \beta\alpha(M) = \pi M \subset M.$$

Secondly we must verify that

$$\beta(K^*) \subset K.$$

But this is equivalent to the condition that the projection of $\alpha\beta(K^*)$ to M^* is zero. This is clear because $\alpha\beta(K^*) = \pi \cdot K^*$.

We may now write $\beta = \beta_{K^*} \oplus \beta_{M^*}$. Let us identify K^* and K via β_{K^*} and M and M^* via α_M. Our diagram takes the form

$$\begin{array}{ccccc} K^* \oplus M & \stackrel{\alpha_{K^*} \oplus id}{\longrightarrow} & K^* \oplus M & \stackrel{id \oplus \beta_M}{\longrightarrow} & K^* \oplus M \\ \| & & \| & & \| \\ N_0 & \longrightarrow & N_{-1} & \longrightarrow & N_0 \end{array} \qquad (A.34)$$

The perfect pairing $\mathcal{B}_0 : N_0 \times N_{-1} \to O_F \otimes R$ is induced by perfect pairings $\mathcal{Q}_{K^*}$ on K^* and $\mathcal{Q}_M$ on M. Since the composition of the two arrows of (A.34) is π we conclude that $\alpha_{K^*} = \pi$ and $\beta_M = \pi$. Hence the isomorphism class of the original polarized chain is determined by the isomorphism classes of the perfect pairings $\mathcal{Q}_M$ and $\mathcal{Q}_{K^*}$, which locally for the étale topology are determined by the ranks of M and K^*. This proves that in the cases $r = 2, a = 1$ respectively $r = 4, a = 1$ two polarized chains of the same type $(\mathcal{G})$ are locally isomorphic for the étale topology.
Finally we treat the case where $r > 4$ is even and $a = 1$. Consider the subchain consisting of the following modules,

$$N_0 \longrightarrow N_{-r+1} \longrightarrow N_{-\frac{r}{2}} \longrightarrow N_{-\frac{r}{2}+1} \longrightarrow N_0. \qquad (A.35)$$

The pairings $\mathcal{B}_0$ and $\mathcal{B}_{-\frac{r}{2}}$ define a polarization on that chain. By the case $r = 4$ and $a = 1$ this chain takes the form (A.31). The modules $N_{-k}, k = 0, \ldots, \frac{r}{2} - 1$ inherit from (A.31) a direct sum decomposition

$$N_{-k} = M_{-k} \oplus U_0^* \oplus V.$$

The M_{-k} form an unpolarized chain of a given type,

$$U_0 = M_{-\frac{r}{2}+1} \subset M_{-\frac{r}{2}+2} \subset \cdots \subset M_0 = U_0. \qquad (A.36)$$

In the same way we get an unpolarized chain

$$U_0^* = M_{-r+1} \subset M_{-r+2} \subset \cdots \subset M_{-\frac{r}{2}} = U_0^*. \qquad (A.37)$$

We see that the data giving a polarized chain $\{N_k\}$ are equivalent to that of a polarized chain of the form (A.31) and to two unpolarized chains of the form (A.36) and (A.37) respectively, which are in duality to each other. This shows that also in the case r even, $a = 1$ two polarized chains of type $(\mathcal{G})$ are locally isomorphic for the étale topology.

A.36 Let us now show the formal smoothness of the functor of isomorphisms in this case. First we look at the case $r = 2$, that is a chain of the form (A.32). We define a *splitting* of such a chain to be an orthogonal decomposition with respect to $\mathcal{Q}_\alpha$

$$V = K \oplus M$$

such that $K \bmod \pi$ is the kernel of $\mathcal{Q}_\alpha \bmod \pi$.
Let us consider a surjection with nilpotent kernel $R \to S$. We claim that any splitting over S

$$V \otimes_R S = K_S \oplus M_S$$

lifts to a splitting over R. Indeed let us denote by a bar the reduction mod π. Consider the diagram

$$\begin{array}{ccc} O_F/\pi \otimes R & \longleftarrow & O_F \otimes R \\ \downarrow & & \downarrow \\ O_F/\pi \otimes S & \longleftarrow & O_F \otimes S \end{array}$$

The kernel $K(\bar{\alpha})$ of $\mathcal{Q}_\alpha \bmod \pi$ is a $O_F/\pi O_F \otimes R$-module. All we need to show is that there is a direct summand K of V, such that $\bar{K} = K(\bar{\alpha})$ and $K \otimes_R S = K_S$. Let us replace in the above diagram $O_F \otimes R$ by a ring T, such that we get a fibre product. Define K_T to be the fibre product of $K(\bar{\alpha})$ and K_S over $\bar{K}_S$. Clearly K_T is a direct summand of $V \otimes T$. Hence it is enough to lift K_T to a direct summand K of V. Since an easy diagram chasing shows that $O_F \otimes R \to T$ is a surjection with nilpotent kernel we have no trouble doing that.

Now consider an automorphism ω of the chain (A.32) tensored with S. Choose a splitting of V as above and set

$$M'_S = \omega(M_S) \quad \text{and} \quad K'_S = \omega(K_S).$$

Then we lift $V \otimes_R S = M'_S \oplus K'_S$ to a splitting over R,

$$V = K' \oplus M'.$$

We need to lift the isomorphisms of perfect quadratic forms

$$\omega : (M_S, \mathcal{Q}_{\alpha,S}) \longrightarrow (M'_S, \mathcal{Q}_{\alpha,S}), \quad \omega : (K_S, \mathcal{Q}_{\beta,S}) \longrightarrow (K'_S, \mathcal{Q}_{\beta,S})$$

to isomorphisms of perfect quadratic forms

$$\omega : (M, \mathcal{Q}_\alpha) \longrightarrow (M', \mathcal{Q}_\alpha),\ \omega : (K, \mathcal{Q}_\beta) \longrightarrow (K', \mathcal{Q}_\beta).$$

This can be done by the propositions (A.13) and (A.14). This finishes the case $r = 2$.

Let us now consider the case of any even period r. We look at the diagram (A.35). We define a *semisplitting* of our polarized chain to be an isotropic subspace $U_0 \subset N_0$ and an isotropic subspace $U_{-\frac{r}{2}} \subset N_{-\frac{r}{2}}$, such that the maps

$$\bar{U}_0 \longrightarrow N_0/N_{-1}, \quad \bar{U}_{-\frac{r}{2}} \longrightarrow N_{-\frac{r}{2}}/N_{-\frac{r}{2}+1}$$

are isomorphisms. Although contrary to the case $a = 0$ the pairings on N_0 and $N_{-\frac{r}{2}}$ are not perfect, it is easy to see that such semisplittings lift with respect to surjections $R \to S$ with nilpotent kernel. Indeed one can use that U_0 and the image of $U_{-\frac{r}{2}}$ in N_0 generate a perfect subspace of N_0. Then one argues by first lifting this perfect subspace.

To lift an automorphism of the polarized chain one is reduced by using a semisplitting, to lift isomorphisms $U_0 \to U_0'$, isomorphisms of a polarized chain (A.32) and isomorphisms of unpolarized chains (A.36) and (A.37). This finishes the case $a = 1$ and r even of our theorem.

The only case left is $a = 1$ and r odd. Fortunately by an index shift this reduces to cases we have already done. Indeed, assume we are given a polarized chain $(\{N_i\}, \mathcal{B}_i)$. Then we define a new polarization by the rule

$$\mathcal{B}'_{-i}(n_{-i}, n_{i+1-r}) = \mathcal{B}_i(n_{-i}, \theta(n_{i+1-r}))$$

where $n_{-i} \in N_{-i}$ $n_{i+1-r} \in N_{i+1-r}$. Then for the new chain $(\{N_i\}, \mathcal{B}'_i)$ we have $a = 1 - r$ even. An index shift $N_i' = N_{i+\frac{1-r}{2}}$ reduces everything to the case $a = 0$, which we already treated above. This finishes the proof of proposition (A.21).

We will now consider the case (III). We assume that $O_B = M_n(O_F)$. The restriction of $*$ to F will be also denoted by $f \longmapsto \bar{f}$.

Let R be a $\mathbf{Z}_p$-algebra. We extend the involution R-linearily to $S = O_F \otimes_{\mathbf{Z}} R$ and denote it by $s \mapsto \bar{s}$. With respect to this involution we may speak of an ε-hermitian form on a S-module M.

Proposition A.37 *There is a hermitian or antihermitian form*

$$\chi : O_F^n \times O_F^n \longrightarrow O_F$$

which is perfect and satisfies the equations

$$\chi(Ax, y) = \chi(x, A^* y), \quad x, y \in O_F^n, A \in M_n(O_F). \qquad \text{(A.38)}$$

The form χ *is uniquely determined up to a unit* $f \in O_{F_0}^\times$.

Proof: Let us remark that there may exist both a hermitian and an antihermitian form meeting the requirements of the proposition; this is how the above uniqueness assertion is to be interpreted. We may write the involution $*$ in the form

$$A^* = U^{-1}\,{}^t\bar{A}U\,.$$

The equation $A^{**} = A$ implies $U^{-1}{}^t\bar{U} \in F$. We write

$${}^t\bar{U} = fU\,.$$

We see that $f\bar{f} = 1$. By Hilbert Satz 90 we conclude that we may assume that ${}^t\bar{U} = U$. Then

$$\chi(x, y) = {}^t\bar{x}Uy$$

is a hermitian form that satisfies (A.38). Since $*$ leaves $M_n(O_F)$ invariant, we see that $U^{-1}O_F^n$ is an $M_n(O_F)$-lattice. By Morita equivalence there is an $u \in F$, such that $U^{-1}O_F^n = u \cdot O_F^n$. We have to find an element $u_1 \in F$, such that u and u_1 have the same order, and $\bar{u}_1 = \pm u_1$. If F/F_0 is unramified or if the order of u is even, we may find $u_1 \in F_0$ having the same order as u. In the ramified case, there is a prime element π of O_F that satisfies $\pi = -\bar{\pi}$. In this case we take u_1 to be a power of π. Clearly $u_1 \cdot \chi$ is the form we are looking for. □

Remark A.38 We may choose a generator ϑ_F of the different ideal of $F/\mathbf{Q}_p$ that satisfies $\vartheta_F = \bar{\vartheta}_F$ or $\vartheta_F = -\bar{\vartheta}_F$. As for O_F-bilinear pairings the equation

$$tr_{F/\mathbf{Q}_p}\vartheta_F^{-1} f\chi(x, y) = \chi_0(x, fy)$$

gives a bijection between hermitian or antihermitian forms and $\mathbf{Z}_p$-bilinear forms χ_0, which are symmetric or antisymmetric and satisfy the relation

$$\chi_0(x, fy) = \chi_0(\bar{f}x, y) .$$

Under this bijection perfect forms correspond to perfect forms.

A.39 Let M be a right $M_n(O_F)$-module. Then we view M as a left module by the equation

$$Am = mA^*, \qquad m \in M, \quad A \in M_n(O_F).$$

With this convention χ_0 becomes an isomorphism of $M_n(O_F)$-modules

$$\chi_0 : O_F^n \longrightarrow \mathrm{Hom}(O_F^n, \mathbf{Z}_p) .$$

Let R be a $\mathbf{Z}_p$-algebra. Let M and M' be $M_n(O_F) \otimes_{\mathbf{Z}_p} R$-modules. Assume we are given a R-bilinear form

$$\mathcal{E} : M \times M' \longrightarrow R ,$$

such that $\mathcal{E}(Am, m') = \mathcal{E}(m, A^* m')$. By Morita equivalence we write $M = O_F^n \otimes_{O_F} N$ and $M' = O_F^n \otimes_{O_F} N'$. Then there is a unique R-bilinear form

$$\mathcal{B} : N \times N' \longrightarrow O_F \otimes R$$

such that $\mathcal{B}(fn, n') = \bar{f}\mathcal{B}(n, n') = \mathcal{B}(n, \bar{f}n')$ for $f \in O_F$. Namely, $\mathcal{B}$ is defined by the equation

$$\mathcal{E}(x_1 \otimes n_1, x_2 \otimes n_2) = \chi_0(x_1, \mathcal{B}(n_1, n_2)x_2). \tag{A.39}$$

More formally we have the map of $M_n(O_F)$-modules induced by $\mathcal{E}$:

$$\begin{array}{rl} O_F^n \otimes_{O_F} N \to \mathrm{Hom}_R(O_F^n \otimes_{O_F} N', R) \cong & \mathrm{Hom}_{O_F \otimes R}(N', \mathrm{Hom}_R(O_F^n \otimes R, R)) \\ & \downarrow \chi_0 \\ & \mathrm{Hom}_{O_F \otimes R}(N', O_F^n \otimes R) \\ & \| \\ & O_F^n \otimes_{O_F} \mathrm{Hom}_{O_F \otimes R}(N', O_F \otimes R) \end{array}$$

Hence again by Morita equivalence we get

$$\mathcal{B}: N \longrightarrow \mathrm{Hom}_{O_F \otimes R}(N', O_F \otimes R).$$

Clearly $\mathcal{B}$ is perfect if and only if $\mathcal{E}$ is perfect, and $\mathcal{B}$ is hermitian if χ_0 is antihermitian and vice versa.

A.40 Assume we are given a selfdual chain of $M_n(O_F)$-lattices $\mathcal{L} = \{\Lambda\}$ in the vectorspace (V, ψ). We write $V = F^n \otimes_F W$. The equation (A.39) applied to this situation yields a hermitian or antihermitian pairing

$$\varkappa : W \times W \longrightarrow F.$$

We have then a selfdual chain of lattices $\mathcal{G} = \{\Gamma\}$ in $(W, \varkappa)$, such that $\{\Lambda = O_F^n \otimes \Gamma\ ;\ \ \Gamma \in \mathcal{G}\}$ is the chain $(\mathcal{L})$ and $\Lambda^* = O_F^n \otimes \Gamma^*$.

Assume we are given a polarized chain of $M_n(O_F) \otimes \mathcal{O}_T$-modules $(M_\Lambda, \mathcal{E}_\Lambda)$ of type $(\mathcal{L})$ on a scheme T. Then there is a chain N_Γ of $O_F \otimes \mathcal{O}_T$-modules of type $(\mathcal{G})$, such that

$$M_{O_F^n \otimes \Gamma} = O_F^n \otimes N_\Gamma .$$

From (A.39) we get perfect pairings

$$\mathcal{B}_\Gamma : N_\Gamma \times N_{\Gamma^*} \longrightarrow O_F \otimes \mathcal{O}_T .$$

They satisfy the equation

$$\mathcal{B}_\Gamma(n, n') = \varepsilon \overline{\mathcal{B}_{\Gamma^*}(n', n)}, \quad n \in N_\Gamma, n' \in N_{\Gamma^*},$$

where the sign ε is opposite to the sign of χ_0.

Definition A.41 *Let $\{N_\Gamma\}$ be a chain of $O_F \otimes \mathcal{O}_T$-modules on a $\mathbf{Z}_p$-scheme T of type $(\mathcal{G})$. A polarization of the chain $\{N_\Gamma\}$ is a set of perfect sesquilinear forms*

$$\mathcal{B}_\Gamma : N_\Gamma \times N_{\Gamma^*} \longrightarrow O_F \otimes \mathcal{O}_T$$

which satisfy the following conditions

1. $\mathcal{B}_\Gamma(\bar{f}n, n') = f\mathcal{B}_\Gamma(n, n') = \mathcal{B}_\Gamma(n, \bar{f}n')$,
   $n \in N_\Gamma,\ n' \in N_{\Gamma^*},\ f \in O_F$

2. $\mathcal{B}_\Gamma(n,n') = \pm\overline{\mathcal{B}_{\Gamma^*}(n',n)}$
Here the sign depends only on the involution with which we started.

3. Let $\Gamma_1 \subset \Gamma_2$ *be lattices in* $\mathcal{G}$. *Then*

$$\mathcal{B}_{\Gamma_1}(m, \varrho_{\Gamma_1^*,\Gamma_2^*}(n)) = \mathcal{B}_{\Gamma_2}(\varrho_{\Gamma_2,\Gamma_1}(m), n),$$

where $m \in N_{\Gamma_1}, n \in N_{\Gamma_2^*}$.

4. There is a periodicity morphism functorial in Γ

$$\theta = \theta_{\pi^{-1}} : N_{\pi\Gamma} \longrightarrow N_\Gamma\,,$$

such that $\pi \cdot \theta = \varrho_{\Gamma,\pi\Gamma}$ *and such that*

$$\mathcal{B}_\Gamma(\theta(n), n') = \frac{\pi}{\overline{\pi}}\mathcal{B}_{\pi\Gamma}(n, \theta(n')), \quad n \in N_{\pi\Gamma}, n' \in N_{\Gamma^*}.$$

In the notation of the definition (3.6) we have $\frac{\pi}{\overline{\pi}}\theta = \theta_{\overline{\pi}^{-1}}$. If we multiply the last equation of condition 4) by $\overline{\pi}$ we get a special case of 3).

We summarize our considerations as follows:

Proposition A.42 *There is an equivalence between the categories of polarized chains of* $M_n(O_F) \otimes \mathcal{O}_T$*-modules on the scheme* T *of type* $(\mathcal{L})$ *and the category of polarized chains of* $O_F \otimes \mathcal{O}_T$*-modules of type* $(\mathcal{G})$. □

The theorem (3.16) may be reformulated as follows in case (III):

Proposition A.43 *Let* $(N_\Gamma, \mathcal{B}_\Gamma)$ *be a polarized chain of* $O_F \otimes \mathcal{O}_T$*-modules of type* $(\mathcal{G})$. *Then locally for the étale topology on* T *there is an isomorphism of polarized chains*

$$(N_\Gamma, \mathcal{B}_\Gamma) \simeq (\Gamma, \varkappa) \otimes \mathcal{O}_T.$$

If r *and* a *(see below) both are even, such an isomorphism exists even locally for the Zariski topology.*
If $(N'_\Gamma, \mathcal{B}'_\Gamma)$ *is a second polarized chain of type* $(\mathcal{G})$, *then the functor on the category of* T*-schemes*

$$T' \longrightarrow Isom((N_\Gamma, \mathcal{B}_\Gamma)_{T'}, (N'_\Gamma, \mathcal{B}'_\Gamma)_{T'})$$

is representable by a smooth affine scheme.

Proof: Since the question is local we will assume that $T = Spec\ R$. Let us choose a prime element $\pi \in F$, such that $\pi = \bar{\pi}$ in the case where F/F_0 is unramified and $\pi = -\bar{\pi}$ in the ramified case. Let us index the selfdual chain $\mathcal{G}$ by $\mathbf{Z}$:

$$\cdots \subset \Gamma_i \subset \Gamma_{i+1} \subset \cdots$$
$$\pi\Gamma_i = \Gamma_{i-r} \quad \Gamma_i^* = \Gamma_{-i+a}\,.$$

If $(\mathcal{N}_\Gamma, \mathcal{B}_\Gamma)$ is a polarized chain of type $(\mathcal{G})$, we will set $N_i = \mathcal{N}_{\Gamma_i}$ and $\mathcal{B}_i = \mathcal{B}_{\Gamma_i}$.

Let us start with the case $r = 1$ and $a = 0$. Then the whole polarized chain $(N_i, \mathcal{B}_i)$ is determined by a perfect ε-hermitian form :

$$\mathcal{B}_0 : N_0 \times N_0 \longrightarrow O_F \otimes R.$$

In the case where $\mathcal{B}_0$ is hermitian or F is unramified over F_0 the proposition (A.14) is applicable. Then the proof is as in case (II). Hence we are left with the situation, where F/F_0 is ramified, and the form $\mathcal{B}_0$ is antihermitian. We claim that Zariski locally N_0 is a direct sum of two isotropic subspaces. Indeed, since $\mathcal{B}_0$ modulo π is an alternating form on $N_0/\pi N_0$, we have locally a decomposition into isotropic subspaces modulo π. By proposition (A.12) we obtain the desired decomposition for N_0. This implies the first part of the proposition for $a = 0$ and $r = 1$. The proof that isomorphisms $(N_0, \mathcal{B}_0) \longrightarrow (N_0', \mathcal{B}_0')$ lift follows from the same proposition.

The case $r = 1, a = 1$ is done in the same way. We have only to use the perfect ε-hermitian pairing

$$\mathcal{Q}(n_0, n_0') = \mathcal{B}_0(n_0, \theta(n_0')), \quad n_0, n_0' \in N_0,$$

which in this case determines the whole chain up to isomorphism.

Next we consider the case $a = 0$ and $r > 1$. There is an obvious analog of the lemma (A.27), which we give without proof:

Lemma A.44 *A polarized chain of $O_F \otimes \mathcal{O}_T$-modules of type $(\mathcal{G})$ of period r and $a = 0$ is given by the following data.*

1. *A circular diagram of T-locally free $O_F \otimes \mathcal{O}_T$-modules*

$$\begin{array}{ccc} N_{-r+1} & \xrightarrow{\varrho} \cdots \xrightarrow{\varrho} & N_{-t}, \\ & {}^{\varrho}\nwarrow \quad \swarrow{}^{\beta} & \\ & N_0 & \end{array}$$

where $t = \left[\frac{r+1}{2}\right]$.

2. $O_{F_0} \otimes \mathcal{O}_T$*-bilinear pairings, which are either hermitian or antihermitian*

$$\begin{aligned} \mathcal{B} &: N_0 \times N_0 \longrightarrow O_F \otimes_{\mathbf{Z}_p} \mathcal{O}_T \\ \mathcal{Q} &: N_{-t} \times N_{-t} \longrightarrow O_F \otimes_{\mathbf{Z}_p} \mathcal{O}_T . \end{aligned}$$

The following conditions are satisfied.

(i) Locally on T *there exist isomorphisms*

$$\begin{array}{rcll} N_i & \simeq & \Gamma_i \otimes_{\mathbf{Z}_p} \mathcal{O}_T & i = 0, -r+1, \ldots -t \\ N_i/N_{i-1} & \simeq & \Gamma_i/\Gamma_{i-1} \otimes_{\mathbf{Z}_p} \mathcal{O}_T & i = -t, \ldots, -r+1 \\ N^*_{-t}/N_{-t} & \cong & \begin{cases} \Gamma_{-t+1}/\Gamma_{-t} \otimes_{\mathbf{Z}_p} \mathcal{O}_T & \text{if } r \text{ is odd} \\ 0 & \text{if } r \text{ is even.} \end{cases} & \end{array}$$

The left hand side of the last isomorphism is a notation for the cokernel of the map induced by $\mathcal{Q}$

$$N_{-t} \longrightarrow N^*_{-t} .$$

The bilinear form $\mathcal{B}$ *is perfect.*
In the case where the extension F/F_0 *is unramified the forms* $\mathcal{B}$ *and* $\mathcal{Q}$ *are hermitian of the same sign as* $\varkappa$. *In the ramified case* $\mathcal{B}$ *has the same sign as* $\varkappa$ *but* $\mathcal{Q}$ *has the opposite sign.*

(ii) Let $\alpha = \varrho^{r-t} : N_0 \longrightarrow N_{-t}$. *Then we have the identity*

$$\mathcal{B}(\beta(m), n) = \mathcal{Q}(m, \alpha(n)), \quad m \in N_{-t}, \quad n \in N_0 .$$

(iii) Going once around the circular diagram is multiplication by π. □

A.45 Let us start with the case where r is even. Assume $r = 2$. Then the circular diagram of lemma (A.44) looks as follows

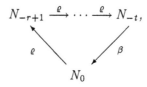

$$N_0 \xrightarrow{\alpha} N_{-1} \xrightarrow{\beta} N_0 \xrightarrow{\alpha} N_{-1},$$

where $\beta\alpha = \alpha\beta = \pi$.

We know that isotropic subspaces of $N_0/\pi N_0$ and $N_{-1}/\pi N_{-1}$ with respect to $\mathcal{B} \bmod \pi$ and $\mathcal{Q} \bmod \pi$ lift, since in the case r even the form $\mathcal{Q}$ is perfect. Hence we find a splitting of our chain by isotropic subspaces (compare (A.18)):

$$\begin{aligned} N_0 &= U_0 \oplus \beta(U_{-1}) \\ N_1 &= U_{-1} \oplus \alpha(U_0). \end{aligned}$$

We have a perfect sesquilinear form, $U_{-1} \times U_0 \to O_F \otimes R$ given by

$$\mathcal{B}(\beta(u_{-1}), u_0) = \mathcal{Q}(u_{-1}, \alpha(u_0)).$$

We see that the $O_F \otimes R$-module U_0 determines the whole chain up to isomorphism. Since $U_0 \bmod \pi$ is isomorphic to the cokernel of β, it follows that U_0 is T-locally free. Hence any two chains of type $(\mathcal{G})$ are locally isomorphic for the Zariski topology. The rest of the argument for r even is exactly as in the case of $M_n(O_F)$ with an involution of the first kind.

The case, where r is odd and $a = 0$ is done by an obvious modification of the arguments in case (II). We therefore skip it.

A.46 We will now consider the case $a = 1$. We consider the sesquilinear form

$$\mathcal{B} : N_0 \times N_0 \longrightarrow O_F \otimes R,$$

which is defined by

$$\mathcal{B}(n_0, n_0') = \mathcal{B}_0(n_0, \varrho(n_0')).$$

$\mathcal{B}$ is ε-hermitian of the same sign as $\varkappa$. We set $\bar{N}_0 = N_0/\pi N_0$. For $0 \leq t \leq r$ we denote by F_{-t} the image of the map $N_{-t} \longrightarrow \bar{N}_0$ obtained from ϱ.

In the same way as lemma (A.26) one proves:

Lemma A.47 *The orthogonal complement of F_{-t} with respect to $\mathcal{B}$ modulo is F_{t+1-r}* □.

We will start with the case $r = 2$. For an $O_F \otimes R$-module N we denote by ${}^cN$ the module obtained from N via restriction of scalars by the conjugation $O_F \otimes R \to O_F \otimes R, f \otimes r \to \bar{f} \otimes r$.
The perfect sesquilinear pairing

$$\begin{array}{rcl} \mathcal{P} : N_0 \times N_{-1} & \longrightarrow & O_F \otimes R \\ (n_0, n_{-1}) & \longmapsto & \mathcal{B}_0(n_0, \theta(n_{-1})) \end{array} \qquad (A.40)$$

allows us to identify N_{-1} with ${}^cN_0^*$. Let α be the map

$$N_0 \underset{\theta^{-1}}{\longrightarrow} N_{-2} \underset{\varrho}{\longrightarrow} N_{-1} \simeq {}^cN_0^* .$$

We get a diagram

$$N_0 \underset{\alpha}{\longrightarrow} {}^cN_0^* \underset{\beta}{\longrightarrow} N_0 , \qquad (A.41)$$

where the map β is induced by ϱ. We have $\alpha\beta = \beta\alpha = \pi$.
Exactly as in case (II) (A.32) this diagram determines the whole polarized chain up to isomorphism. If $\varkappa$ is hermitian, one checks that β is hermitian with respect to $\mathcal{P}$, while α is hermitian in the case where F/F_0 is unramified but antihermitian in the ramified case. If $\varkappa$ is antihermitian, α and β are ε-hermitian with the opposite sign. By corollary (A.3) we get an exact sequence, if we reduce modulo π,

$$\bar{N}_0 \underset{\bar{\alpha}}{\longrightarrow} \overline{{}^cN_0^*} \underset{\bar{\beta}}{\longrightarrow} \bar{N}_0 \underset{\bar{\alpha}}{\longrightarrow} \overline{{}^cN_0^*} .$$

We may rewrite this diagram in the same way as (A.33):

$$K \oplus M \overset{\alpha_K \oplus \alpha_M}{\longrightarrow} {}^cK^* \oplus {}^cM^* \overset{\beta_K \oplus \beta_M}{\longrightarrow} K \oplus M .$$

Here the maps α_M and β_K are isomorphisms. These isomorphisms determine α_K and β_M uniquely by the equations $\beta_K\alpha_K = \pi$ and $\beta_M\alpha_M = \pi$.

Hence our polarized chain is determined by the maps α_M and β_K, which may be interpreted as perfect hermitian respectively antihermitian forms on M and ${}^cK^*$. Hence we finish the case $r = 2$ by applying the proposition (A.43) in the case $r = 1$, $a = 0$ to α_M and β_K.

A.48 We consider now the case $r = 4$. We have a chain of modules:

$$N_0 \underset{\sim}{\xrightarrow{\theta^{-1}}} N_{-4} \xrightarrow{\varrho} N_{-3} \xrightarrow{\varrho} N_{-2} \xrightarrow{\varrho} N_{-1} \xrightarrow{\varrho} N_0. \tag{A.42}$$

Moreover we have perfect sesquilinear forms

$$\mathcal{P}_0 : N_0 \times N_{-3} \longrightarrow O_F \otimes R$$

$$\mathcal{P}_0(n_0, n_{-3}) = \mathcal{B}_0(n_0, \theta(n_{-3}))$$

and

$$\mathcal{P}_{-2} : N_{-2} \times N_{-1} \longrightarrow O_F \otimes R$$

$$\mathcal{P}_{-2}(n_{-2}, n_{-1}) = \mathcal{B}_{-2}(n_{-2}, \theta(n_{-1}))$$

We have the relations

$$\begin{aligned} \mathcal{P}_0(n_0, \theta^{-1}\varrho(n_0')) &= \pm\overline{\mathcal{P}(n_0', \theta^{-1}\varrho(n_0))} \\ \mathcal{P}_{-2}(n_{-2}, \varrho(n_{-2}')) &= \pm\overline{\mathcal{P}_{-2}(n_{-2}', \varrho(n_{-2}))} \\ \mathcal{P}_0(\varrho(n_{-1}), n_{-3}) &= \pm\overline{\mathcal{P}_{-2}(\varrho(n_{-3}), n_{-1})} \end{aligned}$$

Here the signs depend on $\varkappa$ and the ramification of F/F_0. We also introduce the ε-hermitian forms

$$\mathcal{B}(n_0, n_0') = \mathcal{P}(n_0, \theta^{-1}\varrho(n_0')) = \mathcal{B}_0(n_0, \varrho(n_0'))$$

and

$$\mathcal{Q}(n_{-2}, n_{-2}') = \mathcal{P}_{-2}(n_{-2}, \varrho(n_{-2}')) = \mathcal{B}_{-2}(n_{-2}, \theta\varrho(n_{-2}')).$$

We can use $\mathcal{P}_0$ (respectively $\mathcal{P}_{-2}$) to identify N_{-3} with ${}^cN_0^*$ (respectively N_{-1} with ${}^cN_{-2}^*$). Then the diagram (A.42) takes the form:

$$N_0 \xrightarrow{\tilde{\varrho}} {}^cN_0^* \xrightarrow{\tilde{\varrho}} N_{-2} \xrightarrow{\tilde{\varrho}} {}^cN_{-2}^* \xrightarrow{\tilde{\varrho}} N_0. \tag{A.43}$$

The relations between the bilinear forms above signify that the second and the forth arrow in this diagram are anti-dual to each other, while the first is induced by $\mathcal{B}$ and the third is induced by $\mathcal{Q}$.

As in the case (II) we find totally isotropic direct summands $U_0 \subset N_0$ with respect to $\mathcal{B}$ and $U_{-2} \subset N_{-2}$ with respect to $\mathcal{Q}$, such that the maps

$\bar{U}_0 \to N_0/N_{-1}$ and $\bar{U}_{-2} \to N_{-2}/N_{-3}$ are isomorphisms. Since by lemma (A.47) the form $\mathcal{B}$ modulo π induces a perfect pairing

$$\begin{array}{rcccl} \bar{\mathcal{B}} : N_{-2}/N_{-3} & \times & N_0/N_{-1} & \longrightarrow & O_F/\pi O_F \otimes R \\ n_{-2} & \times & n_0 & \longmapsto & \mathcal{B}(\varrho^2(n_{-2}), n_0) \ \text{ modulo } \pi\,, \end{array}$$

we get a perfect pairing

$$U_{-2} \times U_0 \longrightarrow O_F \otimes R\,, \tag{A.44}$$

which maps $u_{-2} \times u_0$ to $\mathcal{B}(\varrho^2(u_{-2}), u_0) = \mathcal{Q}(u_{-2}, \varrho^2\theta^{-1}(u_0))$. From the unpolarized case we know that ϱ^2 maps U_0 isomorphically to a direct summand of N_{-2} and U_{-2} isomorphically to a direct summand of N_0. Hence we may identify U_0 with a direct summand of N_{-2} and U_{-2} with a direct summand of N_0. Then the restrictions of $\mathcal{B}$ to the subspace $U_0 \oplus U_{-2} \subset N_0$ respectively of $\mathcal{Q}$ to the subspace $U_0 \oplus U_{-2} \subset N_{-2}$ are the same perfect pairing given by (A.44).

Taking the orthogonal complements to $U_0 \oplus U_{-2}$ we may write:

$$N_0 = U_0 \oplus U_{-2} \oplus V, \quad N_{-2} = U_0 \oplus U_{-2} \oplus W\,.$$

If we identify $U_0 \oplus U_{-2}$ with ${}^c(U_0 \oplus U_{-2})^*$ by (A.44) we may write

$${}^cN_0^* = U_0 \oplus U_{-2} \oplus {}^cV^*, \quad {}^cN_{-2}^* = U_0 \oplus U_{-2} \oplus {}^cW^*.$$

As in the case of an involution of the first kind, one checks that the map $N_{-1} \to N_0$ takes with these identifications the form

$$U_0 \oplus U_{-2} \oplus {}^cW^* \overset{\pi \oplus id_{U_{-2}} \oplus \gamma}{\longrightarrow} U_0 \oplus U_{-2} \oplus V.$$

The map γ is an isomorphism, which we use to identify V and ${}^cW^*$. Then our original chain (A.42) looks as follows:

$$\begin{array}{l} U_0 \oplus U_{-2} \oplus V \overset{id \oplus \alpha}{\longrightarrow} U_0 \oplus U_{-2} \oplus {}^cV^* \overset{id_{U_0} \oplus \pi \oplus id_{{}^cV^*}}{\longrightarrow} U_0 \oplus U_{-2} \oplus {}^cV^* \overset{id \oplus \beta}{\longrightarrow} \\ U_0 \oplus U_{-2} \oplus V \overset{\pi \oplus id_{U_{-2}} \oplus id_V}{\longrightarrow} U_0 \oplus U_{-2} \oplus V. \end{array}$$

Here $\alpha : V \to {}^cV^*$ and $\beta : {}^cV^* \to V$ are maps, which are either hermitian or antihermitian and $\alpha\beta = \beta\alpha = \pi$. This is exactly the situation of the diagram (A.41). This shows that the ranks of the modules N_i and N_i/N_{i-1}

determine locally for the étale topology the polarized chain (A.42) up to isomorphism.
In the case where $r > 4$ is an even number the proof of this fact may be reduced as in case (II) to the case $r = 4$.
The proof of the formal smoothness of the functor is as usual. The case $a = 1$ and r odd reduces to the case $a = 0$ by a shift of indices.

A.49 We will now consider the case (IV). For the maximal order in B we may take $O_B = M_n(O_D)$, where O_D is the unique maximal order in the quaternion algebra D.
We will fix an unramified extension F_2/F of degree two that is contained in D. We also fix a prime element $\Pi \in D$, such that conjugation by Π induces on F_2 the nontrivial automorphism over F:

$$\bar{e} = \Pi e \Pi^{-1} \quad \text{for} \quad e \in F_2 .$$

Then $\Pi^2 = \pi \in F$ is a prime element of F. The maximal order in D is

$$O_D = O_{F_2}[\Pi].$$

By assumption the involution $X \longmapsto X^*, X \in M_n(D)$ leaves $M_n(O_D)$ invariant. Let us classify the possible involutions of this kind up to isomorphism. In terms of matrices this problem may be rephrased as follows. For $X \in M_n(D)$ we denote by X' the matrix obtained by applying the main involution to the entries of X. Then

$$X \longmapsto {}^tX'$$

is an involution of $M_n(D)$ that leaves $M_n(O_D)$ invariant. Any other involution of the first kind is of the form

$$X \longmapsto C\,{}^tX'C^{-1},$$

where $C \in GL_n(D)$ is a matrix that satisfies one of the equations $C = \pm\,{}^tC'$. Two involutions $X^* = C\,{}^tX'C^{-1}$ and $X^+ = C_1\,{}^tX'C_1^{-1}$ are isomorphic, iff there is an $A \in GL_n(D)$ and $f \in F^\times$, such that

$$AC^tA' = fC_1 .$$

Since the involution $X \longmapsto X'$ fixes $M_n(O_D)$, we see that $CO_D^n \subset D^n$ is a left $M_n(O_D)$-submodule. By Morita equivalence we conclude that

$$CO_D^n = \Pi^k O_D^n \quad \text{for some } k\,.$$

Since we may multiply C by an element of F without changing the involution, we may assume that $k = 0$ or $k = 1$.
Let $U = O_D^n$ considered as a right O_D-module. We have $O_B = \mathrm{End}_{-O_D}(U)$. In the case $k = 0$ we have a perfect hermitian respectively antihermitian form on the right O_D-module U with respect to the main involution

$$H(u,v) = {}^t u' C^{-1} v, \quad u, v \in U\,.$$

In particular we have the equations

$$H(ud, v) = d' H(u,v)$$

and

$$H(u, vd) = H(u,v) d\,.$$

This form H induces the involution $X \longmapsto X^*$ on O_B

$$H(Xu, v) = H(u, X^* v)\,.$$

In the case $k = 1$ we define $A \in GL_n(O_D)$ by the equation $C = A\,\Pi$. We will work with the nebeninvolution on D,

Definition A.50 *The nebeninvolution on D with respect to the prime element Π is defined by $\tilde{d} = \Pi d' \Pi^{-1}$.*

We note that even up to isomorphism this definition depends on the prime element Π.
Hence in the case $k = 1$ we have with the obvious notation

$$X' = A\,{}^t\tilde{X} A^{-1}\,,$$

where $A \in GL_n(O_D)$, ${}^t\tilde{A} = \mp A$. In this case we get a perfect ε-hermitian form on the right O_D-module U with respect to the nebeninvolution

$$H(u,v) = {}^t\tilde{u} A^{-1} v\,, \quad u, v \in U\,.$$

This pairing induces the involution $X \longmapsto X^*$ on $O_B = \mathrm{End}_{-O_D}(U)$. We note that isomorphic ε-hermitian forms on U induce isomorphic involutions on O_B. Let us summarize our conclusions.

Proposition A.51 *Any involution of the first kind on O_B is induced by a perfect ε-hermitian form on the right O_D-module $U = O_D^n$ with respect to the main involution or the nebeninvolution,*

$$H : U \times U \longrightarrow O_D .$$

The reader may verify that up to isomorphism there are the following possibilities for perfect ε-hermitian forms on U.
Let us denote by $d \longmapsto \check{d}$ an involution on D, which is either the main involution or the nebeninvolution. In the table below we represent a ε-hermitian form H on U by its matrix $N \in M_n(O_D)$,

$$H(u, v) = {}^t\check{u}Nv .$$

We also denote by ζ a root of unity of order prime to p that generates F_2/F

Case: H hermitian with respect to the main involution,

$$N = E \quad \text{the unit matrix.}$$

Case: H antihermitian with respect to the main involution ,

$$N = \zeta E.$$

Case: H hermitian with respect to the nebeninvolution,

$$N = E \quad \text{or} \quad N = \begin{pmatrix} E & 0 \\ 0 & \zeta \end{pmatrix}.$$

Case: H antihermitian with respect to the nebeninvolution

$$N = \begin{pmatrix} 0 & E \\ -E & 0 \end{pmatrix},$$

where the blocks are quadratic of the same size.

Instead of H we can consider the perfect (anti-)symmetric O_F-bilinear form

$$h_F : U \times U \longrightarrow O_F ,$$

which is given by the formula

$$h_F(u_1, u_2) = tr^0 \Pi^{-1} H(u_1, u_2) , \qquad \text{(A.45)}$$

where tr^0 denotes the reduced trace form on D. Let us denote by $d \longmapsto d^+$ the involution on D given by $d^+ = \Pi \check{d} \Pi^{-1}$. This is either the main involution or the nebeninvolution. One checks that h_F satisfies the equation

$$h_F(u_1, u_2 d) = h_F(u_1 d^+, u_2). \tag{A.46}$$

Conversely a perfect (anti-)symmetric O_F-bilinear form h_F that satisfies the equation (A.46) comes from a perfect hermitian pairing H on the right O_D-module U with respect to the involution $d \longmapsto \check{d}$.
As usual to h_F there corresponds a $\mathbf{Z}_p$-bilinear perfect form

$$h(u_1, u_2) = tr_{F/\mathbf{Q}_p} \vartheta_F^{-1} h_F(u_1, u_2).$$

Here ϑ_F is any generator of the different ideal of F over $\mathbf{Q}_p$. Again h satisfies the equation (A.46). It induces on $\mathrm{End}_{-O_D}(U) = O_B$ the given involution $X \longmapsto X^*$.
We may reformulate the Proposition (A.51):

Proposition A.52 *Any involution of the first kind on O_B is induced by a perfect (anti-) symmetric $\mathbf{Z}_p$-bilinear form on the right O_D-module $U = O_D^n$*

$$h : U \times U \longrightarrow \mathbf{Z}_p$$

that satisfies the equation

$$h(u_1 \check{d}, u_2) = h(u_1, u_2 d),$$

where $d \longmapsto d^{\vee}$ is either the main involution or the nebeninvolution.

Proposition A.53 *Let $M_1 = O_D^n \otimes_{O_D} N_1$ and $M_2 = O_D^n \otimes_{O_D} N_2$ be left $O_B \otimes R$-modules which are projective and finite.*
Then there is a bijection between perfect R-bilinear forms

$$\mathcal{E} : M_1 \times M_2 \longrightarrow R,$$

that satisfy the equation

$$\mathcal{E}(b^* m_1, m_2) = \mathcal{E}(m_1, b m_2), \quad b \in O_B,$$

and perfect sesquilinear forms with respect to the involution $d \otimes r \longmapsto \check{d} \otimes r$ on $O_D \otimes R$,

$$\mathcal{B} : N_1 \times N_2 \longrightarrow O_D \otimes R.$$

The forms $\mathcal{B}$ and $\mathcal{E}$ determine each other uniquely by the equation

$$\mathcal{E}(u_1 \otimes m_1, u_2 \otimes m_2) = h(u_1, u_2\mathcal{B}(m_1, m_2))$$

Proof: Let us denote by M_1' the left $O_B \otimes R$-module obtained from the natural right $O_B \otimes R$-module $\mathrm{Hom}_R(M_1, R)$ by restriction of scalars with respect to the given involution $b \longmapsto b^*$ on O_B. Then $\mathcal{E}$ becomes a linear map of left O_B-modules

$$\begin{array}{ccc} M_2 & \longrightarrow & M_1' \\ m_2 & \longmapsto & \mathcal{E}(-, m_2) \end{array}$$

Using the Morita equivalence we get a map

$$O_D^n \otimes_{O_D} N_2 \longrightarrow \mathrm{Hom}_R(O_D^n \otimes_{O_D} N_1, R). \tag{A.47}$$

Let us extend h to a perfect form on $O_D^n \otimes R$,

$$h : O_D^n \otimes R \times O_D^n \otimes R \longrightarrow R\,.$$

We see that h defines an isomorphism of left $O_B \otimes R$-modules,

$$\begin{array}{ccc} (O_D^n \otimes R) & \longrightarrow & (O_D^n \otimes R)' = \mathrm{Hom}_R(O_D^n \otimes R, R) \\ u & \longmapsto & h(-, u)\,. \end{array}$$

Using this isomorphism, we may rewrite the right hand side of (A.47),

$$\mathrm{Hom}_R(O_D^n \otimes_{O_D} N_1, R) \cong \mathrm{Hom}_{O_D \otimes R-}(N_1, \mathrm{Hom}_R(O_D^n \otimes R, R))$$
$$\overset{h}{\simeq} \mathrm{Hom}_{O_D \otimes R-}(N_1, O_D^n \otimes R) \simeq O_D^n \otimes_{O_D} \mathrm{Hom}_{O_D \otimes R-}(N_1, O_D \otimes R).$$

We note that the left O_D-module structure on $\mathrm{Hom}_R(O_D^n \otimes R, R)$ used to form the second Hom is the one induced by the natural right O_D-module structure on O_D^n. The left $O_D \otimes R$-module structure on $O_D^n \otimes R$ respectively on $O_D \otimes R$ used to form the Hom in the second line is induced from the natural right module structure by restriction of scalars with respect to the involution $d \longmapsto \check{d}$ on O_D.

Inserting what we did in (A.47) we get by Morita equivalence an O_D-module homomorphism

$$N_2 \longrightarrow \mathrm{Hom}_{O_D\otimes R-}(N_1, O_D \otimes R). \tag{A.48}$$

The O_D-module structure on the Hom is induced by the natural left O_D-module structure on $O_D \otimes R$. Clearly (A.48) defines the sesquilinear form $\mathcal{B}$ we are looking for. Moreover $\mathcal{B}$ is perfect, iff $\mathcal{E}$ is perfect. □

Consider the given symplectic B-module $(V, (\ ,\ ))$ with respect to the involution $b \longmapsto b^*$ on B

$$(bv, w) = (v, b^* w)\,.$$

We write V as a left B-module in the form $V = D^n \otimes_D W$. Then by proposition (A.53) we find a sesquilinear form

$$\chi : W \times W \longrightarrow D\,,$$

that satisfies the equation

$$\begin{aligned} h(u_1, u_2\chi(w_1, w_2)) &= (u_1 \otimes w_1, u_2 \otimes w_2) \\ \chi(dw_1, w_2) &= \chi(w_1, w_d)\check{d} \\ \chi(w_1, dw_2) &= d\chi(w_1, w_2) \\ \chi(w_1, w_2)^{\vee} &= \pm\chi(w_2, w_1). \end{aligned}$$

The sign in the last equation is $+$ if h is antisymmetric and $-$ if h is symmetric.

Now assume that we are given a selfdual chain of lattices $\{\Lambda_i\}_{i\in\mathbf{Z}}$ in V, indexed as in corollary (3.7). We write $\Lambda_i = O_D^n \otimes_{O_D} \Gamma_i$, where $\mathcal{G} = \{\Gamma_i\}_{i\in\mathbf{Z}}$ is a chain of O_D-lattices in W. If the lattices Λ_i and $\Lambda_{i'}$ are dual with respect to $(\ ,\ )$, then by proposition (A.53) the sesquilinear form

$$\chi : \Gamma_i \times \Gamma_{i'} \longrightarrow O_D$$

is perfect, i.e. $\{\Gamma_i\}$ is selfdual with respect to χ. There is an integer a such that

$$\Gamma_{i'} = \Gamma_{-i+a}.$$

If the chain $\{\Lambda_i\}$ has period r we get

$$\Gamma_{i-r} = \Pi\Gamma_i\,.$$

Definition A.54 *Let T be a scheme, where p is locally nilpotent. A polarized chain of $O_D \otimes \mathcal{O}_T$-modules of type $\mathcal{G} = \{\Gamma_i\}$ on T is given by the following data:*

1. *A sequence of left $O_D \otimes \mathcal{O}_T$-modules*

$$\cdots \longrightarrow N_{i-1} \xrightarrow{\varrho} N_i \xrightarrow{\varrho} N_{i+1} \longrightarrow \cdots, \qquad i \in \mathbf{Z}\,.$$

2. *A set of periodicity isomorphisms*

$$\theta : N_i^{\Pi} \longrightarrow N_{i-r}.$$

3. *A set of perfect sesquilinear forms with respect to the involution $d \longmapsto d^{\vee}$ on D*

$$\mathcal{B}_i : N_i \times N_{-i+a} \longrightarrow O_D \otimes \mathcal{O}_T\,.$$

These data are subject to the following conditions :

(i) Locally on T there exist isomorphisms of $O_D \otimes \mathcal{O}_T$-modules

$$N_i \cong \Gamma_i \otimes \mathcal{O}_T, \quad N_i/N_{i-1} \cong \Gamma_i/\Gamma_{i-1} \otimes \mathcal{O}_T.$$

(ii)

$$\theta\varrho = \varrho\theta, \quad \varrho^r\theta = \Pi.$$

(iii) The forms $\mathcal{B}_i$ satisfy the following relations:

$$\begin{aligned}
\mathcal{B}_i(m_1, dm_2) &= d\mathcal{B}_i(m_1, m_2) \\
\mathcal{B}_i(dm_1, m_2) &= \mathcal{B}_i(m_1, m_2)\check{d} \\
\mathcal{B}_i(m_1, m_2) &= \pm\mathcal{B}_{-i+a}(m_2, m_1)^{\vee} \\
\mathcal{B}_i(\varrho(m_3), m_2) &= \mathcal{B}_{i-1}(m_3, \varrho(m_2)) \\
\mathcal{B}_{i-r}(\theta(m_1), m_4) &= -\Pi\mathcal{B}_i(m_1, \theta(m_4)),
\end{aligned}$$

where $m_1 \in N_i$, $m_2 \in N_{-i+a}$, $m_3 \in N_{i-1}$, $m_4 \in N_{-i+r+a}$, $d \in O_D$.

Let $\{M_i\}_{i\in\mathbf{Z}}, \mathcal{E}$ be a polarized chain of modules of type $(\mathcal{L})$ on a $\mathbf{Z}_p$-scheme T. The polarization form $\mathcal{E}$ consists of a set of perfect pairings

$$\mathcal{E}_i : M_i \times M_{-i+a} \longrightarrow \mathcal{O}_T,$$

where a is some fixed integer.
By Morita equivalence we get a chain $\{N_i\}_{i\in\mathbf{Z}}$ of $O_D \otimes \mathcal{O}_T$-modules of type $(\mathcal{G})$,

$$M_i \cong O_D^n \otimes_{O_D} N_i.$$

By proposition (A.53) we have perfect sesquilinear forms

$$\mathcal{B}_i : N_i \times N_{-i+a} \longrightarrow \mathcal{O}_T.$$

Proposition A.55 *The chain* $\{N_i\}_{i\in\mathbf{Z}}$ *with the forms* $\mathcal{B}_i$ *is a polarized chain of* $O_D \otimes \mathcal{O}_T$*-modules of type* $(\mathcal{G})$. *The functor which associates to a polarized chain* $(\{M_i\}, \mathcal{E})$ *of type* $(\mathcal{L})$ *the polarized chain* $(\{N_i\}, \mathcal{B})$ *is an equivalence of categories.*

Proof: For any element b in the normalizer of O_B we have an isomorphism

$$\theta_b : M_\Lambda^b \longrightarrow M_{b\Lambda}.$$

Let us begin with the definition of the isomorphisms $\theta : N_i^\Pi \to N_{i-r}$, and show how the θ_b may be recovered from θ.
We view θ_b as an isomorphism of $\mathcal{O}_T$-modules

$$\theta_b : M_\Lambda \longrightarrow M_{b\Lambda} \tag{A.49}$$

that satisfies the relation

$$\theta_b(am) = bab^{-1}\theta_b(m), \quad a \in O_B, m \in M_\Lambda.$$

We note that any element b of the normalizer of O_B may be written

$$b = \Pi^m u \qquad u \in GL_n(O_D). \tag{A.50}$$

We denote the exponent m by $\mathrm{ord}_D b$. By the equation $(\Pi^*)^2 = \pi$ we see that

$$\mathrm{ord}_D b = \mathrm{ord}_D b^* \,.$$

If $\mathrm{ord}_D b = 0$ the map (A.49) is just multiplication by b. For a general b we have

$$\theta_b = (\theta_\Pi)^m \cdot u \,.$$

We note that there is a natural isomorphism

$$M_i^\Pi = (O_D^n)^\Pi \otimes_{O_D} N_i \xrightarrow{\sim} O_D^n \otimes_{O_D} N_i^\Pi$$
$$x \otimes n \longmapsto \Pi x \Pi^{-1} \otimes n.$$

Hence the map $\theta_\Pi : M_i^\Pi \longrightarrow M_{i-r}$ induces by Morita equivalence a map

$$\bar{\theta}_\Pi : N_i^\Pi \longrightarrow N_{i-r} \,.$$

Conversely the map θ_Π may be written

$$\begin{aligned} \theta_\Pi : \ O_D^n \otimes_{O_D} N_i \quad &\longrightarrow O_D^n \otimes_{O_D} N_{i-r} \\ x \otimes n_i \quad &\longmapsto \Pi x \Pi^{-1} \otimes \bar{\theta}_\Pi(n_i). \end{aligned}$$

Next we need to verify that the relations (iii) of definition (A.54) hold for $\mathcal{B}$. Only the last relation needs a verification.
We have $\Pi^* = u\Pi$ for some $u \in O_B^\times$. Then the map θ_{Π^*} is

$$\begin{aligned} \theta_{\Pi^*} : \ O_D^n \otimes N_i \quad &\longrightarrow \quad O_D^n \otimes_{O_D} N_{i-r} \\ x \otimes n \quad &\longmapsto \quad u\Pi x \Pi^{-1} \otimes \bar{\theta}_\Pi(n). \end{aligned}$$

By definition the periodicity isomorphisms are compatible with the polarization $\mathcal{E}$ in the following sense:

$$\mathcal{E}_{i-r}(\theta_\Pi(m_1), m_2) = \mathcal{E}_i(m_1, \theta_\Pi^*(m_2)) \,. \tag{A.51}$$

Let us compute, what these equations say about the sesquilinear forms $\mathcal{B}_i$:

$$\begin{aligned}\mathcal{E}_{i-r}(\theta_\Pi(x_1 \otimes (n_1)), x_2 \otimes n_2) &= \mathcal{E}_{i-r}(\Pi x_1 \Pi^{-1} \otimes \bar{\theta}_\Pi n_1, x_2 \otimes n_2) = \\ &= h(\Pi x_1 \Pi^{-1}, x_2 \mathcal{B}_{i-r}(\bar{\theta}_\Pi(n_1), n_2)) \\ &= h(x_1, \Pi^* x_2 \mathcal{B}_{i-r}(\bar{\theta}_\Pi(n_1), n_2)\check{\Pi}^{-1}) \\ &= -h(x_1, u\Pi x_2 \mathcal{B}_{i-r}(\bar{\theta}_\Pi(n_1), n_2)\Pi^{-1}).\end{aligned}$$

Computing the other side of the equality (A.51) we get

$$\mathcal{E}_i(x_1 \otimes n_1, \theta_{\Pi^*}(x_2 \otimes n_2)) = \mathcal{E}_i(x_1 \otimes n_1, u\Pi x_2 \Pi^{-1} \otimes \theta_{\bar{\Pi}}(n_2))$$

$$= h(x_1, u\Pi x_2 \Pi^{-1} \mathcal{B}_i(n_1, \theta_{\bar{\Pi}}(n_2))) .$$

Comparing both results, we get the last equation of the proposition. □

The theorem (3.16) may now be reformulated in the case at hand.

Proposition A.56 *The polarized chain $\{N_i\}$ of type $\{\mathcal{G}\}$ is locally for the étale topology on T isomorphic to $\{\Gamma_i \otimes \mathcal{O}_T\}$. If r and a both are even, such an isomorphism exists even locally for the Zariski topology.*
Moreover if $\{N_i'\}$ is a second chain of type $(\mathcal{G})$ on T, then the functor of isomorphisms of polarized chains on the category of T-schemes,

$$T' \longrightarrow \mathrm{Isom}(\{N_i\} \otimes_{\mathcal{O}_T} \mathcal{O}_{T'}, \{N_i'\} \otimes_{\mathcal{O}_T} \mathcal{O}_{T'}),$$

is representable by a smooth affine scheme.

The proof of this proposition will be the rest of this appendix. We may assume that $T = Spec\, R$ is affine. Let us start with the case where $a = 0$ and $r = 1$. Then we have a perfect sesquilinear form

$$\mathcal{B}_0 : N_0 \times N_0 \longrightarrow O_D \otimes R .$$

From $\mathcal{B}_0$ we recover the whole chain of modules by the rules:

$$N_{-i} = N_0^{\Pi^i}, \varrho = \Pi, \theta = id, \mathcal{B}_{-i}(n_1, n_2) = \Pi^{-i}\mathcal{B}_0(n_1, n_2)\check{\Pi}^i .$$

Consider first the cases, where either $\mathcal{B}_0$ is hermitian or $\check{}$ is the main involution. Then the modules N_0 and N_0' admit locally for the Zariski topology an orthogonal basis by proposition (A.14). The smoothness of the functor of automorphisms follows by proposition (A.13). To show the first assertion of proposition (A.56) in the case where $\mathcal{B}_0$ is hermitian with respect to the main involution it is enough to show that there exists an orthonormal base. Since an orthogonal basis exists we may assume that N_0 is a free $O_D \otimes R$-module of rank 1. In this case the assertion is that we have a surjection of étale sheaves in R,

$$(O_D \otimes R)^\times \longrightarrow (O_F \otimes R)^\times . \tag{A.52}$$

This follows because the last map is a smooth morphism of algebraic groups over $\mathbf{Z}_p$, for $p \neq 2$. Next we have the case, where $\mathcal{B}_0$ is antihermitian with respect to the main involution. This reduces to the case before, by multiplying $\mathcal{B}_0$ with a root of unity ζ of order prime to p, which generates F_2/F. In the case where $\mathcal{B}_0$ is hermitian with respect to the nebeninvolution we consider the reduction of $\mathcal{B}_0$ modulo Π,

$$\bar{\mathcal{B}} : N_0/\Pi N_0 \times N_0/\Pi N_0 \longrightarrow \kappa_2 \otimes R\,,$$

where κ_2 denotes the residue class field of F_2. To show the existence of an orthonormal basis we may replace by proposition (A.13) $\mathcal{B}_0$ by $\bar{\mathcal{B}}$. Since the nebeninvolution is the identity on κ_2 the form $\bar{\mathcal{B}}$ is a perfect symmetric bilinear form, which has an orthonormal basis locally for the étale topology.

Finally let $\mathcal{B}_0$ be antihermitian for the nebeninvolution. Then the reduction $\bar{\mathcal{B}}$ modulo Π is an antisymmetric perfect form, which even with respect to the Zariski topology has a standard symplectic basis. Then the same is true for the form $\mathcal{B}_0$ by proposition (A.12). The smoothness of the functor of isomorphisms follows from the same proposition.

Hence we have treated the case $a = 0$ and $r = 1$. The case $a = 1$ and $r = 1$ is similar. In this case

$$\mathcal{P}(n_0, n_0') = \mathcal{B}_1(\theta^{-1}(n_0), n_0')$$

is a perfect hermitian or antihermitian form with respect to the involution $d \longmapsto \Pi d \Pi^{-1}$:

$$\mathcal{P} : N_0 \times N_0 \longrightarrow O_D \otimes R\,.$$

Again $\mathcal{P}$ determines the whole polarized chain uniquely. Hence in this case the proposition (A.56) is valid too.

We may now assume $r > 1$. Let us start with the case where $a = 0$. Then $\mathcal{B}_0$ is a perfect ε-hermitian form on N_0 with respect to the given involution $d \longmapsto \check{d}$. We will denote this pairing also by $\mathcal{B}$. The lemma (A.26) holds with the same proof.

Lemma A.57 *For $0 \le t \le r$ let F_{-t} be the image of N_{-t} in $N_0/\Pi N_0$ under the map ϱ^t. Then F_{-r+t} is the orthogonal complement of F_{-t} with respect to $\mathcal{B}$ modulo Π.* □

From now on let $t = \left[\frac{r+1}{2}\right]$. Then we define a form

$$\mathcal{Q} : N_{-t} \times N_{-t} \longrightarrow O_D \otimes R$$

$$\mathcal{Q}(n, n') = \begin{cases} \mathcal{B}_t(\theta^{-1}\varrho(n), n') & \text{if } r \text{ is odd} \\ \mathcal{B}_t(\theta^{-1}(n), n') & \text{if } r \text{ is even.} \end{cases}$$

Then $\mathcal{Q}$ is hermitian or antihermitian with respect to the involution $d \longmapsto \Pi \check{d} \Pi^{-1} = d^+$.

The definition A.54 may be rephrased in terms of the sesquilinear forms $\mathcal{P}$ and $\mathcal{Q}$.

Before doing this we need to fix a notation. Let N be a left $O_D \otimes R$-module. Then $N^* = \mathrm{Hom}_{O_D \otimes R-}(N, O_D \otimes R)$ is in a natural way a right $O_D \otimes R$-module. We denote the left $O_D \otimes R$-modules obtained by restriction of scalars with respect to the involutions $d \longmapsto \check{d}$ respectively $d \longmapsto d^+$ by $N^{\check{}}$ respectively N^+. There is a natural isomorphism of left $O_D \otimes R$-modules

$$(N_0^{\Pi})^{\check{}} \cong N_0^+ \cong (N_0^{\check{}})^{\Pi}.$$

It is given by conjugation with Π:

$$\mathrm{Hom}_{O_D \otimes R-}(N_0^{\Pi}, O_D \otimes R) \longrightarrow \mathrm{Hom}_{O_D \otimes R-}(N_0, O_D \otimes R)$$

$$\varphi \longmapsto \psi : \psi(n) = \Pi^{-1}\varphi(n)\Pi\,.$$

Proposition A.58 *To give a polarized chain of $O_D \otimes R$-modules on $T = \operatorname{Spec} R$ of type $(\mathcal{G})$ with period r and $a = 0$ is equivalent to giving the following data*

1. *A circular diagram of locally on T free $O_D \otimes R$-modules:*

$$\begin{array}{ccc} N_{-r+1} \xrightarrow{\varrho} \cdots \xrightarrow{\varrho} N_{-t} & & \\ {\scriptstyle\gamma}\nwarrow \qquad \swarrow{\scriptstyle\beta} & \qquad & t = \left[\dfrac{r+1}{2}\right] \\ N_0 & & \end{array} \tag{A.53}$$

The maps ϱ and β are $O_D \otimes R$-linear. The map γ satisfies the relation

$$\gamma(an) = \Pi a \Pi^{-1} \gamma(n), \quad a \in O_D \otimes R, \ n \in N_0.$$

2. *Two R-bilinear forms*

$$\begin{array}{rcccl} \mathcal{B}: & N_0 & \times & N_0 & \longrightarrow O_D \otimes R \\ \mathcal{Q}: & N_{-t} & \times & N_{-t} & \longrightarrow O_D \otimes R. \end{array}$$

$\mathcal{B}$ is perfect and ε-hermitian with respect to the involution $d \longmapsto d^{\smile}$ and $\mathcal{Q}$ is ε-hermitian with respect to the involution $d \longmapsto \Pi d^{\smile} \Pi^{-1}$. The forms $\mathcal{B}$ and $\mathcal{Q}$ are ε-hermitian of the opposite sign.

The following conditions are required.

(i) Going once around the diagram (A.53) is multiplication by Π.

(ii) There exists locally on T isomorphisms of $O_D \otimes R$-modules

$$\begin{array}{rcll} N_i & \simeq & \Gamma_i \otimes R & \\ N_i/\varrho(N_{i-1}) & \simeq & \Gamma_i/\Gamma_{i-1} \otimes R, & i = -r+2, \ldots, -t \\ N_{-r+1}/\gamma(N_0) & \simeq & \Gamma_{-r+1}/\Gamma_{-r} \otimes R. & \end{array}$$

The cokernel of the map

$$\begin{array}{ccc} N_{-t} & \longrightarrow & N_{-t}^{+} \\ n & \longmapsto & \mathcal{Q}(n, -) \end{array}$$

is locally isomorphic to $\Gamma_{-t+1}/\Gamma_{-t} \otimes R$.

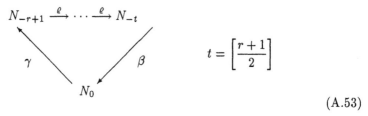

(iii) Let $\alpha = \varrho^{r-t-1}\gamma : N_0^\Pi \longrightarrow N_{-t}$. *Then we have the relation*

$$\mathcal{Q}(\alpha(n), n') = \mathcal{B}(n, \beta(n')), \quad n \in N_0, n' \in N_{-t}.$$

Remark: Because we are in a non commutative situation some care is needed:

$$\mathcal{Q}(n', \alpha(n)) = -\Pi\mathcal{B}(\beta(n'), n)\Pi^{-1}.$$

Proof: It is obvious that a polarized chain of modules of type $(\mathcal{G})$ of the definition (A.54) gives rise to a circular diagram with the properties described in the proposition. Let us prove the opposite assertion.
We note that β is uniquely determined by α and the equation

$$\mathcal{Q}(\alpha(n), n') = \mathcal{B}(n, \beta(n')), \quad n \in N_0, \ n' \in N_{-t}.$$

We denote the map $n' \longmapsto \mathcal{Q}(n', -)$ by $q : N_{-t} \longrightarrow N_{-t}^+$. For $t \leq i \leq r-1$ let N_{-i}^+ be the dual of N_{i-r} with respect to the involution $+$. We denote the corresponding sesquilinear forms by

$$< , >: N_{-i}^+ \times N_{i-r} \longrightarrow O_D.$$

If we denote by ϱ^+ the dual map to ϱ defined by these pairings, we get a diagram

$$N_0^\Pi \xrightarrow{\gamma} N_{-r+1} \xrightarrow{\varrho} \cdots \xrightarrow{\varrho} N_{-t} \xrightarrow{q} N_{-t}^+ \xrightarrow{\varrho^+} \cdots \xrightarrow{\varrho^+} N_{-r+1}^+ \xrightarrow{\delta} N_0. \tag{A.54}$$

Here the map δ is given by the equation

$$\mathcal{B}(n_0, \delta(m)) = \varepsilon_{\mathcal{Q}} < m, \gamma(n_0) >^+,$$

or equivalently

$$-\Pi^{-1}\mathcal{B}(\delta(m), n_0)\Pi = < m, \gamma(n_0) > . \tag{A.55}$$

Here $\varepsilon_{\mathcal{Q}}$ denotes the sign of the ε–hermitian form $\mathcal{Q}$.
Let $k = r - t - 1$. We claim that the map $\delta(q^+)^k q$ coincides with β. Indeed

$$\mathcal{B}(n_0, \delta(\varrho^+)^k q(n)) = \varepsilon_{\mathcal{Q}} < (\varrho^+)^k q(n), \gamma(n_0) >^+$$
$$= \varepsilon_{\mathcal{Q}} < q(n), \varrho^k \gamma(n_0) >^+ = \varepsilon_{\mathcal{Q}} \mathcal{Q}(n, \varrho^k \gamma(n_0))^+$$
$$= \mathcal{Q}(\varrho^k \gamma(n_0), n) = \mathcal{Q}(\alpha(n_0), n) = \mathcal{B}(n_0, \beta(n)).$$

Next we consider the chain (A.54) modulo Π. Then the maps

$$\bar{N}_0^{\Pi} \xrightarrow{\gamma} \bar{N}_{-r+1} \quad \text{and} \quad \bar{N}^+_{-r+1} \xrightarrow{\delta} \bar{N}_0$$

are dual to each other by the equation (A.55). Since the first map has by our condition a $O_D/\Pi O_D \otimes R$-projective cokernel which is free locally on T, we conclude that the second map has a cokernel with the same properties and of the same rank.

Finally we have to show that going once around the circle (A.54) starting at any point is multiplication by Π. Indeed, this is a requirement except if we start in a point N^+_{-i}. In this case let $(\varrho^+)^a$ and $(\varrho^+)^b$ be the maps

$$(\varrho^+)^a : N^+_{-i} \longrightarrow N^+_{-r+1}, \quad (\varrho^+)^b : N^+_{-t} \longrightarrow N^+_{-i}.$$

For $n_{-i} \in N^+_{-i}$ and $n_{i-r} \in N_{i-r}$ we get

$$< (\varrho^+)^b q\alpha\delta(\varrho^+)^a(n_{-i}), n_{i-r} >=$$
$$< q\alpha\delta(\varrho^+)^a(n_{-i}), \varrho^b(n_{i-r}) >= \mathcal{Q}(\alpha\delta(\varrho^+)^a(n_{-i}), \varrho^b(n_{i-r})) =$$
$$= \mathcal{B}(\delta(\varrho^+)^a(n_{-i}), \beta\varrho^b(n_{i-r})) = -\Pi^{-1} < (\varrho^+)^a n_{-i}, \beta\varrho^b n_{i-r} > \Pi =$$
$$= -\Pi^{-1} < n_{-i}, \Pi n_{i-r} > \Pi = - \quad < n_{-i}, n_{i-r} > \Pi$$
$$=< \Pi n_{-i}, n_{i-r} > .$$

This proves the proposition. According to the proof we will set in the following $N_{-i} = N^+_{-r+i}$ for $1 \leq i \leq t-1$. □

A.59 In the case $r = 2$ and $a = 0$ we have a diagram

$$N_0 \xrightarrow{\alpha} N_{-1} \xrightarrow{\beta} N_0 \xrightarrow{\alpha} N_{-1} \qquad \alpha\beta = \beta\alpha = \pi . \tag{A.56}$$

The forms $\mathcal{B}$ and $\mathcal{Q}$ are perfect. Then the chain arises as follows:

Let $\mathcal{P} : U_0 \times U_{-1} \longrightarrow O_D \otimes R$ be a perfect sesquilinear form with respect to the involution $d \longmapsto \check{d}$, where U_0 and U_{-1} are locally on T free $O_D \otimes R$-modules. Then the diagram (A.56) above is given as follows

$$N_0^\Pi = U_0^\Pi \oplus U_{-1}^\Pi \stackrel{id \oplus \Pi}{\longrightarrow} U_0^\Pi \oplus U_{-1} \stackrel{\Pi \oplus id}{\longrightarrow} U_0 \oplus U_{-1}\,.$$

The forms $\mathcal{B}$ and $\mathcal{Q}$ are uniquely defined by the condition that the spaces U_0, U_{-1} respectively U_0^Π are totally isotropic and by the equations

$$\begin{aligned} \mathcal{B}(u_0, u_{-1}) &= \mathcal{P}(u_0, u_{-1}) & \mathcal{B}(u_{-1}, u_0) &= \pm\mathcal{P}(u_0, u_{-1})^{\vee} \\ \mathcal{Q}(u_0, u_{-1}) &= \mathcal{P}(u_0, u_{-1}) & \mathcal{Q}(u_{-1}, u_0) &= \mp\mathcal{P}(u_0, u_{-1})^{+}. \end{aligned}$$

One verifies the equation $\mathcal{Q}(\Pi u_{-1}, u_0) = \mathcal{B}(u_{-1}, \Pi u_0)$, which shows that the condition (iii) of the proposition (A.58) holds.
This shows that in the case $r = 2$ and $a = 0$ any two polarized chains of type $(\mathcal{G})$ are locally isomorphic for the Zariski topology. The same is true for any even r by the argument given in the case (II). Also the formal smoothness can be seen as in the previous cases because isotropic direct summands lift.

A.60 We consider now the case, where r is odd. Let us start with $r = 3$. In this case the circle looks as follows

$$N_0^\Pi \stackrel{\alpha}{\longrightarrow} N_{-2} \stackrel{\beta}{\longrightarrow} N_0$$

where α and β are $O_D \otimes R$-linear maps.
The given ε-hermitian forms on $\mathcal{B}$ and $\mathcal{Q}$ allow us to factor β as follows:

$$N_0^\Pi \stackrel{\alpha}{\longrightarrow} N_{-2} \stackrel{q}{\longrightarrow} N_{-2}^{+} \stackrel{\delta}{\longrightarrow} N_0\,. \tag{A.57}$$

The map q is the one defined by the form $\mathcal{Q}$ (Prop. A.58) (ii)), and the map δ is given by the equation

$$\mathcal{B}(n_0, \delta(m)) = \varepsilon_{\mathcal{Q}}(m(\alpha(n_0)))^{+}\,,$$

where $\varepsilon_{\mathcal{Q}}$ denotes the sign of the ε-hermitian form $\mathcal{Q}$.
Let us denote by $\bar{U}_0$ an isotropic complement of the image F_{-2}^0 of N_{-2} in $\bar{N}_0 = N_0/\Pi N_0$. If F_{-1}^0 denotes the image of N_{-1} in $\bar{N}_0$ we find a direct sum decomposition

$$\bar{N}_0 = F_{-1}^0 \oplus \bar{U}_0\,.$$

Let $U_0 \subset N_0$ be an isotropic direct summand that lifts $\bar{U}_0$. Next let F_0^{-2} be the image of α in $\bar{N}_{-2} = N_{-2}/\Pi N_{-2}$ and $F_{-1}^{-2} \subset F_0^{-2}$ be the image of $\alpha\delta$. By the lemma (A.3) we know that F_{-1}^{-2} is the kernel of $q \bmod \Pi$. It is easily checked that the orthogonal complement of F_0^{-2} with respect to $\mathcal{Q} \bmod \Pi$ is F_0^{-2}. We find an isotropic subspace $\bar{U}_{-2}$ in $\bar{N}_{-2}$ such that its image in $\bar{N}_{-2}/F_{-1}^{-2}$ is an isotropic complement of F_0^{-2}/F_{-1}^{-2}. We get

$$\bar{U}_{-2} \oplus F_0^{-2} = \bar{N}_{-2}\,.$$

Let $U_{-2} \subset N_{-2}$ be an isotropic subspace that lifts $\bar{U}_{-2}$. The spaces U_0 and U_{-2} are part of a splitting of the unpolarized chain (A.57). Hence by the proof of proposition (A.4) $\beta(U_{-2}) \oplus U_0 \subset N_0$ is a direct summand. Since $\beta(U_{-2}) \bmod \Pi$ and $U_0 \bmod \Pi$ are isotropic complements, the same holds for $\beta(U_{-2})$ and U_0. Hence we get an orthogonal decomposition

$$N_0 = (U_0 \oplus \beta(U_{-2})) \perp N\,.$$

It is clear that $U_{-2} \subset N_{-2}$ and $U_0 \oplus N \subset N_0$ represents a splitting of the unpolarized chain

$$N_0 \xrightarrow{\alpha} N_{-2} \xrightarrow{\beta} N_0\,.$$

Hence we get $N_{-2} = U_{-2} \oplus \alpha(U_0 \oplus N)$.
We have a perfect sesquilinear pairing with respect to the involution $d \longmapsto d^{\sim}$:

$$\mathcal{P}:\ U_0 \times U_{-2} \longrightarrow O_D \otimes R$$

$$\mathcal{P}(u_0, u_{-2}) = \mathcal{B}(u_0, \beta u_{-2})$$

and moreover a perfect $\varepsilon_{\mathcal{B}}$-hermitian form induced by $\mathcal{B}$:

$$\mathcal{B}:\ N \times N \longrightarrow O_D \otimes R\,. \tag{A.58}$$

Our whole situation is uniquely determined by $(N, \mathcal{B})$ and the dual spaces $(U_0, U_{-2}, \mathcal{P})$.
Indeed the circle takes the form

$$N_0^{\Pi} = U_0^{\Pi} \oplus N^{\Pi} \oplus U_{-2}^{\Pi} \xrightarrow{id \oplus \Pi_{U_{-2}}} U_0^{\Pi} \oplus N^{\Pi} \oplus U_{-2} = N_{-1} \xrightarrow{\Pi \oplus id_{U_{-2}}} U_0 \oplus N \oplus U_{-2}. \tag{A.59}$$

The bilinear forms $\mathcal{B}$ and $\mathcal{Q}$ are uniquely given by the rules

$$\begin{array}{lcl} \mathcal{B}(u_0, u_{-2}) & = & \mathcal{Q}(u_0, u_{-2}) = \mathcal{P}(u_0, u_{-2}), \quad u_{-2} \in U_{-2},\ u_0 \in U_0 \\ \mathcal{B}(v, v') & & \text{coincides with (A.58) for } v, v' \in N \\ \mathcal{Q}(v, v') & = & \Pi\mathcal{B}(v, v'). \end{array}$$

Hence the question whether two polarized chains of the same type ($\mathcal{G}$) are locally isomorphic for the étale topology is reduced to the question whether two perfect ε-hermitian spaces $(N, \mathcal{B})$ and $(N', \mathcal{B}')$ of the same rank and sign are isomorphic. But this was settled in the case $r = 1$. The smoothness of the functor follows in the usual way from the lifting property of isotropic direct summands.
To treat the more general case where r is odd and $a = 0$ one proceeds exactly as in the case (II).

A.61 Let us now consider the case $a = 1$. If $r = 2t + 1$ is an odd number we define a new polarization on our chain by the rule

$$\mathcal{B}'_{-i}(n_{-i}, n_{i+1-r}) = \mathcal{B}_{r-i}(\theta^{-1}(n_{-i}), n_{i+1-r}).$$

This is a polarization for the involution $d \longmapsto d^+$ with $a = 1 - r$ even. Since we have already done this case we may restrict our attention to the case $r = 2t$ and $a = 1$. Let us start with the case $r = 2$,

$$N_0^{\Pi} \xrightarrow{\theta} N_{-2} \xrightarrow{\varrho} N_{-1} \longrightarrow N_0. \tag{A.60}$$

We consider the sesquilinear form with respect to the involution $+$,

$$\mathcal{P} : N_{-1} \times N_0 \longrightarrow O_D \otimes R$$

$$\mathcal{P}(n_{-1}, n_0) = \mathcal{B}_1(\theta^{-1}(n_{-1}), n_0).$$

The form $\mathcal{P}$ determines the polarization $\mathcal{B}_i$ uniquely. We define a form $\mathcal{B}$ on N_0 and a form $\mathcal{Q}$ on N_{-1} by the equations

$$\begin{array}{lcl} \mathcal{B}(n_0, n'_0) & = & \mathcal{P}(\varrho\theta(n_0), n'_0) \\ \mathcal{Q}(n_{-1}, n'_{-1}) & = & \mathcal{P}(n_{-1}, \varrho(n'_{-1})). \end{array}$$

Then $\mathcal{B}$ is a $\varepsilon_{\mathcal{B}}$-hermitian form for the involution $d \longrightarrow d^{\vee}$ and $\mathcal{Q}$ is a $\varepsilon_{\mathcal{Q}}$-hermitian form for the involution $d \longmapsto d^{+}$.

Let us denote $\varrho\theta$ by α and θ by β. We look first at the situation modulo Π:

$$\bar{N}_0^{\Pi} \xrightarrow{\bar{\alpha}} \bar{N}_{-2} \xrightarrow{\bar{\beta}} \bar{N}_0 \xrightarrow{\bar{\alpha}} \bar{N}_0^{\Pi^{-1}} . \tag{A.61}$$

We note that $\mathcal{P}$ induces a perfect pairing

$$\mathcal{P} : \bar{\alpha}(\bar{N}_0^{\Pi}) \times \bar{N}_0/\bar{\beta}(\bar{N}_{-1}) \longrightarrow O_D/\Pi \otimes R. \tag{A.62}$$

Indeed, assume that $\mathcal{P}(\alpha n_0', n_0) \equiv 0 \bmod \Pi$ for all $n_0' \in N_0$. We have $\mathcal{P}(\alpha(n_0'), n_0) = \varepsilon_{\mathcal{B}}\mathcal{P}(\alpha(n_0), n_0')$. Hence $\alpha(n_0) \equiv 0 \bmod \Pi$ and by the exactness of (A.61) $n_0 \bmod \Pi$ is in the image of $\bar{\beta}$.

Let $\bar{C}_0$ be a direct summand of $\bar{N}_0$ projecting isomorphically to $\bar{N}_0/\beta(\bar{N}_{-1})$. Let $\bar{C}_1$ be the orthogonal complement of $\bar{C}_0$ with respect to $\mathcal{P}$. Then by (A.62) $\bar{C}_1$ is complementary to $\bar{\alpha}(\bar{N}_0^{\Pi})$ and hence projects isomorphically onto $\bar{N}_{-1}/\mathrm{Im}\,\bar{\alpha}$. Again from (A.62) we deduce that $\mathcal{B}$ is perfect on $\bar{C}_0$ and by symmetry $\mathcal{Q}$ is perfect on $\bar{C}_1$.

Let us lift $\bar{C}_0$ to a direct summand C_0 of N_0. Then $\mathcal{B}$ is perfect on C_0 and hence we find an orthogonal decomposition with respect to $\mathcal{B}$:

$$N_0 = C_0 \oplus C_0^{\perp} .$$

Let $C_{-1} \subset N_{-1}$ be the orthogonal complement of C_0 with respect to $\mathcal{P}$. Then $C_{-1} \bmod \Pi = \bar{C}_{-1}$ and $\mathcal{Q}$ is perfect on C_{-1}. We get an orthogonal decomposition with respect to $\mathcal{Q}$:

$$N_{-1} = C_{-1} \oplus C_{-1}^{\perp} .$$

We claim that the map β induces an isomorphism $\beta : C_{-1} \longrightarrow C_0^{\perp}$. For $c_{-1} \in C_{-1}$ and $c_0 \in C_0$ we have the equation

$$\mathcal{B}(\beta(c_{-1}), c_0) = \mathcal{P}(\alpha\beta(c_{-1}), c_0) = -\Pi\mathcal{P}(c_{-1}, c_0) = 0 .$$

Hence $\beta(C_1) \subset C_0^{\perp}$. On the other hand $C_0^{\perp} \bmod \Pi = \mathrm{Ker}\,\bar{\alpha} = \mathrm{Im}\,\bar{\beta}$. We conclude by the lemma of Nakayama. By symmetry $\alpha : C_0^{\Pi} \longrightarrow C_{-1}^{\perp}$ is an isomorphism. We obtain

$$\alpha(C_0^{\perp}) = \alpha\beta(C_{-1}) = \Pi C_{-1}$$

$$\beta(C_{-1}^{\perp}) = \Pi C_0$$

$$\mathcal{P}(C_{-1}^{\perp}, C_0^{\perp}) = \mathcal{P}(C_{-1}^{\perp}, \beta(C_{-1})) = \mathcal{Q}(C_{-1}, C_{-1}^{\perp}) = 0\,.$$

It follows that the couple of ($\pm$)-hermitian perfect spaces $(C_0, \mathcal{B})$ and $(C_{-1}, \mathcal{Q})$ for the involution $^{\vee}$ and $^{+}$ respectively determine the polarized chain uniquely.
Indeed, using the isomorphisms $1\oplus\beta : C_0 \oplus C_{-1} \simeq N_0$ and $\alpha\oplus 1 : C_0^{\Pi} \oplus C_{-1} \simeq N_{-1}$ our polarized chain becomes

$$N_0^{\Pi} = C_0^{\Pi} \oplus C_{-1}^{\Pi} \overset{id\oplus\Pi}{\longrightarrow} N_{-1} = C_0^{\Pi} \oplus C_{-1} \overset{\Pi\oplus id}{\longrightarrow} N_0 = C_0 \oplus C_{-1}\,.$$

The polarization $\mathcal{P}$ is uniquely determined by the properties

$$\begin{aligned} \mathcal{P}(c_0', c_0) &= \mathcal{B}(c_0', c_0) \quad c_0, c_0' \in C_0 \\ \mathcal{P}(c_{-1}', c_{-1}) &= \mathcal{Q}(c_{-1}', c_{-1}). \end{aligned}$$

The spaces C_{-1} and C_0 (respectively C_0^{Π} and C_{-1}) are orthogonal to each other.
Two chains are locally isomorphic for the étale topology because by proposition (A.56) in the case $r = 0, a = 0$ the corresponding spaces $(C_0, \mathcal{B})$ resp. $(C_{-1}, \mathcal{Q})$ are locally isomorphic.
We call a *splitting of the chain* (A.60) a pair of direct summands $C_0 \subset N_0$ and $C_{-1} \subset N_{-1}$, such that C_0 modulo Π maps isomorphically to $N_0/\beta(N_{-1})$ and C_{-1} is the orthogonal complement of C_0 with respect to $\mathcal{P}$. The smoothness of the functor of proposition (A.56) follows from the fact that splittings lift with respect to surjections $R \longrightarrow S$ with nilpotent kernel.

In the case where $a = 1$ and r is any even number, we consider the $\varepsilon_{\mathcal{B}}$-hermitian form

$$\mathcal{B} : \; N_0 \times N_0 \longrightarrow O_D \otimes R$$

$$\mathcal{B}(n_0, n_0') = \mathcal{B}_1(\varrho(n_0), n_0').$$

We denote by a bar its reduction modulo Π

$$\bar{\mathcal{B}} : \; \bar{N}_0 \times \bar{N}_0 \longrightarrow O_D/\Pi \otimes R.$$

Let F_{-i} be the image of $N_{-i} \longrightarrow \bar{N}_0$.

$$(0) = F_{-r} \subset F_{-r+1} \subset \cdots \subset F_0 = \bar{N}_0$$

One verifies as usual

Lemma A.62 *The orthogonal complement of F_{-t} with respect to the form $\bar{\mathcal{B}}$ is F_{t+1-r}.* □

Let $\mathcal{Q} : N_{-\frac{r}{2}} \times N_{-\frac{r}{2}} \longrightarrow O_D \otimes R$ be the $\varepsilon_{\mathcal{Q}}$-hermitian form with respect to the involution $+$ given by

$$\mathcal{Q}(n, n') = \mathcal{B}_{\frac{r}{2}+1}(\varrho\theta^{-1}(n), n'), \quad n, n' \in N_{-\frac{r}{2}}\,.$$

Then the lemma holds for $\bar{\mathcal{Q}}$ with the obvious modifications.

A.63 Next we consider the case $r = 4$ and $a = 1$. We get a chain

$$N_{-4} \xrightarrow{\sim} N_0 \xrightarrow{\varrho\theta} N_{-3} \xrightarrow{\varrho} N_{-2} \xrightarrow{\varrho} N_{-1} \xrightarrow{\varrho} N_0. \tag{A.63}$$

The polarization $\mathcal{B}_i$ is uniquely determined by the following perfect sesquilinear forms with respect to the involution $+$

$$\mathcal{P}_{-1} :\ N_{-1} \times N_{-2} \longrightarrow O_D \otimes R,$$

$$\mathcal{P}_{-1}(n_{-1}, n_{-2}) = \mathcal{B}_3(\theta^{-1}(n_{-1}), n_{-2})$$

$$\mathcal{P}_{-3} :\ N_{-3} \times N_0 \longrightarrow O_D \otimes R,$$

$$\mathcal{P}_{-3}(n_{-3}, n_0) = \mathcal{B}_1(\theta^{-1}(n_{-3}), n_0)\,.$$

We have the relations:

$$\mathcal{B}(n_0, n_0') = \mathcal{P}_{-3}(\theta\varrho(n_0), n_0') = \mathcal{B}_1(\varrho(n_0), n_0')$$

$$\mathcal{Q}(n_{-2}, n'_{-2}) = \mathcal{P}_{-1}(\varrho(n_{-2}), n'_{-2}) = \mathcal{B}_3(\theta^{-1}\varrho(n_{-2}), n'_{-2})\,.$$

With the notations of lemma (A.62) one checks that $(\bar{N}_0/F_{-3}, \bar{\mathcal{B}})$ is a perfect $\varepsilon_{\mathcal{B}}$-hermitian space. Let $\hat{C}_0 \subset \bar{N}_0/F_{-3}$ be an isotropic complement of F_{-2}/F_{-3}. Then we have by the lemma that $\hat{C}_0 \oplus F_{-1}/F_{-3} = \bar{N}_0/F_{-3}$. We lift $\hat{C}_0$ to a direct summand $\bar{C}_0$ of $\bar{N}_0$ and F_{-2}/F_{-3} to a direct summand $\bar{M}_{-2}$ of $\bar{N}_0$. Then $\bar{C}_0$ and $\bar{M}_{-2}$ are isotropic subspaces and $\bar{C}_0 \oplus \bar{M}_{-2}$ is perfect. We lift $\bar{C}_0 \oplus \bar{M}_{-2}$ to a direct summand M of N. Then $\mathcal{B}$ restricted

to N is perfect and hence $\bar{C}_0$ may be lifted to a direct summand C_0 of M which is isotropic.
Hence we have shown that there is an isotropic direct summand C_0 of N_0 with respect to $\mathcal{B}$, whose reduction modulo Π maps isomorphically to N_0/F_{-1}. There is an isotropic direct summand C_{-2} of N_{-2}, whose reduction modulo Π maps isomorphically to $N_{-2}/\mathrm{Im}N_{-3}$.
Let L_0 be the orthogonal complement with respect to $\mathcal{B}$ of the perfect direct summand $C_0 \oplus \varrho^2(C_{-2})$ of N_0. We get the decomposition

$$N_0 = (C_0 \oplus \varrho^2(C_{-2})) \oplus L_0 .$$

Similarly we have an orthogonal decomposition with respect to $\mathcal{Q}$:

$$N_{-2} = (\varrho^2\theta(C_0) \oplus C_{-2}) \oplus L_{-2} .$$

We define $L_{-1} \subset N_{-1}$ to be the orthogonal complement of $(\varrho^2\theta(C_0) \oplus C_{-2})$ with respect to $\mathcal{P}_{-1}$.
From the definition of $\mathcal{Q}$ we have $\varrho(L_{-2}) \subset L_{-1}$. We claim that ϱ induces an isomorphism $L_{-1} \longrightarrow L_0$. Let us verify that $\varrho(L_{-1}) \subset L_0$. Indeed this is equivalent to $\mathcal{B}(\varrho(l), x) = 0$ for any $l \in L_{-1}$ and $x \in C_0 \oplus \varrho^2(C_{-2})$. But we have

$$\begin{aligned}\mathcal{B}(\varrho(l), x) &= \mathcal{B}_1(\varrho^2(l), x) = \mathcal{B}_{-1}(l, \varrho^2(x)) \\ &= -\Pi\mathcal{B}_3(\theta^{-1}(l), \theta\varrho^2(x))\Pi = -\Pi\mathcal{P}_{-1}(l, \theta\varrho^2(x))\Pi = 0 .\end{aligned}$$

Since $L_0 \oplus \varrho^2(C_{-2})$ is the orthogonal complement of $\varrho^2(C_{-2})$ with respect to $\mathcal{B}$, it follows from the lemma (A.62) that $L_0 \oplus \varrho^2(C_{-2})$ reduces modulo Π to F_{-1}. Hence $L_{-1} \oplus \varrho(C_{-2}) \xrightarrow{\varrho} L_0 \oplus \varrho^2 C_{-2}$ is an isomorphism modulo Π and we get our assertion.
Similarly we define L_{-3} and get that ϱ induces an isomorphism $L_{-3} \longrightarrow L_{-2}$. Hence our original polarized chain

$$N_0 \xrightarrow{\varrho\theta} N_{-3} \longrightarrow N_{-2} \longrightarrow N_{-1} \longrightarrow N_0$$

splits into two orthogonal parts, one formed by the modules C_0 and C_{-2} and one formed by the modules L_i. On the first part the polarization is given by a perfect pairing between C_0 and C_{-2},

$$\mathcal{Q}(\varrho^2\theta(c_0), c_{-2}) = \mathcal{B}(c_0, \varrho^2(c_{-2})) . \tag{A.64}$$

Hence locally for the Zariski topology this part is determined by the rank of C_0.
On the part

$$L_0 \xrightarrow{\varrho\theta} L_{-3} \underset{\sim}{\xrightarrow{\varrho}} L_{-2} \xrightarrow{\varrho} L_{-1} \underset{\sim}{\xrightarrow{\varrho}} L_0 \tag{A.65}$$

the forms $\mathcal{P}_{-1}$ and $\mathcal{P}_{-3}$ that define the polarization are given by a single sesquilinear form with respect to the involution $+$,

$$\mathcal{P}: \ L_{-2} \times L_0 \longrightarrow O_D \otimes R.$$

It is given by one of the following equivalent equations:

$$\mathcal{P}(l_{-2}, l_0) = \mathcal{P}_{-3}(\varrho^{-1}(l_{-2}), l_0) = -\varepsilon_{\mathcal{B}}\mathcal{P}_{-1}(\varrho^{-1}(l_0), l_{-2})^{+}.$$

We note that the forms $\mathcal{P}$ and $\mathcal{Q}$ are given by

$$\mathcal{B}(l_0, l'_0) = \mathcal{P}(\theta\varrho^2(l_0), l'_0), \quad \mathcal{Q}(l_{-2}, l'_2) = \mathcal{P}(l_{-2}, \varrho^2(l'_{-2}))\,.$$

Hence the polarized chain (A.65) is determined by the chain

$$L_0 \xrightarrow{\varrho^2\theta} L_{-2} \xrightarrow{\varrho^2} L_0\,,$$

which is polarized by the form $\mathcal{P}$. The last chain is exactly of the type $r = 2$ and $a = 1$, already considered (A.60). It follows that the whole polarized chain $\{N_i\}$ (A.63) is determined up to isomorphism locally for the étale topology by its type $\{\Gamma_i\}$, and that the functor of isomorphisms is smooth.

A.64 In the case where $r > 4$ is any even number and $a = 1$ we consider the following subchain of modules

$$N_0 \longrightarrow N_{-r+1} \longrightarrow N_{-\frac{r}{2}} \longrightarrow N_{-\frac{r}{2}+1} \longrightarrow N_0.$$

The pairings $\mathcal{B}_0$ and $\mathcal{B}_{-\frac{r}{2}}$ define a polarization. Hence we are in the situation $r = 4$. We conclude that the map $N_{-\frac{r}{2}+1} \longrightarrow N_0$ looks as follows,

$$L_{-1} \oplus C_0^{\Pi} \oplus C_{-2} \longrightarrow L_0 \oplus C_0 \oplus C_{-2}\,.$$

This map has the form $\varrho \oplus \Pi \oplus id_{C_{-2}}$, where $\varrho : L_{-1} \longrightarrow L_0$ is an isomorphism and $\Pi : C_0^{\Pi} \longrightarrow C_0$ is multiplication by Π.

Hence the part $N_{-\frac{r}{2}+1} \longrightarrow N_{-\frac{r}{2}+2} \longrightarrow \cdots \longrightarrow N_{-1} \longrightarrow N_0$ of our chain is given by an unpolarized chain

$$C_0^{\Pi} = N'_{-\frac{r}{2}+1} \longrightarrow N'_{-\frac{r}{2}} \longrightarrow \cdots N'_{-1} \longrightarrow N'_0 = C_0. \tag{A.66}$$

Similarly we get a chain

$$C_{-2}^{\Pi} = N'_{-r+1} \longrightarrow N'_{r+2} \longrightarrow \cdots \longrightarrow N'_{-\frac{r}{2}} = C_{-2},$$

which is dual to the chain above, with respect to the pairing (A.64). From this we obtain the theorem for r even and $a = 1$, which was the last case to be treated.

This completes the proof of proposition (A.56) and also of theorem (3.16).

4. The formal Hecke correspondences

In this chapter we shall define the Hecke correspondences. They will be self–correspondences of any one of the formal schemes constructed in the previous chapter. We shall first explain the relative position of lattices and lattice chains and then pass to the corresponding concepts for isogenies of p-divisible groups. The case of a moduli problem of type (PEL) will be reduced to the case of type (EL).

4.1 Let us recall the notion of a Hodge polygon. Let D be a finite dimensional division algebra over $\mathbf{Q}_p$. We denote by O_D the ring of integers in D, and by Π a prime element. We consider an injection of finite torsion free O_D-modules of the same rank,

$$\varphi : M \longrightarrow N. \tag{4.1}$$

Then there exists a basis $\{v_i\}_{i=1,\ldots,r}$ of the O_D-module N, such that there is a basis of M of the form $\{\Pi^{e(i)}v_i\}_{i=1,\ldots,r}$. where the $e(i)$ are nonnegative integers. We define nonnegative integers t_k:

$$t_k = f(D/\mathbf{Q}_p)\text{card}\{i; e(i) = k\}$$

Here $f(D/\mathbf{Q}_p) = \dim_{\mathbf{F}_p} O_D/\Pi O_D$ denotes the index of inertia of D over $\mathbf{Q}_p$. For negative integers k we set $t_k = 0$. In a more invariant way the numbers t_k for nonnegative k may be expressed as follows:

$$t_k = \text{length}_{\mathbf{Z}_p}(M \cap \Pi^k N + \Pi M)/(M \cap \Pi^{k+1} N + \Pi M).$$

From this definition we conclude

$$\sum_{k\in\mathbf{Z}} t_k = f(D/\mathbf{Q}_p) rank_{O_D} M = e(D/\mathbf{Q}_p)^{-1} dim_{\mathbf{Q}_p} M \otimes \mathbf{Q}_p.$$

Sometimes it is convenient to work with the non-decreasing function

$$t(k) = \sum_{l<k} l t_l + k \sum_{l\geq k} t_l. \tag{4.2}$$

For nonnegative k this function may be written:

$$t(k) = \text{length}_{\mathbf{Z}_p} ker(\Pi^k : N/M \longrightarrow N/M).$$

Let $\{t_k\}_{k\in\mathbf{Z}}$ be any sequence of nonnegative integers, such that $h = \sum_{k\in\mathbf{Z}} t_k$ is bounded. We call such a sequence *finite.*

Definition 4.2 *The Hodge function associated to the sequence t_k is the unique nondecreasing continuous function in the real interval $[0, h]$, which is linear of slope k on the interval $[\sum_{l<k} t_l, \sum_{l\leq k} t_l]$ and vanishes at the origin. The graph of this function is called the Hodge polygon. If the numbers arise from the situation as in (4.1), we will speak of the Hodge polygon of the injection φ.*

Clearly this function takes the following values:

$$H(\sum_{l\leq k} t_l) = \sum_{l\leq k} l t_l.$$

Lemma 4.3 *Suppose we are given two sets of nonnegative integers:*

$$\{t_k\}_{k\in\mathbf{Z}}, \quad \{t_k^0\}_{k\in\mathbf{Z}}.$$

Assume that $h = \sum_{k\in\mathbf{Z}} t_k = \sum_{k\in\mathbf{Z}} t_k^0$ is bounded. Denote by HP and HP^0 the associated Hodge polygons. The condition that the Hodge polygon HP^0 lies above the Hodge polygon of HP is equivalent to the following inequality for the t-functions (4.2).

$$t(k) \leq t^0(k)$$

We content ourselves with giving the geometric reason for this elementary lemma. Consider two convex continuously differentiable functions $y = H(x)$ and $y = H^0(x)$ on the real interval $[x_1, x_2]$, i.e. the derivatives of the functions are nondecreasing. Then the graph of the function H lies below the graph of the function H^0 if and only if the tangent line of slope λ to the graph of H lies below the tangent line of slope λ to the graph of H^0. For the proof of the lemma it is enough to note that $t(k)$ is the t-coordinate of the intersection point of the tangent line of slope k to HP with the vertical line $t = h$.

4.4 The Hodge polygon of an O_D-morphism (4.1) is an invariant that expresses the relative position of O_D-lattices. To explain this, consider a D-vectorspace V. We consider an O_D-lattice $M \subset V$, i.e. a finitely generated O_D-submodule of V such that $M \otimes_{\mathbf{Z}} \mathbf{Q} = V$. Given two O_D-lattices M and N of V we associate to them a Hodge polygon $HP(M, N)$ as follows. There is a power Π^m such that $\Pi^m M \subset N$. Let $t_k^{(m)}$ be the sequence associated to this inclusion. Then we associate to the pair of lattices (M, N) the sequence $t_k = t_{k+m}^{(m)}$. This is independent of the number m chosen.

Definition 4.5 *The Hodge polygon $HP(M, N)$ of the pair (M, N) is the Hodge polygon of the sequence t_k.*

4.6 More generally we will consider the situation where B is a simple algebra of finite dimension over $\mathbf{Q}_p$. Let O_B be a maximal order of B. Then there is an isomorphism $B = M_n(D)$ with a matrix algebra over a division algebra D such that $O_B = M_n(O_D)$ under this isomorphism. Assume we are given a B-vectorspace V. The Hodge polygon of two O_B-lattices M and N in V is by definition the Hodge polygon $HP(M, N)$ of M and N viewed as O_D-lattices. We will denote the corresponding Hodge function by $H(M, N)$. It is a real function on the interval $[0,h]$, where $h = e(D/\mathbf{Q}_p)^{-1} dim_{\mathbf{Q}_p} V$. Here $e(D/\mathbf{Q}_p)$ denotes the ramification index. There is a natural submodule W of V such that $V = W^n$ as an $M_n(D)$-module. Any O_B-lattice $M \subset V$ is of the form $\bar{M}^n$ for the natural O_D-lattice $\bar{M} \subset W$. The Hodge functions are related:

$$H(M, N)(nt) = nH(\bar{M}, \bar{N})(t).$$

Consider the algebraic group $G = GL_B(V)$ over $\mathbf{Q}_p$. We fix an O_B-lattice $\Lambda \subset V$, which we call the standard lattice. The stabilizer of Λ is a maximal

open compact subgroup $K \subset G(\mathbf{Q}_p)$. With the notations introduced above we have $G = GL_D(W)$.
To any pair of O_B-lattices M and N in V one associates a double coset in $K\backslash G(\mathbf{Q}_p)/K$. To do this we take an element $x \in G(\mathbf{Q}_p)$ such that $N = x\Lambda$. Then there is a $g \in G(\mathbf{Q}_p)$ such that $M = xg\Lambda$. One verifies that the double coset KgK is independent of the choice of x and g.

Definition 4.7 *The relative position of the pair (M, N) is the double coset KgK. We will use the notation :* $\mathrm{pos}(M, N) = g$.

Example 4.8 Let M and N be lattices in a D-module V. Let $\{u_i\}_{i=1\dots r}$ be a basis of Λ as an O_D-module. With respect to this basis we have an isomorphism $G = GL_r(D^{opp})$. Assume there is a basis $\{v_i\}_{i=1\dots r}$ of the O_D-module N, such that $\{\Pi^{e(i)} v_i\}_{i=1\dots r}$ is a basis of M. Then $\mathrm{pos}(M, N)$ is given by the diagonal matrix

$$\begin{pmatrix} \Pi^{e(1)} & 0 & \dots & 0 \\ 0 & \Pi^{e(2)} & \dots & 0 \\ \dots & \dots & \dots & \dots \\ 0 & 0 & \dots & \Pi^{e(r)} \end{pmatrix}.$$

Lemma 4.9 *The map*

$$H : K \backslash G(\mathbf{Q}_p)/K \longrightarrow \mathcal{C}[0, h]$$

that associates to a double coset KgK the Hodge function $H(g\Lambda, \Lambda)$ on the interval $[0, h], h = e(D/\mathbf{Q}_p)^{-1} dim_{\mathbf{Q}_p} V$, is an injection.

We omit the verification. Hence the relative position and the Hodge function determine each other by the relation:

$$H(\mathrm{pos}(M, N)) = H(M, N).$$

4.10 We next consider the extension of these concepts to periodic lattice chains. Let us first consider the case where D is a division algebra and V is a D-vector space. Let $\mathcal{L} = \{\Lambda_i\}_{i \in \mathbf{Z}}$ be a lattice chain of V (cf. 3.1) which we index by the integers. We put

$$m_i = \mathrm{length}_{\mathbf{Z}_p}(\Lambda_i/\Lambda_{i-1}).$$

Then $m_{i+r} = m_i$, where r denotes the period of $\mathcal{L}$, (cf. 3.2).

Lemma 4.11 *Let $\mathcal{L} = \{\Lambda_i\}$ and $\mathcal{L}' = \{\Lambda'_i\}$ be indexed lattice chains. There exists $g \in G(\mathbf{Q}_p) = GL_D(V)$ with $g\Lambda'_i = \Lambda_i$, all i, if and only if $\mathcal{L}$ and $\mathcal{L}'$ have the same period r and if $m_i = m'_i$, $i \in \mathbf{Z}$.*

Proof: Only the if-direction is non-trivial. Let $g' \in G(\mathbf{Q}_p)$ with $g'\Lambda'_0 = \Lambda_0$. Then also $g\Lambda'_{-r} = \Lambda_{-r}$ and $g'\mathcal{L}'$ and $\mathcal{L}$ induce two filtrations of the same length on the $O_D/\Pi O_D$-vector space Λ_0/Λ_{-r}. By assumption the successive quotients have identical dimensions. Therefore we find $g'' \in G(\mathbf{Q}_p)$ with $g''\Lambda_0 = \Lambda_0$ which carries one filtration into the other. The element $g = g''g'$ therefore takes Λ'_i into Λ_i, all i. □

4.12 Let $\mathcal{L} = \{\Lambda_i\}$ and $\mathcal{L}' = \{\Lambda'_i\}$ be two conjugate indexed lattice chains in V, i.e. there exists $g \in G(\mathbf{Q}_p)$ with $g\Lambda'_i = \Lambda_i$, all i. We introduce new lattices

$$\Lambda_{ij} = \Lambda_i \cap \Lambda'_j + \Lambda'_{j-1}.$$

Then for fixed j these form an increasing sequence of O_D-lattices between Λ'_{j-1} and Λ'_j which for small values of i all coincide with Λ'_{j-1} and for large values of i all coincide with Λ'_j. We introduce the non-negative integers

$$t_{ij} = \mathrm{length}_{\mathbf{Z}_p}(\Lambda_{ij}/\Lambda_{i-1,j}).$$

Lemma 4.13 *The integers t_{ij} have the following properties.*

(i) $\sum_i t_{ij} = m_j$, $\sum_j t_{ij} = m_i$.

(ii) $t_{i+r,j+r} = t_{i,j}$

(iii) $\Lambda_{ij} = \Lambda'_j \Longleftrightarrow t_{l,j} = 0$ for $l > i$

$\Lambda_{ij} = \Lambda'_{j-1} \Longleftrightarrow t_{l,j} = 0$ for $l \leq i$.

(iv) $\Lambda'_j \subset \Lambda_i \Longleftrightarrow t_{l,k} = 0$ for $l > i, k \leq j$

$\Lambda_i \subset \Lambda'_j \Longleftrightarrow t_{l,k} = 0$ for $l \leq i, k > j$.

(v)

$$\forall i\ \exists j_1, j_2 \text{ with } t_{l,k} = 0 \quad \text{for } l > i \text{ and } k \leq j_1 \text{ and for } l \leq i \text{ and } k > j_2$$

$$\forall j \ \exists i_1, i_2 \ \text{with} t_{l,k} = 0 \quad \text{for } l > i_1 \text{ and } k \leq j \text{ and}$$
$$\text{for } l \leq i_2 \text{ and } k > j.$$

Proof: The first assertion of (i) is trivial and the second follows from the isomorphisms $Gr_i^{\mathcal{L}} Gr_j^{\mathcal{L}'} \simeq Gr_j^{\mathcal{L}'} Gr_i^{\mathcal{L}}$. The assertions (ii) and (iii) are obvious. Let us prove the first statement of (iv). The one implication is obvious, so assume $\Lambda_{i,k} = \Lambda_k, k \leq j$. But then $\Lambda_i \cap \Lambda'_j + \Lambda'_{j-1} = \Lambda'_j, \Lambda_i \cap \Lambda'_{j-1} + \Lambda'_{j-2} = \Lambda'_{j-1}$, etc. From these equalities we inductively deduce that

$$\Lambda_i \cap \Lambda'_j + \Lambda'_k = \Lambda'_j, \quad k \leq j.$$

However, if k is small enough that $\Lambda'_k \subset \Lambda_i$ we deduce that $\Lambda'_j \subset \Lambda_i$. The second statement of (iv) is similar. Finally, using (iv), the first statement of (v) says that

$$\forall i \ \exists j_1, j_2 \ \text{with} \ \Lambda'_{j_1} \subset \Lambda_i \subset \Lambda'_{j_2},$$

and similarly for the second statement of (v).

Example 4.14 (i) Let $r = 1$ and let $M = \Lambda_0$ and $N = \Lambda'_0$. Because of the periodicity condition (ii) of (4.13) the integers t_{ij} are determined by the integers

$$t_{i,0} = \text{length}_{\mathbf{Z}_p}(\Pi^{-i} N \cap M + \Pi M)/(\Pi^{-(i-1)} N \cap M + \Pi M).$$

However, it is obvious that the sequence of integers t_k associated in (4.4.) to the pair (M, N) is given by $t_k = t_{-k,0}$. In this sense the collection of integers t_{ij} generalizes the definition given in (4.4.).
(ii) Let $r = \dim_D V$, i.e. $m_i = m'_i = 1$, all i. Then $\forall j \ \exists! i = w(j)$ with $t_{ij} = 1$. For $i \neq w(j)$ we have $t_{ij} = 0$. Lemma (4.13.), (i) and (ii) easily imply that the map $w : \mathbf{Z} \to \mathbf{Z}$ belongs to the *affine Weyl group* $W^{aff} = \{w : \mathbf{Z} \to \mathbf{Z} \text{ bijective; } \ w(j+d) = w(j) + d\}$.

4.15 The integers t_{ij} serve to determine the relative position of two indexed lattice chains. To explain this we fix a indexed lattice chain $\mathcal{L}^0$ which we call the standard chain. The subgroup

$$K = K_{\mathcal{L}^0} = \{g \in G(\mathbf{Q}_p) \ ; \ g\Lambda = \Lambda, \Lambda \in \mathcal{L}^0\}$$

is an open compact subgroup of $G(\mathbf{Q}_p)$. If $\mathcal{L}$ and $\mathcal{L}'$ are two indexed lattice chains conjugate to $\mathcal{L}^0$ one associates to them in a way completely analogous to (4.7.) above a double coset KgK. We will again use the notation

$$\text{pos}(\mathcal{L}, \mathcal{L}') = g \Leftrightarrow \exists x \in G(\mathbf{Q}_p) : \mathcal{L}' = x\mathcal{L}^0, \mathcal{L} = xg\mathcal{L}^0.$$

The analogue of lemma (4.9.) above is the following assertion.

Lemma 4.16 *The map which associates to the double coset KgK the collection of integers $t(g) = (t_{ij}(g)) = (t_{ij}(g\mathcal{L}^0, \mathcal{L}^0))$ is injective. In other words, if $\mathcal{L}, \mathcal{L}', \mathcal{L}_1, \mathcal{L}'_1$ are indexed lattice chains all conjugate to $\mathcal{L}^0$, then there exists $x \in G(\mathbf{Q}_p)$ with*

$$x\Lambda_i = \Lambda'_i,\ x\Lambda_{i,1} = \Lambda'_{i,1}, \quad \textit{all } i$$

if and only if $t_{ij}(\mathcal{L}, \mathcal{L}') = t_{ij}(\mathcal{L}_1, \mathcal{L}'_1)$. Put yet another way, if $\mathcal{L}$ and $\mathcal{L}'$ are indexed lattice chains conjugate to $\mathcal{L}^0$, then there exists $k \in K$ with

$$k\Lambda_i = \Lambda'_i$$

if and only if $t_{ij}(\mathcal{L}, \mathcal{L}^0) = t_{ij}(\mathcal{L}', \mathcal{L}^0)$.

Proof: We prove the lemma in its last form. We consider a segment of the chain $\mathcal{L}^0$,

$$\Lambda_0^0 \subset \Lambda_1^0 \subset \ldots \subset \Lambda_r^0 .$$

A set of direct summands M_j of Λ_r^0 is called a splitting of this segment if the image of M_i in Λ_r^0/Λ_0^0 is equal to Λ_j^0/Λ_0^0. The existence of a splitting shows the surjectivity of the map

$$K \longrightarrow \prod_{j=1}^{r} GL_{O_D}(\Lambda_j^0/\Lambda_{j-1}^0).$$

Therefore, replacing $\mathcal{L}$ by $k\mathcal{L}, k \in K$, we may and will assume that

$$\Lambda_i \cap \Lambda_j^0 + \Lambda_{j-1}^0 = \Lambda'_i \cap \Lambda_j^0 + \Lambda_{j-1}^0.$$

We use the following statement.

Sublemma 4.17 *Let $s \geq 1$. Assume that*

$$\Lambda_i \cap \Lambda_j^0 + \Lambda_{j-s}^0 = \Lambda'_i \cap \Lambda_j^0 + \Lambda_{j-s}^0.$$

Then there exists $k \in K$ with $(k-1)\Lambda_j^0 \subset \Lambda_{j-s}^0$ such that

$$k\Lambda_i \cap \Lambda_j^0 + \Lambda_{j-s-1}^0 = \Lambda'_i \cap \Lambda_j^0 + \Lambda_{j-s-1}^0.$$

Proof: We consider the submodules

$$\begin{aligned} G_{ij} &= \Lambda_i \cap \Lambda_j^0 / \Lambda_i \cap \Lambda_{j-s}^0 \subset \Lambda_j^0/\Lambda_{j-s}^0 \\ \tilde{G}_{ij} &= \Lambda_i \cap \Lambda_j^0 / \Lambda_i \cap \Lambda_{j-s-1}^0 \subset \Lambda_j^0/\Lambda_{j-s-1}^0. \end{aligned}$$

and

$$\begin{aligned} V_{ij} &= \Lambda_i \cap \Lambda_{j-s+1}^0 / \Lambda_i \cap \Lambda_{j-s}^0 \subset \Lambda_{j-s+1}^0/\Lambda_{j-s}^0 \\ \tilde{V}_{ij} &= \Lambda_i \cap \Lambda_{j-s+1}^0 / \Lambda_i \cap \Lambda_{j-s-1}^0 \subset \Lambda_{j-s+1}^0/\Lambda_{j-s-1}^0. \end{aligned}$$

We have a cartesian diagram with exact rows

$$\begin{array}{ccccccccc} 0 & \to & V_{i,j-1} & \to & \tilde{G}_{ij} & \to & G_{ij} & \to & 0 \\ & & \| & & \cup & & \cup & & \\ 0 & \to & V_{i,j-1} & \to & \tilde{V}_{ij} & \to & V_{ij} & \to & 0. \end{array}$$

If we replace $\mathcal{L}$ by $\mathcal{L}'$ only the middle terms change and $G_{ij}/V_{ij} = G'_{ij}/V'_{ij}$. Let $y \in V_{ij}$ and let $x, x' \in \Lambda_{j-s+1}^0/\Lambda_{j-s-1}^0$ be liftings to the middle terms of the lower rows in the above diagram for $\mathcal{L}$ resp. $\mathcal{L}'$. Then the residue class of $x - x'$ in

$$(\Lambda_{j-s}^0/\Lambda_{j-s-1}^0)/\Lambda_{i,j-1}^0$$

is independent of the choices of x and x'. Hence the difference between the diagrams is measured by a homomorphism

$$\alpha_{ij} : V_{ij} \to (\Lambda_{j-s}^0/\Lambda_{j-s-1}^0)/\Lambda_{i,j-1}^0.$$

Now let $k \in K$ with $(k-1)\Lambda_j^0 \subset \Lambda_{j-s}^0$. Then the difference between the extension for $\mathcal{L}$ and for $k\,\mathcal{L}$ is given by the maps induced as follows from $k-1$:

$$\begin{array}{ccc} V_{ij} & \dashrightarrow & (\Lambda^0_{j-s}/\Lambda^0_{j-s-1})/\Lambda_{i,j-1} \\ \cap & & \uparrow \\ \Lambda^0_j/\Lambda^0_{j-s} & \xrightarrow{k-1} & \Lambda^0_{j-s}/\Lambda^0_{j-s-1} \end{array}$$

Therefore, to prove the sublemma, we have to find $k \in K$ with $(k-1)\Lambda^0_j \subset \Lambda^0_{j-s}$ which induces the given maps α_{ij}, all i, j. Clearly any set of homomorphisms

$$\zeta_j : \Lambda^0_{j-s+1}/\Lambda^0_{j-s} \longrightarrow \Lambda^0_{j-s}/\Lambda^0_{j-s-1}$$

$(j = 1, \ldots, r)$ is induced by some $k-1$. Therefore we must prove that for fixed j there is a homomorphism ζ_j which induces α_{ij}, all i. However, this is an exercise in linear algebra which we may formulate as follows.

Let V and W be vector spaces with finite separating and exhaustive increasing filtrations $V_\bullet$ and $W_\bullet$. Suppose we are given linear maps $\alpha_i : V_i \longrightarrow W/W_i$ such that the following diagrams are commutative.

$$\begin{array}{ccc} V_i & \longrightarrow & W/W_i \\ \uparrow & & \uparrow \\ V_{i-1} & \longrightarrow & W/W_{i-1} \end{array}$$

Then the maps α_i are induced by a homomorphism

$$V \longrightarrow W$$

The proof is by induction, constructing at the i-th stage a dotted arrow so as to make following diagram commutative.

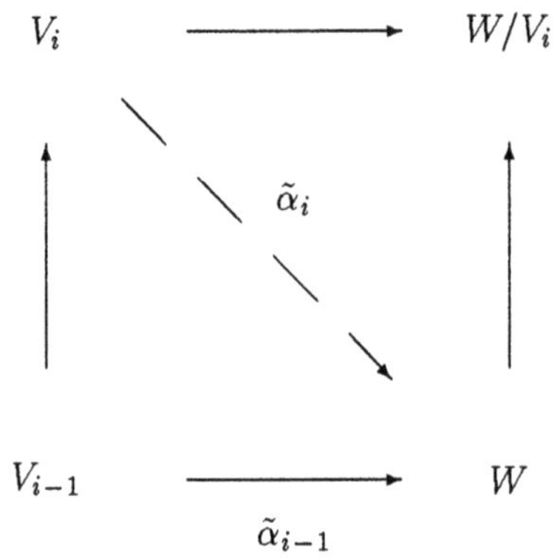

End of the proof of lemma (4.16). By an obvious induction we construct elements $k_s \in K$ with $(k_s - 1)\Lambda_j^0 \subset \Lambda_{j-s}^0 (s = 2, 3, \ldots)$ such that putting $k^{(s)} = k_s \cdots k_2$ we have $(k^{(s)} - 1)\Lambda_j^0 \subset \Lambda_{j-1}^0$ and

$$k^{(s)}\Lambda_j \cap \Lambda_j^0 + \Lambda_{j-s} = \Lambda'_j \cap \Lambda_j^0 + \Lambda_{j-s}.$$

However, $k^{(s)}$ converges to $k \in K$ with $(k - 1)\Lambda_j^0 \subset \Lambda_{j-1}^0$ and

$$k\Lambda_j \cap \Lambda_j^0 = \Lambda'_j \cap \Lambda_j^0,$$

i.e., $k\mathcal{L} = \mathcal{L}'$.

4.18 In the setting of (4.15) we define the Hecke correspondence associated to the double coset KgK (with $K = K_{\mathcal{L}^0}, \mathcal{L}^0 = \{\Lambda_i^0\}$). It is the correspondence on the set of indexed lattice chains conjugate to $\mathcal{L}^0$ defined by

$$T_g(\mathcal{L}) = \{\mathcal{L}'; \ \text{pos}(\mathcal{L}, \mathcal{L}') = g\}.$$

Alternatively, by lemma (4.16.), this set may be described as follows. Let

$$t(g) = (t_{ij}(g)) = (t_{ij}(g\mathcal{L}^0, \mathcal{L}^0)).$$

Then

$$T_g(\mathcal{L}) = \{\mathcal{L}'; \ t_{ij}(\mathcal{L}, \mathcal{L}') = t_{ij}(g), \ \text{all } i, j\}.$$

Lemma 4.19 *The set $T_g(\mathcal{L})$ is finite.*

Proof: We may assume $\mathcal{L} = \mathcal{L}^0$. The set $T_g(\mathcal{L})$ may then be identified with the orbit under K of $g\mathcal{L}^0$. Being discrete and compact it is therefore finite.

4.20 More generally we will consider the situation where B is a simple algebra of finite dimension over $\mathbf{Q}_p$ and O_B is a maximal order of B. As in (4.6.) we write $B = M_n(D)$ for a division algebra D and $O_B = M_n(O_D)$. We consider indexed lattice chains $\mathcal{L} = \{\Lambda_i\}$ in a B-vector space V, cf.(3.2). Let $G = GL_B(V)$. For two indexed lattice chains $\mathcal{L}$ and $\mathcal{L}'$ we introduce the numbers $t_{ij}(\mathcal{L}, \mathcal{L}')$ as above by considering $\mathcal{L}$ and $\mathcal{L}'$ as O_D-lattice chains. The obvious analogue of lemma (4.16) is true and may in fact be reduced to this lemma by Morita equivalence, cf. (3.2) and (4.6). In fact, under this equivalence we may write any O_B-lattice Λ in the form $\bar{\Lambda}^n$ for an O_D-lattice in the D-vector space W with $V = W^n$. Corresponding to the indexed lattice chains $\mathcal{L}$ and $\mathcal{L}'$ there are indexed lattice chains $\bar{\mathcal{L}}$ and $\bar{\mathcal{L}}'$ and

$$t_{ij}(\mathcal{L}, \mathcal{L}') = n \cdot t_{ij}(\bar{\mathcal{L}}, \bar{\mathcal{L}}')$$

Similarly, $G = GL_B(V) = GL_D(W)$ and $K_{\mathcal{L}} = K_{\bar{\mathcal{L}}}$, which proves the claim.

4.21 We now turn to p-divisible groups. Let B be a simple algebra of finite dimension over $\mathbf{Q}_p$ and O_B a maximal order of B. We consider p-divisible groups X with an O_B-action:

$$O_B \longrightarrow EndX\ .$$

We require that morphisms respect the O_B-action.
Let X and Y be p-divisible O_B-groups over a perfect field L of characteristic p. Assume we are given an O_B-isogeny

$$\alpha : X \longrightarrow Y.$$

Let $\varphi : M \to N$ be the morphism of the corresponding crystals. This is a morphism of $O_B \otimes_{\mathbf{Z}_p} W(L)$-modules. For nonnegative integers k we define:

$$t_k = \text{length}_{W(L)}(M \cap \Pi^k N + \Pi M)/(M \cap \Pi^{k+1} N + \Pi M).$$

For negative integers k we set $t_k = 0$.

Definition 4.22 *The Hodge function of the isogeny α is the Hodge function associated to the sequence t_k. We will denote the Hodge function respectively the Hodge polygon of α by $H(\alpha)$ respectively $HP(\alpha)$.*

In contrast to the situation considered in (4.1) there is no reason for the integers t_k to be divisible by $f(D/\mathbf{Q}_p)$. We have the relation:

$$\sum_{k \in \mathbf{Z}} t_k = e(D/\mathbf{Q}_p)^{-1} \text{ height } X = \text{ height } X(\Pi)$$

Here $X(\Pi)$ denotes the kernel of the isogeny $\Pi : X \rightarrow X$. The kernel of the isogeny α is a finite group scheme A with an O_B-action. We denote by $A(\Pi^k)$ the kernel of the multiplication by Π^k on A. The function $t(k)$ associated to the Hodge polygon of α (cf. 4.2) is for nonnegative k given by the formula:

$$t(k) = \text{ height } A(\Pi^k).$$

We note that the numbers t_k for nonnegative k and height X determine the Hodge function of α.

4.23 We denote by $Nilp$ the category of schemes where p is locally nilpotent (cf. chapter 2). Let us consider an isogeny of p-divisible O_B-groups $\alpha : X \rightarrow Y$ over a general base $S \in Nilp$. Then A is by definition a finite locally free group scheme. We denote by $\tilde{A}(\Pi^k)$ the quasicoherent $\mathcal{O}_S$-algebra associated to $A(\Pi^k)$. It is locally of finite representation.
Before we proceed, we need to recall a definition. Let R be a commutative ring. Let M be a finitely generated R-module. One associates to M a sequence of ideals of R,

$$\vartheta_0(M) \subset \vartheta_1(M) \subset \vartheta_2(M) \subset \ldots,$$

the determinantal or Fitting ideals of M (Bourbaki, Algebre commutative, Exerc.10 Chapt.VII, 4). These ideals commute with arbitrary base change, i.e. for any ring homomorphism $R \rightarrow S$, we have

$$\vartheta_i(M)S = \vartheta_i(M \otimes_R S).$$

The variety $V(\vartheta_i(M))$ consists of all points $s \in Spec\, R$, such that the rank $rk_{\kappa(s)}(M \otimes_R \kappa(s)) > i$. If M is of finite presentation the Fitting ideals are finitely generated.

Definition 4.24 *Let $\{a_k\}_{k\in\mathbf{Z}}$ be any sequence of nonnegative integers such that $a_k = 0$ for $k < 0$. We say that the Hodge polygon of an isogeny of p-divisible O_B-groups $\alpha : X \to Y$ over $S \in Nilp$ lies above the Hodge polygon associated to the numbers a_k, if the $(p^{a(k)}-1)$ Fitting ideal of the $\mathcal{O}_S$-module $\tilde{A}(\Pi^k)$ is zero.*

If S is the spectrum of a perfect field this means simply that the Hodge polygon associated to α (cf. 4.22)) lies above the Hodge polygon $HP(a)$ associated to the sequence a_k. If S is reduced the Hodge polygon of α lies above $HP(a)$, if and only if for any geometric point $\bar{s}$ of S the Hodge polygon of $\alpha_{\bar{s}}$ lies above $HP(a)$.

4.25 We make the well known fact that the Hodge polygon rises under specialization a little more precise. For a given isogeny $\alpha : X \to Y$ over S, we define the functor $HP^{\geq a}$ on $Nilp$. The T-valued points $HP^{\geq a}(T)$ consists of morphisms $f : T \to S$ such that the Hodge polygon of $f^*(\alpha)$ lies above the Hodge polygon $HP(a)$.

Proposition 4.26 *The functor $HP^{\geq a}$ is representable by a closed subscheme of S.*

4.27 We want to extend these considerations to quasi-isogenies of p-divisible groups with O_B-action. We recall that any $b \in B^\times$ normalizing O_B defines an isogeny (cf. 3.20)

$$b : X^b \longrightarrow X.$$

Let $a = \{a_k\}_{k\in\mathbf{Z}}$ be any sequence of nonnegative integers. We denote by $a^{(m)}$ the sequence given by $a_k^{(m)} = a_{k-m}$.

Definition 4.28 *Let $a = \{a_k\}_{k\in\mathbf{Z}}$ be a finite sequence of nonnegative integers. We denote by m the smallest nonnegative integer, such that $a_k^{(m)} = 0$ for $k < 0$. We say that the Hodge polygon of a quasi-isogeny $\alpha : X \to Y$ lies above $HP(a)$, if $\Pi^m\alpha : X^{\Pi^m} \to Y$ is an isogeny, whose Hodge polygon lies above $HP(a^{(m)})$.*

Remark 4.29 Assume that $a_k = 0$ for $k < 0$. Assume that the Hodge polygon of α lies above $HP(a)$. Then the Hodge polygon of $\Pi\alpha$ lies above $HP(a^{(1)})$, but the converse need not be true. Indeed, if A denotes the kernel of α and A' the kernel of $\Pi\alpha$, we have an exact sequence of group schemes for any $k \geq 1$:

$$0 \to X(\Pi) \to A'(\Pi^k) \to A(\Pi^{k-1}) \to 0$$

This implies that locally for the Zariski topology on S the algebra $\tilde{A}'(\Pi^k)$ associated to $A'(\Pi^k)$ is as an $\mathcal{O}_S$-module a direct sum of p^h copies of $\tilde{A}(\Pi^{k-1})$, where $h =$ height $X(\Pi)$. Let $\vartheta_u(\tilde{A}(\Pi^{k-1}))$ be the first nonvanishing Fitting ideal of $\tilde{A}(\Pi^{k-1})$. We conclude by Bourbaki loc.cit. that $\vartheta_i(\tilde{A}'(\Pi^k)) = 0$ for $i < p^h u$ and that $\vartheta_{p^h u}(\tilde{A}'(\Pi^k)) = \vartheta_u(\tilde{A}(\Pi^{k-1}))^{p^h}$. Hence $u \geq p^{a(k-1)}$ implies $p^h u \geq p^{a(k-1)h} = p^{a^{(1)}(k)}$. This proves the assertion of the remark.

Proposition 4.30 *Let $HP^{\geq a}$ be the subfunctor of S, where the Hodge polygon of the quasi-isogeny α lies above $HP(a)$. Then $HP^{\geq a}$ is representable by a closed subscheme of S.*

Proof: This follows from proposition (4.26) since the subfunctor of S where $\Pi^m\alpha$ is an isogeny is representable by a closed subscheme of S, (cf. 2.9).

4.31 We also introduce the functor $HP^{=a}$. It is the subfunctor of S which consists of points $f : T \to S$, such that $\Pi^m f^*(\alpha)$ is an isogeny and moreover if A' denotes the kernel of $\Pi^m f^*(\alpha)$ then $A'(\Pi^k)$ is a locally free group scheme of height $a^{(m)}(k)$.

Proposition 4.32 *$HP^{=a}$ is an open subfunctor of $HP^{\geq a}$.*

This is a consequence of the following elementary fact.

Lemma 4.33 *Let $\tilde{M}$ be a quasicoherent module on a scheme S, which is locally of finite type. Assume that the $(t-1)$ Fitting ideal of $\tilde{M}$ is zero. If for a point $s \in S$ the inequality $rk_{\kappa(s)}\tilde{M} \otimes \kappa(s) \leq t$ holds, then $\tilde{M}$ is a free module of rank t locally around s.*

Proof: Let $n = rk_{\kappa(s)}\tilde{M} \otimes \kappa(s)$. By the lemma of Nakayama there is an affine neighbourhood $Spec\, R \subset S$ of s, such that the R-module $M = H^0(Spec\, R, \tilde{M})$ admits a resolution

$$0 \longrightarrow K \longrightarrow R^n \longrightarrow M \longrightarrow 0\,.$$

Let $e_1, \ldots, e_n$ be the standard basis of R^n and $e_1^*, \ldots, e_n^*$ a dual basis. The i-th Fitting ideal ϑ_i is by definition generated by the $(n-i) \times (n-i)$ minors of the following (possibly infinite) matrix

$$a_{k,i} = \langle k, e_i^* \rangle, \quad k \in K, \quad i = 1, \ldots, n,$$

where $\vartheta_i = R$ for $i \geq n$.

If M is free of rank t, we have $\vartheta_{t-1} = 0$.

For the converse assume $\vartheta_{t-1} = 0$. Since $\vartheta_n = R$ we conclude $n > t-1$. Hence $n = t$, since by assumption $n \leq t$. But then ϑ_{t-1} is spanned by the element $a_{k,i}$. It follows that $K = 0$.

□

4.34 Propositions (4.30) and (4.32) justify the following definition. The *quasi-isogeny* $\alpha : X \to Y$ *puts* X *and* Y *in relative position* a if the locally closed subscheme $HP^{=a}$ of S is all of S. In fact, we shall not use this definition in the sequel.

4.35 We now return to the definition of the formal Hecke correspondences. We wish to transpose the definition of (4.18) to the context of chains of p-divisible groups with O_B-action of a fixed type $(\mathcal{L})$, (cf. 3.21). We number $\mathcal{L}$ once and for all,

$$\mathcal{L} = \{\Lambda_i\}_{i \in \mathbf{Z}}.$$

Therefore a chain of p-divisible groups of type $(\mathcal{L})$ over some base scheme $S \in Nilp$ inherits a natural numbering,

$$\longrightarrow X_{i-1} \xrightarrow{\alpha_i} X_i \xrightarrow{\alpha_{i+1}} \ldots .$$

Let

$$\alpha : X_\bullet \to Y_\bullet$$

be a O_B-quasi-isogeny of chains of p-divisible groups of type $(\mathcal{L})$ over S. This means that we are given O_B-quasi-isogenies $\alpha_i : X_i \to Y_i$ commuting with the transition morphisms. Obviously, giving α is equivalent to giving α_i for one $i \in \mathbf{Z}$. Let S be the spectrum of a perfect field L of characteristic p. Then the quasi-isogeny α allows us to identify the rational Dieudonné modules of X_i and of Y_i. We denote by V this $B \otimes K_0(L)$-module. The

Dieudonné modules of $(X_i)_i$ define a indexed chain of lattices in an obvious sense (these are $O_B \otimes W(L)$-modules)

$$\ldots \subset \Lambda_i \subset \Lambda_{i+1} \subset \ldots$$

and similarly for $Y_\bullet$, which defines $\{\Lambda'_j\}$. We put

$$t_{ij}(\alpha) = \text{length}_{W(L)}(\Lambda_i \cap \Lambda'_j + \Lambda'_{j-1})/(\Lambda_{i-1} \cap \Lambda'_j + \Lambda'_{j-1}).$$

There is the relation

$$\sum_i t_{ij} = \text{height Ker}(\alpha_j : Y_{j-1} \to Y_j).$$

We shall use this relation to treat the case of a general base scheme $S \in Nilp$.

4.36 We fix a collection of integers $t = (t_{ij})$. One example we have in mind is when $t = t(g)$, cf. (4.16). We wish to say what it means for the quasi-isogeny α to be of type t. Let i, j, k be integers such that $i < k$ and that α induces a true isogeny $X_i \to Y_j$. Let $A_{ik}(j)$ be the intersection of the finite group schemes arising as the kernels of isogenies,

$$A_{ik}(j) = \text{Ker}(X_i \to X_k) \cap \text{Ker}(X_i \to Y_j).$$

Definition 4.37 *The* O_B-quasi-isogeny α is of type $t = (t_{ij})$ *if, for all integers* i, j, k *as above,* $A_{ik}(j)$ *is a finite locally free group scheme of rank* p^n *with*

$$n = \sum_{i \leq r \leq k} \sum_{l \leq i} t_{lr}.$$

4.38 Let S be the spectrum of a perfect field L and (Λ_i) and (Λ'_j) the $O_B \otimes W(L)$-lattice chains associated to $X_\bullet$ and $Y_\bullet$ in the $B \otimes K_0(L)$-module V. We suppose that $t_{ij}(\alpha) = t_{ij}$ (cf. (4.35)). Then $A_{ik}(j)$ is a finite group scheme of rank p^n with

$$n = \text{length}_{W(L)}(\Lambda_k \cap \Lambda'_j / \Lambda_i).$$

Here $\Lambda_i \subset \Lambda_k$ and $\Lambda_i \subset \Lambda'_j$. However,

$$\text{length}\,(\Lambda_k \cap \Lambda'_j / \Lambda_i) \quad = \quad \sum_{r=i+1}^{k} \text{length}\,(\Lambda'_j \cap \Lambda_r / \Lambda'_j \cap \Lambda_{r-1})$$

$$= \sum_{r=i+1}^{k} \text{length}\,(\Lambda'_j \cap \Lambda_r + \Lambda_{r-1}/\Lambda_{r-1})$$
$$= \sum_{j<r\leq k} \sum_{l\leq i} t_{l,r} \ .$$

Therefore α is of type (t_{ij}) in the sense of definition (4.37). Conversely, if α is of type (t_{ij}) in the sense of (4.37) then $t_{ij}(\alpha) = t_{ij}$. Indeed, the numbers length $(\Lambda_k \cap \Lambda'_j/\Lambda_i)$ for all i, j, k with $i < k$ and $\Lambda_i \subset \Lambda'_j$ determine uniquely all integers $t_{ij}(\alpha)$. In this sense the definition (4.37) is the correct definition of the relative position of α in the case of a perfect field L.
We now analyze the definition (4.37) over a general base scheme $S \in Nilp$.

Proposition 4.39 *The subfunctor of S where the quasi-isogeny α is of type (t_{ij}) is representable by a locally closed subscheme of S.*

Proof: Due to the periodicity condition there is a positive integer c such that the quasi-isogenies

$$X_{j-c} \longrightarrow Y_j \longrightarrow X_{j+c}$$

are isogenies for all j. Furthermore, for any i, j, k as in (4.37) there is an exact sequence of group schemes

$$0 \to X_{i-r}(\Pi) \longrightarrow A_{i-r,k}(j) \longrightarrow A_{ik}(j) \to 0 \ .$$

Therefore, locally for the Zariski topology on S, the affine algebra associated to $A_{i-r,k}(j)$ is a direct sum of p^h copies of the affine algebra of $A_{ik}(j)$, where $h =$ height $X_i(\Pi)$. Hence the condition in (4.36) for i, j, k is equivalent to the condition for $i-r, j, k$. Furthermore, for fixed i, j, k as in (4.36) we have if $k \geq j + c$,

$$A_{ik}(j) = \text{Ker}(X_i \to Y_j)$$

since the isogeny $X_i \to X_k$ factors through Y_j. Therefore for such indices the condition that $A_{ik}(j)$ be locally free is automatic and the condition on the rank follows from the condition for i, j, k_0 with $k_0 = \max(i, j + c)$. Finally, multiplication by Π induces isomorphisms of group schemes

$$A_{ik}(j) \xrightarrow{\sim} A_{i+r,k+r}(j + r).$$

Summarizing, we see that the subfunctor of S where α is of type (t_{ij}) is defined by *finitely many* conditions of the type that a certain finite S-group scheme be locally free of a given rank. By (4.30) and (4.33) this is representable by a locally closed subscheme of S.

Remark 4.40 We have reduced here the proof of proposition (4.39) to (4.30) and (4.33) which use the Fitting ideals. In fact, the assertion that the condition that a certain finite S-group scheme be locally free of a given rank defines a locally closed subscheme of S is more elementary.

4.41 We finally remark that everything generalizes to the case where B is a finite - dimensional semi-simple algebra over $\mathbf{Q}_p$ and O_B a maximal order in B and where we consider indexed multichains of lattices and of p-divisible groups and O_B-quasi-isogenies between them. In this case B, O_B, and G decompose as a product and we associate to each $g \in G(\mathbf{Q}_p)$ the function $t(g) = (t_{ij})$ of each factor and to a pair of conjugate indexed multichains of lattices $\mathcal{L}$ and $\mathcal{L}'$ the function $t(\mathcal{L}, \mathcal{L}')$ of each factor. Similarly, if a collection of integers $t = (t_{ij})$ is given for each simple factor, it makes sense to say of a O_B-quasi-isogeny between multichains of p-divisible groups to be of type t.

4.42 We now return to the set-up of chapter 3 and consider a moduli problem of type (EL), corresponding to $(F, B, O_B, V, b, \mu, \mathcal{L})$ (cf (3.21)) relative to an algebraically closed field L of characteristic p. We denote by $\breve{\mathcal{M}} = \breve{\mathcal{M}}_{\mathcal{L}}$ the formal scheme over $Spf(O_{\breve{E}})$ representing the functor (3.21). Let $S \in Nilp_{O_{\breve{E}}}$ and let $(X_{\mathcal{L}}, \varrho) \in \breve{\mathcal{M}}(S)$ and $(X'_{\mathcal{L}}, \varrho') \in \breve{\mathcal{M}}(S)$. Then $\varrho'\varrho^{-1}$ defines a quasi-isogeny from $X_{\mathcal{L}} \times_S \bar{S}$ to $X_{\mathcal{L}'} \times_S \bar{S}$ which extends in a unique way to a quasi-isogeny

$$\varrho'\varrho^{-1} : X_{\mathcal{L}} \longrightarrow X'_{\mathcal{L}}.$$

Definition 4.43 *Let t be as in (4.41). The formal Hecke correspondence associated to t is the functor on $Nilp$ which to S associates the isomorphism classes of objects*

$$(X_{\mathcal{L}}, \varrho), (X'_{\mathcal{L}}, \varrho') \in \breve{\mathcal{M}}(S) \times \breve{\mathcal{M}}(S)$$

such that the resulting quasi-isogeny $\varrho'\varrho^{-1}$ from $X_{\mathcal{L}}$ to $X'_{\mathcal{L}}$ is of type t, cf. (4.41).

From (4.39) we deduce immediately the following statement.

Proposition 4.44 *The preceding functor is representable by a locally closed formal subscheme*

$$\mathcal{C}orr(t) \subset \breve{\mathcal{M}} \times_{Spf\, O_{\breve{E}}} \breve{\mathcal{M}}.$$

We remark that the projection morphisms

$$\breve{\mathcal{M}} \longleftarrow \mathcal{C}orr(t) \longrightarrow \breve{\mathcal{M}}$$

very often are *not proper* (comp. however (5.42)). In fact, $\mathcal{C}orr(t)$ is in general not Zariski–closed in $\breve{\mathcal{M}} \times_{Spf\, O_{\breve{E}}} \breve{\mathcal{M}}$. The reason for this is that the Hodge polygon of a family of quasi-isogenies may vary with the point of the base scheme.
If $g \in G(\mathbf{Q}_p)$ and if $t = t(g)$, cf. (4.41), we use the notation

$$\mathcal{C}orr(g) = \mathcal{C}orr(t).$$

In general, the collection of integers need not arise in this way.

4.45 We now pass to the polarized case which will be reduced to the previous case. Let then $(F, B, O_B, *, V, (\ ,\ ))$ be data of case (PEL). We denote by G the associated reductive algebraic group over $\mathbf{Q}_p$. Let F_0 be the invariants under the involution $*$ in F.
Let $\mathcal{L}$ be a selfdual multichain of lattices in V. Let $K_{\mathcal{L}}$ be its fix group,

$$K_{\mathcal{L}} = \{g \in G(\mathbf{Q}_p);\ g\Lambda = \Lambda,\ \Lambda \in \mathcal{L}\}.$$

Then, as in (4.15), one associates a double coset $K_{\mathcal{L}} g K_{\mathcal{L}}$ to a pair of indexed selfdual multichains of lattices, conjugate to $\mathcal{L}$ under $G(\mathbf{Q}_p)$.
Let

$$i : G \longrightarrow \tilde{G} = GL_B(V)$$

be the canonical embedding, and let $\tilde{K}_{\mathcal{L}} \subset \tilde{G}(\mathbf{Q}_p)$ be the fix group of $\mathcal{L}$ in $\tilde{G}(\mathbf{Q}_p)$.
The proof of the following theorem was communicated to us by Waldspurger. It is the generalization of a lemma of Kottwitz [Ko3], 7.4 which is valid for hyperspecial open compact subgroups $K_{\mathcal{L}}$.

Theorem 4.46 *The embedding i induces an injective map on sets of double cosets,*

$$K_{\mathcal{L}} \setminus G(\mathbf{Q}_p)/K_{\mathcal{L}} \longrightarrow \tilde{K}_{\mathcal{L}} \setminus \tilde{G}(\mathbf{Q}_p)/\tilde{K}_{\mathcal{L}}.$$

To present Waldspurger's proof we need some preparations. Any self-dual multichain of lattices $\mathcal{L}$ may be refined to a maximal self-dual multichain of lattices $\mathcal{L}'$ (i.e. $\Lambda \in \mathcal{L} \Rightarrow \Lambda \in \mathcal{L}'$) and any two maximal self-dual multichains are conjugate by an element $g \in G(\mathbf{Q}_p)$ with $c(g) = 1$. We fix one such maximal self-dual multichain $\mathcal{L}^0$ and denote the corresponding fix group in $G(\mathbf{Q}_p)$ by K^0. It is an Iwahori subgroup of $G(\mathbf{Q}_p)$ and we may assume that $\mathcal{L}^0$ is a refinement of $\mathcal{L}$, i.e. $K^0 \subset K_{\mathcal{L}}$.

4.47 We shall first assume that F is a field. Let $B = M_n(D)$ and $O_B = M_n(O_D)$ for a division algebra D. We consider D^n as a D–module from the right and a B–module from the left. Let $* : D \to D$ be a main involution if D is a quaternion algebra. In the case that the involution on B is of second kind, we have $D = F$ and we take for $*$ the restriction of the involution to F. Then there exists a D–valued sesquilinear form

$$H : D^n \times D^n \longrightarrow D$$

such that

$$\begin{aligned} H(xd, yd') &= d^* H(x, y)d', \quad d, d' \in D \\ H(bx, y) &= H(x, b^* y), \quad b \in B. \end{aligned}$$

Let

$$h(x, y) = \mathrm{Tr}_{F/\mathbf{Q}_p} \mathrm{Tr}^0(H(x, y)).$$

By Morita equivalence we may write $V = D^n \otimes_D \bar{V}$. Furthermore, there is a uniquely defined sesquilinear form

$$<,>: \bar{V} \times \bar{V} \longrightarrow D$$

such that

$$(x \otimes v, x' \otimes v') = h(x, x' < v, v' >), \quad x, x' \in D^n, \quad v, v' \in \bar{V}.$$

Then

$$< dv, d'v' >= d' < v, v' > d^*, \quad d, d' \in D.$$

and $<,>$ is hermitian or anti-hermitian. Let G' be the algebraic group over $\mathbf{Q}_p$ with

$$G'(\mathbf{Q}_p) = \{g \in GL_D(\bar{V});\ < gv, gv' >= c(g) < v, v' >, c(g) \in F_0^{\times}\}.$$

We write $\bar{V}$ as an orthogonal sum of an isotropic space and an anisotropic one,

$$\bar{V} = \bar{V}' \oplus \bar{V}''.$$

Here $\bar{V}'$ posesses a maximal Witt basis. In $\bar{V}''$ there is a unique O_D-lattice $L'' \subset \bar{V}''$ such that

$$(L'')^* \supset L'' \supset \Pi(L'')^*.$$

We introduce the integers

$$\begin{aligned} \alpha &= \dim_{\kappa(D)}(L''/\Pi(L'')^*) \\ \beta &= \dim_{\kappa(D)}((L'')^*/L'') \\ 2r &= \dim_D \bar{V}'. \end{aligned}$$

Then the dimension d of $\bar{V}$ is $2r + \alpha + \beta$. Let

$$\mathbf{Z}_\alpha = \begin{cases} \mathbf{Z} & \text{if } \alpha \text{ is odd} \\ \mathbf{Z} + 1/2 & \text{if } \alpha \text{ is even.} \end{cases}$$

We choose a basis $\{e_{\frac{1-\alpha}{2}-r}, \ldots, e_{\frac{\alpha-1}{2}+r+\beta}\}$ of $\bar{V}$ such that L'' has as O_D-basis

$$\{e_{\frac{1-\alpha}{2}}, \ldots, e_{\frac{\alpha-1}{2}}\} \cup \{e_{\frac{\alpha-1}{2}+r+1}, \ldots, e_{\frac{\alpha-1}{2}+r+\beta}\},$$

$(L'')^*$ has as O_D-basis

$$\{e_{\frac{1-\alpha}{2}}, \ldots, e_{\frac{\alpha-1}{2}}\} \cup \{\Pi^{-1} e_{\frac{\alpha-1}{2}+r+1}, \ldots, \Pi^{-1} e_{\frac{\alpha-1}{2}+r+\beta}\},$$

and such that

$$\{e_{\frac{1-\alpha}{2}-r}, \ldots, e_{\frac{1-\alpha}{2}-1}\} \cup \{e_{\frac{\alpha-1}{2}+1}, \ldots, e_{\frac{\alpha-1}{2}+r}\}$$

is a basis of $\bar{V}'$ such that $< e_i, e_j >= 0$ if e_i and e_j both are in the same of the two sets above and such that

$$< e_i, e_j >= \delta_{i,-j}$$

if e_i lies in the first set and e_j in the second set. Before proceeding we make a simple remark. Let

$$c_0' = \min\{c > 0;\ \exists g \in G'(\mathbf{Q}_p) : c(g) \in \Pi^c O_D^\times\}.$$

Assume that c_0' is odd. Then $D = F$ and hence $c_0' = 1$. Let $g_0 \in G'(\mathbf{Q}_p)$ with $c(g_0) \in \Pi O_D^\times$. Then $g_0(\bar{V}')$ has again a Witt basis and hence by Witt's theorem there exists h with $c(h) = 1$ which takes $\bar{V}'$ into $g_0(\bar{V}')$. Replacing g_0 by $h^{-1}g_0$ we may therefore assume that $g_0(\bar{V}') = \bar{V}', g_0(\bar{V}'') = \bar{V}''$. However, since $c(g_0) \in \Pi O_D^\times$, by the uniqueness of the lattice L'' we have

$$g_0(L'')^* = L'', g_0 L'' = \Pi(L'')^*.$$

and g_0 induces an isomorphism

$$(L'')^*/L'' \xrightarrow{\sim} L''/\Pi(L'')^*.$$

In particular, $\alpha = \beta$ and d is even in this case. We therefore have proved the following statement.

Let $c_0 = min\{\operatorname{ord} c(g) > 0; g \in G(\mathbf{Q}_p)\}$. *Then* $c = c_0 d/2 \in \mathbf{Z}$.

Indeed, if c_0 is even there is nothing to prove. If c_0 is odd, then since c_0 is a multiple of c'_0 the latter is odd and hence, by the above, d is even.
We now extend the definition of e_i to all of $\mathbf{Z}_\alpha$ by putting

$$e_{i+jd} = \Pi^j e_j.$$

The basis $\{e_i\}$ defines a maximal split torus S of G, namely $S(\mathbf{Q}_p)$ consists of those elements $g \in G(\mathbf{Q}_p)$ such that

$$g e_i = \lambda_i e_i, \quad \lambda_i \in \mathbf{Q}_p^\times, \text{all } i.$$

Furthermore, for $e_i \in \bar{V}''$, the scalar λ_i is independent of e_i and satisfies

$$\lambda_i^2 = c(g), \quad e_i \in \bar{V}''.$$

The centralizer H of S is therefore defined by

$$ge_i = \lambda_i e_i, \quad \lambda_i \in D^\times, e_i \in \bar{V}'.$$

In particular, any $g \in H$ preserves the subspace $\bar{V}''$. The maximal compact subgroup of $H(\mathbf{Q}_p)$ is defined by the conditions

$$\begin{aligned} ge_i &= \lambda_i e_i, \quad \lambda_i \in O_D^\times, \, e_i \in \bar{V}', \\ c(g) &\in \mathbf{Z}_p^\times. \end{aligned}$$

The last condition is automatic if $\bar{V}' \neq (0)$. The normalizer N of S is formed by the elements $g \in G$ which permute the lines $D\,e_i$ in $\bar{V}'$ and which preserve the subspace $\bar{V}''$.

Lemma 4.48 *Recall the integer $c = c_0 d/2$. The affine Weyl group W_G^{aff} of G may be identified with the group of permutations $w : \mathbf{Z}_\alpha \to \mathbf{Z}_\alpha$ such that*

(i) $w(i+d) = w(i) + d, \; i \in \mathbf{Z}_\alpha$.

(ii) There exists $\gamma(w) \in \mathbf{Z}$ such that

$$w(i) + w(-i) = 2c\gamma(w), \; i \in \mathbf{Z}_\alpha$$

$$w(i) = i + c\gamma(w), \; i \in \{\tfrac{1-\alpha}{2}, \ldots, \tfrac{\alpha-1}{2}\} \cup \{\tfrac{\alpha-1}{2}+r+1, \ldots, \tfrac{\alpha-1}{2}+r+\beta\}.$$

Proof: By definition the affine Weyl group W_G^{aff} is the factor group of $N(\mathbf{Q}_p)$ by the maximal compact subgroup of $H(\mathbf{Q}_p)$. Let $g \in N(\mathbf{Q}_p)$. If $e_i \in \bar{V}'$ we define $w(i)$ through the following identity

$$ge_i = \lambda_i e_{w(i)}, \quad \lambda_i \in O_D^\times, \; e_{w(i)} \in \bar{V}'.$$

Let $\gamma(w) \in \mathbf{Z}$ such that

$$\text{ord}\; c(g) = \gamma(w) c_0.$$

Then w is a permutation of

$$\mathbf{Z}_\alpha \setminus ((\{\frac{1-\alpha}{2},\ldots,\frac{\alpha-1}{2}\}\cup\{\frac{\alpha-1}{2}+r+1,\ldots,\frac{\alpha-1}{2}+r+\beta\})+d\mathbf{Z})$$

which satisfies (i) and the first half of (ii). We extend w to the remaining elements of $\mathbf{Z}_\alpha$ by the second half of (ii). It is then easy to see that the map $g \mapsto w$ induces an isomorphism of W_G^{aff} with the group of permutations of $\mathbf{Z}_\alpha$ satisfying (i) and (ii). □

4.49 Let Λ_i^0 denote the O_D-lattice with basis $\{e_{-i},\ldots,e_{-i+d-1}\}$. Let $\mathbf{Z}'_\alpha =$

$$\mathbf{Z}_\alpha\setminus\{i+jd;\ i\in\{\frac{1-\alpha}{2}+1,\ldots,\frac{\alpha-1}{2}\}\cup\{\frac{\alpha-1}{2}+r+2,\ldots,\frac{\alpha-1}{2}+r+\beta\}\}.$$

Then $\mathcal{L}^0 = (\Lambda_i^0)_{i\in\mathbf{Z}'_\alpha}$ is a maximal self-dual chain of lattices. Let $\mathcal{L}$ be a self-dual chain extracted from $\mathcal{L}^0$. Let

$$W_{\mathcal{L}} = W_G^{aff}\cap K_{\mathcal{L}},$$

i.e., $W_{\mathcal{L}}$ is the factor group of $K_{\mathcal{L}}\cap N(\mathbf{Q}_p)$ by the maximal compact subgroup of $H(\mathbf{Q}_p)$.

Lemma 4.50 *There is an equality*

$$K_{\mathcal{L}} = K^0\, W_{\mathcal{L}}\, K^0.$$

Furthermore, there is an identification of sets of double cosets,

$$K^0\setminus G(\mathbf{Q}_p)/K^0 = W_G^{aff}$$

and

$$K_{\mathcal{L}}\setminus G(\mathbf{Q}_p)/K_{\mathcal{L}} = W_{\mathcal{L}}\setminus W_G^{aff}/W_{\mathcal{L}}.$$

More generally, if $\mathcal{L}'$ is another selfdual chain extracted from $\mathcal{L}^0$ with associated subgroups $K'_{\mathcal{L}}$ and $W_{\mathcal{L}'}$,

$$K_{\mathcal{L}}\setminus G(\mathbf{Q}_p)/K_{\mathcal{L}'} \simeq W_{\mathcal{L}}\setminus W_G^{aff}/W_{\mathcal{L}'}.$$

Proof: This is a general fact about reductive groups over local fields, although a comprehensive reference does not seem to be available. If G were semi-simple and simply connected, then $(G(\mathbf{Q}_p), K^0, N(\mathbf{Q}_p))$ would form a Tits system and the statements would follow from corresponding statements valid for general Tits systems (Bourbaki, groupes et algebrès de Lie, chap. IV). Since this is not applicable, the most expedient way seems to verify that $(G(\mathbf{Q}_p), K^0, N(\mathbf{Q}_p))$ is a *generalized Tits system* in the sense of Iwahori [I] and to transpose the proofs of the corresponding statements in Iwahori–Matsumoto [IM], (2.27) and (2.34), compare also [T], (3.31). □

We now return to the proof of theorem (4.46). The lemma (4.50) transposes in an obvious way to the group $\tilde{G}$. The affine Weyl group $W^{aff}_{\tilde{G}}$ of $\tilde{G}$ is identified with the group of permutations $w : \mathbf{Z}_\alpha \to \mathbf{Z}_\alpha$ satisfying (i) in Lemma (4.48). To the chain $\mathcal{L}$ extracted from $\mathcal{L}^0$ we associate the subgroup $\tilde{W}_\mathcal{L} \subset W^{aff}_{\tilde{G}}$ and we have

$$\begin{aligned} \tilde{K}_\mathcal{L} &= \tilde{K}^0 \; \tilde{W}_\mathcal{L} \; \tilde{K}^0 \\ \tilde{K}_\mathcal{L} \setminus \tilde{G}(\mathbf{Q}_p)/\tilde{K}_{\mathcal{L}'} &\simeq \tilde{W}_\mathcal{L} \setminus W^{aff}_{\tilde{G}}/\tilde{W}_{\mathcal{L}'}. \end{aligned}$$

Therefore theorem (4.46) follows from the case $\mathcal{L} = \mathcal{L}'$ of the following proposition.

Proposition 4.51 *The natural map*

$$W_\mathcal{L} \setminus W^{aff}_G/W_{\mathcal{L}'} \longrightarrow \tilde{W}_\mathcal{L} \setminus W^{aff}_{\tilde{G}}/\tilde{W}_{\mathcal{L}'}$$

is injective.

Proof: Let $\mathbf{Z}_\mathcal{L} \subset \mathbf{Z}'_\alpha$ be the corresponding subset, i.e. $i \in \mathbf{Z}_\mathcal{L} \Leftrightarrow \Lambda_i^0 \in \mathcal{L}$. If $i \in \mathbf{Z}_\mathcal{L}$, let i^+ be the smallest element of $\mathbf{Z}_\mathcal{L}$ with $i^+ > i$ and let I_i be the interval $\{i, i+1, \ldots, i^+ - 1\}$. Then the subgroups $W_\mathcal{L}$ of W^{aff}_G resp. $\tilde{W}_\mathcal{L}$ of $W^{aff}_{\tilde{G}}$ are the common stabilizers of the intervals I_i, $i \in \mathbf{Z}_\mathcal{L}$.
Every double coset $\tilde{W}_\mathcal{L} \; w \; \tilde{W}_{\mathcal{L}'}$ contains a unique element $\tilde{w}$ of minimal length (Kostant representative). It may be characterized as follows. Let the chain $\mathcal{L}'$ define the intervals $I'_j = \{j, j+1, \ldots, j^+ - 1\}$, $j \in \mathbf{Z}_{\mathcal{L}'}$. Then $\tilde{w}$ is monotone increasing on each interval $I_i, i \in \mathbf{Z}_\mathcal{L}$, and $\tilde{w}^{-1}$ is monotone increasing on each interval $I'_j, j \in \mathbf{Z}_{\mathcal{L}'}$.
We are going to show that if such a double coset contains a element $w \in W^{aff}_G$, then $\tilde{w} \in W_\mathcal{L} w W_{\mathcal{L}'}$. Let

$$d_{ij} = |I_i \cap w^{-1}(I'_j)|.$$

The selfduality of $\mathcal{L}$ and $\mathcal{L}'$ imply that if $i \in \mathbf{Z}_\mathcal{L}$, $j \in \mathbf{Z}_{\mathcal{L}'}$ then

$$-i^+ + 1 \in \mathbf{Z}_{\mathcal{L}}, \quad -j^+ + 1 \in \mathbf{Z}_{\mathcal{L}'}.$$

Furthermore, condition (ii) in (4.10) on w implies easily that

$$\begin{aligned} -(I_i \cap w^{-1}(I'_j)) &= I_{-i^++1} \cap w^{-1}(I'_{-j^++1+2c\gamma(w)}), \\ d_{-i^++1,j} &= d_{i,-j^++1+2c\gamma(w)}. \end{aligned}$$

We now define permutations u, u' of $\mathbf{Z}_\alpha$ by the following rules.
Let $i \in \mathbf{Z}_{\mathcal{L}}, j \in \mathbf{Z}_{\mathcal{L}'}$, write $I_i \cap w^{-1}(I'_j) = \{k_1, \ldots, k_{d_{ij}}\}$ with $k_1 < k_2 < \ldots < k_{d_{ij}}$. Then

$$\begin{aligned} u(k_l) &= i - 1 + \textstyle\sum_{j'<j} d_{ij'} + l \\ u'^{-1}w(k_l) &= j - 1 + \textstyle\sum_{i'<i} d_{i'j} + l. \end{aligned}$$

Claim: $u \in W_G^{aff}$ *with* $\gamma(u) = 0$, *and similarly for* $u'^{-1}w$.

We indicate the proof for u. The condition (i) of (4.48) is trivial. Let us check the condition (ii), a). By the above relations we have

$$\begin{aligned} u(-k_l) &= (-i^+ + 1) - 1 + \textstyle\sum_{j'<-j^++1+2c\gamma(w)} d_{-i^++1,j'} \quad + d_{ij} - l \\ &= -i^+ + \textstyle\sum_{j'<-j^++1+2c\gamma(w)} d_{i,-j^++1+2c\gamma(w)} \quad + d_{ij} - l. \end{aligned}$$

The inequality in the index of the last sum is equivalent to

$$-j' + 1 + 2c\gamma(w) > j^+.$$

We therefore obtain

$$\begin{aligned} u(-k_l) &= -i^+ + \textstyle\sum_{j'\geq j} d_{ij'} - l \\ &= -(i-1) - \textstyle\sum_{j'<j} d_{ij'} - l \\ &= u(k_l). \end{aligned}$$

This proves condition (ii) a) and condition (ii) b) follows. Indeed, let us for instance show that $u|\{\frac{1-\alpha}{2}, \ldots, \frac{\alpha-1}{2}\}$ is the identity. By condition (ii), b) for w there exists a unique $(i,j) \in \mathbf{Z}_{\mathcal{L}} \times \mathbf{Z}_{\mathcal{L}'}$ such that $\{\frac{1-\alpha}{2}, \ldots, \frac{\alpha-1}{2}\} \subset I_i \cap w^{-1}(I'_j)$. Hence the set $\{\frac{1-\alpha}{2}, \ldots, \frac{\alpha-1}{2}\}$ is mapped under u to an interval, in a monotone increasing manner. Hence $u|\{\frac{1-\alpha}{2}, \ldots, \frac{\alpha-1}{2}\}$ is a translation and therefore, taking into account (ii) a), the identity.
Since u stabilizes each interval I_i, in fact $u \in W_{\mathcal{L}}$. Similarly $u' \in W_{\mathcal{L}'}$.

We now define the permutation v of $\mathbf{Z}'_\alpha$ by

$$v(i-1+\sum_{j'<j} d_{ij'}+k)=j-1+\sum_{i'<i} d_{i'j}+k,\quad k\in\{1,2,\ldots,d_{ij}\}.$$

Then obviously

$$w=u'vu$$

and all factors lie in W_G^{aff}. Furthermore, v is the representative of shortest length since v is monotone on I_i $(i\in\mathbf{Z}_\mathcal{L})$ and v^{-1} is monotone on I'_j $(j\in\mathbf{Z}'_\mathcal{L})$. The proposition follows.

Remark 4.52 (i) Given a self-dual lattice chain $\mathcal{L}$ there exists a unique chain extracted from $\mathcal{L}^0$ and conjugate to $\mathcal{L}$ by an element $g\in G(\mathbf{Q}_p)$ with $c(g)=1$. However, two self-dual chains extracted from $\mathcal{L}^0$ may be conjugate under $G(\mathbf{Q}_p)$. In fact, this occurs if and only if these two chains are conjugate under $w_0\in W_G^{aff}$ where

$$w_0(i)=i+c\ ,\ \ i\in\mathbf{Z}_\alpha.$$

It is easy to give a criterion for two self-dual lattice chains $\mathcal{L}$ and $\mathcal{L}'$ to be conjugate by an element $g\in G(\mathbf{Q}_p)$ with $c(g)=1$. (An analogous criterion for conjugacy under $G(\mathbf{Q}_p)$ is complicated by the above phenomenon). Namely, we index the chain $\mathcal{L}$ in such a way that

$$\Lambda_0\subseteq\Lambda_0^*=\Lambda_a$$

and that there is no member of $\mathcal{L}$ strictly between these two lattices. The integer a is 0 or 1 and is an invariant of $\mathcal{L}$. We furthermore have as before (cf. (4.10)) as invariants the period r and the integers $m_i=\mathrm{length}_{\mathbf{Z}_p}(\Lambda_i/\Lambda_{i-1})$. Then the self-dual chains $\mathcal{L}$ and $\mathcal{L}'$ are conjugate by $g\in G(\mathbf{Q}_p)$ with $c(g)=1$ if and only if $a=a', r=r'$ and $m_i=m'_i,\ \ i\in\mathbf{Z}$.
On the other hand it is easy to give examples of self-dual lattice chains which are conjugate under $\tilde{G}(\mathbf{Q}_p)$ but not under $G(\mathbf{Q}_p)$.
(ii) The analogue of theorem (4.46) where instead of the *fix groups* $K_\mathcal{L}$ resp. $\tilde{K}_\mathcal{L}$ we take the *stabilizer groups*, e.g.

$$\{g\in G(\mathbf{Q}_p);\ \ \forall\Lambda\in\mathcal{L}\ \ \exists\Lambda'\in\mathcal{L}:g\Lambda=\Lambda'\}$$

is *false*.

4.53 We now prove theorem (4.46) in the special case where F_0 is a field and where F is the direct sum of two copies of F_0 which are interchanged by the involution $*$. The decomposition of F induces a decomposition of B and V,

$$B = B^1 \times B^2, \quad V = V^1 \times V^2$$

and an identification $V^2 = V^{1*}$, $B^2 = B^{1opp}$ and finally an identification of the natural embedding i with the inclusion

$$\{(g_1, g_2) \in GL_{B^1}(V^1) \times GL_{B^1}(V^1); \quad g_2 = cg_1, \quad c \in \mathbf{Q}_p^\times\} \subset GL_{B^1}(V^1) \times GL_{B^1}(V^1).$$

The claim of the theorem is in this case a trivial exercise which is left to the reader.

4.54 We now prove theorem (4.46) in general. We may decompose our data of type (PEL) into a product,

$$(F, B, O_B, *, V, (\ ,\ )) = \prod_{i=1}^{m} (F_i, B_i, O_{B_i}, *_i, V_i, (\ ,\ )_i),$$

where each factor is of one of the two types already treated, i.e. either F_i is a field or is the direct sum of two fields which are permuted under the involution $*_i$. Let G_i resp. $\tilde{G}_i = GL_{B_i}(V_i)$ be the groups corresponding to the i–th factor. We obtain a commutative diagram of natural inclusions,

$$\begin{array}{ccc} G & \subset & \prod_{i=1}^{m} G_i \\ \cap & & \cap \\ \tilde{G} & = & \prod_{i=1}^{m} \tilde{G}_i \end{array} .$$

For each $i = 1, \ldots, m$, there is a unique multichain of O_{B_i}–lattices $\mathcal{L}_i$ in V_i such that $\mathcal{L}$ consists of the O_B–lattices for which the projection into V_i lies in $\mathcal{L}_i$. (If F_i is a field, $\mathcal{L}_i$ is a chain of lattices). Let $K_{\mathcal{L}_i} \subset G_i(\mathbf{Q}_p)$ resp. $\tilde{K}_{\mathcal{L}_i} \subset \tilde{G}_i(\mathbf{Q}_p)$ be the corresponding fix groups. We have $\tilde{K}_{\mathcal{L}} = \prod \tilde{K}_{\mathcal{L}_i}$. We put

$$G' = \prod G_i, \quad K'_{\mathcal{L}} = \prod K_{\mathcal{L}_i}.$$

The above commutative diagram induces a map on sets of double cosets,

$$\begin{array}{ccc} K_{\mathcal{L}} \setminus G(\mathbf{Q}_p)/K_{\mathcal{L}} & \xrightarrow{j} & K'_{\mathcal{L}} \setminus G'(\mathbf{Q}_p)/K'_{\mathcal{L}} \\ {\scriptstyle i} \searrow & & \swarrow {\scriptstyle i'} \\ & \tilde{K}_{\mathcal{L}} \setminus \tilde{G}(\mathbf{Q}_p)/\tilde{K}_{\mathcal{L}} & \end{array}$$

The map i' is the product of the corresponding maps for the individual factors, for $i = 1, \ldots, m$. By what has already been proved it is therefore injective. Theorem (4.46) therefore follows from the following statement.

Proposition 4.55 *The map j above is injective.*

Proof: We introduce the kernels of the multiplicator homomorphism c in the various groups,

$$G^{(o)} \subset G, \quad G_i^{(o)} \subset G_i, \quad G'^{(o)} \subset G'.$$

Then

$$G^{(o)} = G'^{(o)} = \prod G_i^{(o)}.$$

We make a compatible choice of a maximal $\mathbf{Q}_p$–split torus in all these groups adapted to the maximal chain $\mathcal{L}^\circ$. Let $W^{aff}_{G^{(o)}}$, W^{aff}_G, $W^{aff}_{G'}$ etc. be the corresponding affine Weyl groups. The multichain $\mathcal{L}$ defines corresponding subgroups $W^{aff}_{G^{(o)}} \cap K_{\mathcal{L}}, W^{aff}_G \cap K_{\mathcal{L}}, W^{aff}_{G'} \cap K'_{\mathcal{L}}$, etc. Now the map j may be identified with the map on double cosets of affine Weyl groups induced by the inclusion $W^{aff}_G \subset W^{aff}_{G'}$,

$$W^{aff}_G \cap K_{\mathcal{L}} \setminus W^{aff}_G / W^{aff}_G \cap K_{\mathcal{L}} \longrightarrow W^{aff}_{G'} \cap K'_{\mathcal{L}} \setminus W^{aff}_{G'} / W^{aff}_{G'} \cap K'_{\mathcal{L}}.$$

The assertion therefore follows from the following statement.

Claim:

$$\begin{aligned} W^{aff}_{G'} \cap K'_{\mathcal{L}} &= W^{aff}_{G^{(o)}} \cap K_{\mathcal{L}}, \quad \textit{or equivalently} \\ W^{aff}_{G_i} \cap K_{\mathcal{L}_i} &= W^{aff}_{G_i^{(o)}} \cap K_{\mathcal{L}_i}, \quad i = 1, \ldots, m. \end{aligned}$$

It suffices to prove the last statement factor by factor. The case that F_i is the direct sum of two fields which are permuted by the involution is easy and left to the reader.

We now consider the case where F_i is a field. We drop the index from the notations and place ourselves in the situation considered in (4.47) – (4.52). In the notation of (4.48) the subgroup $W^{aff}_{G(0)}$ of W^{aff}_G is defined by $\gamma(w) = 0$ so that all that has to be shown is that

$$w \in W^{aff}_G \cap K_{\mathcal{L}} \Longrightarrow \gamma(w) = 0.$$

However, any $w \in W^{aff}_G \cap K_{\mathcal{L}}$ stabilizes each interval $I_i, i \in \mathbf{Z}_{\mathcal{L}}$. On the other hand, condition (ii) in (4.48) implies

$$\begin{aligned} w(-I_i) &= -I_i + 2c\gamma(w), \quad \textit{i.e.} \\ w(I_{-i^{+}+1}) &= I_{-i^{+}+1} + 2c\gamma(w), \end{aligned}$$

hence $\gamma(w) = 0$ as required.

Corollary 4.56 *Let $\mathcal{L}$ and $\mathcal{L}'$ be indexed self-dual multichains of lattices in V which are conjugate under $G(\mathbf{Q}_p)$. Let $g \in G(\mathbf{Q}_p)$. Then*

$$\begin{aligned} \mathrm{pos}(\mathcal{L}, \mathcal{L}') &= g \;\; (cf.(4.45)) \;\; \textit{if and only if} \\ t(\mathcal{L}, \mathcal{L}') &= t(g) \;\; (cf.(4.41)). \end{aligned}$$

Proof: This is just the conjunction of (4.46) and (4.16), as generalized to the composite case in (4.41). □

The preceding considerations justify the following definition relative to the case (PEL) which is the analogue of definition (4.43) for the case (EL). Let $\breve{\mathcal{M}} = \breve{\mathcal{M}}_{\mathcal{L}}$ be the formal scheme over $Spf\, O_{\breve{E}}$ representing the moduli problem (3.21) relative to the data $(F, B, O_B, V, (\ ,\ ), b, \mu, \mathcal{L})$ of type (PEL).

Definition 4.57 *Let t be as in (4.41). The formal Hecke correspondence associated to t is the functor on $Nilp_{O_{\breve{E}}}$ which to S associates the isomorphism classes of objects*

$$(X_{\mathcal{L}}, \varrho),\ (X'_{\mathcal{L}}, \varrho') \in \breve{\mathcal{M}}(S) \times \breve{\mathcal{M}}(S)$$

such that the resulting quasi-isogeny $\varrho'\varrho^{-1}$ from $X_{\mathcal{L}}$ to $X'_{\mathcal{L}}$ is of type t (cf.(4.41)). This functor is representable by a locally closed formal subscheme

$$Corr(t) \subset \breve{\mathcal{M}} \times_{Spf\, O_{\breve{E}}} \breve{\mathcal{M}}.$$

5. The period morphism and the rigid–analytic coverings

In this chapter we first explain the Berthelot–Raynaud functor which associates to one of the formal schemes $\breve{\mathcal{M}}$ encountered in the earlier chapters a smooth rigid–analytic space $\breve{\mathcal{M}}^{rig}$. We then construct the period morphism, a rigid–analytic morphism from $\breve{\mathcal{M}}^{rig}$ to one of the p–adic period domains introduced in chapter 1 and investigate its properties. Finally we construct the tower of rigid–analytic coverings of $\breve{\mathcal{M}}^{rig}$.

5.1 In the beginning of this chapter we will change our notation. We shall denote by (F, O, κ, π) a complete discrete valuation ring. Our aim is to describe Berthelot's functor which associates to a formal scheme $\mathcal{X}$ formally locally of finite type over $Spf\, O$ (cf. (2.3)) a rigid–analytic space. Our reference is chapter 0 of Berthelot's projected Astérisque volume [Ber], comp. also the appendix to [dJ2]. Let us start with the case of a π–adic formal scheme $\mathcal{X}$, i.e. $\mathcal{X}$ is locally of finite type over $Spf\, O$ (cf. (2.2)) (the topology on $\mathcal{O}_{\mathcal{X}}$ is the π–adic topology). In this case the construction is due to Raynaud [Ra1]. To describe this construction we start with the affine case. If $\mathcal{X} = Spf\, A$ is an affine π–adic formal scheme of finite type, then $A \otimes F$ is a Tate algebra and we put $\mathcal{X}^{rig} = Spm(A \otimes F)$. In general, define the set $\mathcal{X}^{rig}$ to be the set of closed formal subschemes $\mathcal{Z}$ which are irreducible and reduced and finite and flat over O. This definition coincides with the previous one in case $\mathcal{X}$ is affine (associate to $x \in Spm(A \otimes F)$ the formal spectrum of the image of A in the residue field of x). The support of such a subscheme $\mathcal{Z}$ is a closed point of $\mathcal{X}$, called the *specialization* of the point $x \in \mathcal{X}^{rig}$ corresponding to $\mathcal{Z}$. This construction defines a map

$$sp : \mathcal{X}^{rig} \longrightarrow \mathcal{X}.$$

For any affine open $\mathcal{U} = Spf(A) \subset \mathcal{X}$, $sp^{-1}(\mathcal{U})$ can be identified with $Spm(A \otimes F)$.

Examples 5.2 (i) Let $\mathcal{X} = Spf\, O\{T_1, \ldots, T_n\}$ (restricted power series ring). Then $\mathcal{X}^{rig}$ is the closed unit ball. A point x of $\mathcal{X}^{rig}$ is defined by $(\xi_1, \ldots, \xi_n) \in O(x)^n$ where $O(x)$ is a discrete valuation ring finite over O. Then $sp(x)$ is the point $(\bar{\xi}_1, \ldots, \bar{\xi}_n)$ in the affine space $\mathbf{A}^n_\kappa$, where $\bar{\xi}_i$ are the residue classes of ξ_i.

(ii) Let $\mathcal{X} = \hat{\mathbf{P}}^n_O$ be the formal projective space. Then a point x of $\mathcal{X}^{rig}$ can be represented in homogeneous coordinates $(\xi_0, \ldots, \xi_n) \in O(x)^{n+1}$ where at least one ξ_i is a unit. The point $sp(x)$ is the point of $\mathbf{P}^n_\kappa$ with homogeneous coordinates $(\bar{\xi}_0, \ldots, \bar{\xi}_n)$.

Proposition 5.3 [Ber, 0.2.3.]: *Let $\mathcal{X}$ be a π-adic formal scheme locally of finite type.*

(i) There exists on $\mathcal{X}^{rig}$ a unique structure of a rigid-analytic space over F with the following properties.

(a) The inverse image under $sp : \mathcal{X}^{rig} \longrightarrow \mathcal{X}$ of an open subscheme (resp. of an open covering) of $\mathcal{X}$ is an admissible open subset (resp. an admissible covering) of $\mathcal{X}^{rig}$.

(b) For any affine open subscheme $\mathcal{U} = Spf\, A \subset \mathcal{X}$ the structure on $\mathcal{U}^{rig} = sp^{-1}(\mathcal{U})$ induced from $\mathcal{X}^{rig}$ coincides with the one on $Spm(A \otimes F)$.

(ii) The map sp defines a morphism of ringed sites $\mathcal{X}^{rig} \longrightarrow \mathcal{X}$ with $sp_*(\mathcal{O}^{rig}_{\mathcal{X}}) = \mathcal{O}_{\mathcal{X}} \otimes F$. This morphism has the following universal property. Let $\mathcal{Y}$ be any rigid-analytic space and let $u : \mathcal{Y} \to \mathcal{X}$ be a morphism of ringed sites. Then u factors in a unique way through sp.

(iii) The functor $\mathcal{X} \mapsto \mathcal{X}^{rig}$ has the following properties.

(a) If $\mathcal{X}$ is of finite type, then $\mathcal{X}^{rig}$ is quasi-compact.

(b) It commutes with products and transforms open resp. closed immersions into open resp. closed immersions.

Examples 5.4 (i) Let $f \in O\{T_1, \dots, T_n\}$ and let $\mathcal{U} = D(f)$. Then $\mathcal{U}^{rig}$ is the open subspace of the closed unit ball defined by $|f(x)| = 1$.

(ii) Let $\hat{\Omega}^d_F$ be Deligne's formal scheme. (cf. [Dr2], comp. (3.61)) Then $(\hat{\Omega}^d_F)^{rig}$ is the complement of the union of all F-rational hyperplanes in $\mathbf{P}^{d-1}_F$. Here $\mathbf{P}^{d-1}_F$ is the rigid space defined either by completing projective space over O along its special fibre or by applying the GAGA functor to projective space over F. This identification is obtained as follows. A point of $(\hat{\Omega}^d)^{rig}$ with values in a finite field extension K of F is given by a diagram (3.12) over $Spf\, O_K$. For a given index i_k we tensor φ_{i_k} with F and obtain a morphism $\varphi^{rig} : F^d \to L$, where L is a K–vector space of dimension 1. This morphism is easily checked to be injective and independent of the choice of i_k. Hence φ^{rig} is a point of $\mathbf{P}^{d-1}_F$ in the complement of all rational hyperplanes. Conversely assume we are given an F–linear injection $\varphi^{rig} : F^d \to L$ into a K–vector space L of dimension 1. Consider the chain $\{\Lambda\}$ of all O_K–lattices in L. Then $\eta_\Lambda = \Lambda \cap F^d$ form a chain of O_F–lattices in F^d. The morphisms $\varphi_\Lambda : \eta_\Lambda \to \Lambda$ provide a section of $\hat{\Omega}^d_F$ over $Spf\, O_K$.

5.5 We now consider the general case of a formal scheme $\mathcal{X}$, formally locally of finite type over $Spf\, O$. As before we start with the affine case.
Let $\mathcal{X} = Spf\, A$ and let $f_1, \dots, f_r$ be a system of generators of a defining ideal. For each n put

$$B_n = A\{T_1, \dots, T_r\}/(f_1^n - \pi T_1, \dots, f_r^n - \pi T_r)$$

where $A\{T_1, \dots, T_r\}$ is the π–adic completion of $A[T_1, \dots, T_r]$. The hypothesis implies that B_n is topologically of finite type over O, hence $B_n \otimes F$ is a Tate algebra. For $n' \geq n$ we have a homomorphism

$$B_{n'} \longrightarrow B_n : T_i' \mapsto f_i^{n'-n} \cdot T_i.$$

The corresponding morphism

$$Spm(B_n \otimes F) \longrightarrow Spm(B_{n'} \otimes F)$$

identifies $Spm(B_n \otimes F)$ with the special domain defined by $|f_i(x)| \leq |\pi|^{1/n}$. The rigid space $\mathcal{X}^{rig}$ is then defined as the union of $Spm(B_n \otimes F)$, with the

$\mathit{Spm}(B_n \otimes F)$ as an admissible open covering. One shows easily (cf. [Ber]) that this definition is independent of the choice of the defining ideal, and of the set of generators, and that this definition coincides with the usual one in case where $\mathcal{X}$ is π-adic. The morphisms

$$\mathit{Spm}(B_n \otimes F) \xrightarrow{sp} \mathit{Spf}(B_n) \longrightarrow \mathcal{X}$$

define by passage to the limit the morphism of ringed sites

$$\mathcal{X}^{rig} \to \mathcal{X}.$$

Let $x \in \mathcal{X}^{rig}$ be represented by a maximal ideal $x_n \in \mathit{Spm}(B_n \otimes F)$. It is easy to see that the image R of A in the residue field of x_n is independent of this representative and is an integral domain which is a finite and flat O-algebra. Conversely, if R is a factor algebra of A with these properties it is a local ring and it is easily seen that for large n the images of the elements f_i^n/π lie in its maximal ideal. It follows that R arises in the way described above from a point in $\mathit{Spm}(B_n \otimes F)$.
Now consider the general case. By the preceding remarks we may define as before $\mathcal{X}^{rig}$ to be the set of closed subschemes $\mathcal{Z}$ of $\mathcal{X}$ which are integral and finite and flat over O. The support of such a subscheme $\mathcal{Z}$ is a closed point of $\mathcal{X}$, the specialization of $\mathcal{Z}$. We obtain a map

$$sp : \mathcal{X}^{rig} \longrightarrow \mathcal{X}.$$

For any affine open $\mathcal{U} \subset \mathcal{X}$, $sp^{-1}(\mathcal{U})$ is in bijective correspondence with $\mathcal{U}^{rig}$. *With these remarks, proposition (5.3) carries over* ([Ber, 0.2.6.]). (In ((i), b) the given structure on $\mathcal{U}^{rig}$ is of course supposed to be the one defined above; unless A is π-adic, $Spm(A \otimes F)$ makes no sense).
We call $\mathcal{X}^{rig}$ the *generic fibre* of the formal scheme $\mathcal{X}$ over $Spf\, O$.

Example 5.6 Let $\mathcal{X} = \mathit{Spf}O[[T_1, \ldots, T_n]]$ with ideal of definition $(\pi, T_1, \ldots, T_n)$. Then $\mathcal{X}^{rig}$ is the open unit ball, regarded as the increasing union of closed balls of radius $|\pi|^{1/n}$.

Proposition 5.7 [Ber, 0.2.7] *Let $\mathcal{X}$ be a formal scheme locally of finite type over O. Let I be an ideal of definition and $X_0 = V(I)$. Let $Z \subset X_0$ be a closed subscheme and let $\hat{\mathcal{X}}$ be the completion of $\mathcal{X}$ along Z. Then $sp^{-1}(Z)$ is an open subspace of $\mathcal{X}^{rig}$ and the canonical morphism of rigid* *spaces $\hat{\mathcal{X}}^{rig} \to \mathcal{X}^{rig}$ arising from the canonical morphism $\hat{\mathcal{X}} \to \mathcal{X}$ induces an isomorphism*

$$\hat{\mathcal{X}}^{rig} \xrightarrow{\sim} sp^{-1}(\mathcal{Z}).$$

Example 5.8 Let $\mathcal{X}$ be a π–adic formal scheme and let $f_1, \ldots, f_r$ be elements of $\Gamma(X, O_{\mathcal{X}})$ with reductions $\bar{f}_1, \ldots, \bar{f}_r$ *mod* π such that the closed set $Z = V(\bar{f}_1, \ldots, \bar{f}_r)$. Then $sp^{-1}(\mathcal{Z}) = \{x \in \mathcal{X}^{rig};\ |f_i(x)| < 1,\ i = 1, \ldots, r\}$. This allows in general a more direct description of the generic fibre $\hat{\mathcal{X}}^{rig}$, comp. e.g. the example in (5.6).

5.9 A morphism of rigid spaces $f : Y \to X$ is *smooth (resp. étale)* if there exist admissible affinoid coverings $(Y_i)_i$ and $(X_i)_i$ of Y and X such that

(i) $f(Y_i) \subset X_i$

(ii) if $A_i = \Gamma(X_i, \mathcal{O}_X)$, $B_i = \Gamma(Y_i, \mathcal{O}_Y)$, there exists an isomorphism

$$B_i = A_i\{T_1, \ldots, T_n\}/(f_1, \ldots, f_r)$$

with $det(\partial f_k/\partial T_\ell)_{k,\ell=1,\ldots,r}$ invertible in B_i (resp. and $r = n$). We shall need the following analogue of Grothendieck's infinitesimal criterion for a morphism to be étale.

Proposition 5.10 [Ro, 3.1.]: *The morphism of rigid spaces $f : X \to Y$ is étale if and only if the following condition is satisfied. Let Z be rigid–analytic space with only one point and let $Z_0 \subset Z$ be a closed subspace. Then any commutative diagram of morphisms below with solid arrows can be completed in a unique way by a dotted arrow into a commutative diagram.*

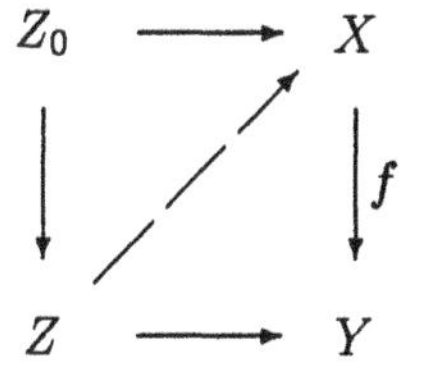

Proof (reduction to the statement of loc. cit.): Both conditions are local on X and Y, hence we may assume that X and Y are affinoid which is the general setting of loc.cit. In loc.cit. the rigid–analytic space Z is any affinoid space and Z_0 is definied by a nilpotent ideal; however a glance at the proof shows that it suffices to consider only spaces appearing in the statement of the proposition (which are, of course, affinoid).

Remark 5.11 The Berthelot functor may be viewed in a natural way from the standpoint of Huber's *adic spaces*, cf. [Hu]. The category of adic spaces over the adic space $Spec(F)^a$ associated to $Spec\, F$ contains as a full subcategory the category of rigid spaces and there is a functor which associates to a locally noetherian formal scheme an adic space. The image under the Berthelot functor is the generic fibre of this associated adic space. We mention without proof the following fact which clarifies the difference between the Raynaud functor and its extension of Berthelot.

Proposition 5.12 (R. Huber) *Let $\mathcal{X}$ be a formal scheme formally locally of finite type and flat over Spf O. Then $\mathcal{X}^{rig}$ is quasicompact if and only if $\mathcal{X}$ is of finite type over Spf O, in particular $\mathcal{X}$ is a noetherian π–adic formal scheme.*

5.13 We now revert to the notations used elsewhere in this paper. Let us start with the data $(B, F, O_B, V, b, \mu, \mathcal{L})$ in the case (EL) relative to $L = \bar{\mathbf{F}}_p$. In the case (PEL) we have in addition a non–degenerate alternating $\mathbf{Q}_p$–pairing $(\ ,\ )$ on V. We denote by $\breve{\mathcal{M}}$ the solution of our moduli problem (3.21). This is a formal scheme, formally locally of finite type over $SpfO_{\breve{E}}$. Here $\breve{E}$ denotes the completion of the maximal unramified extension of the Shimura field E. We denote by $(X_{\mathcal{L}}, \varrho)$ the universal object over $\breve{\mathcal{M}}$. We denote as usual by M_Λ the Lie algebra of the universal extension of X_Λ $(\Lambda \in \mathcal{L})$. The isogenies $\tilde{\varrho}_{\Lambda',\Lambda} : X_\Lambda \to X_{\Lambda'}$ $(\Lambda \subset \Lambda')$ (cf. (3.21)) induce morphisms of coherent locally free $O_{\breve{\mathcal{M}}}$–modules $M_\Lambda \to M_{\Lambda'}$ which induce isomorphisms between the corresponding modules over the structure sheaf of the rigid space $\breve{\mathcal{M}}^{rig}$ associated to $\breve{\mathcal{M}}$. Let us denote the common value of this $O_{\breve{\mathcal{M}}^{rig}}$–module by M_X^{rig}. We are going to exhibit a canonical isomorphism

$$M_X^{rig} = V \otimes O_{\breve{\mathcal{M}}^{rig}} \tag{5.1}$$

5.14 The construction of this isomorphism is best done in a somewhat more general context. Let (F, O, κ, π) be a complete discrete valuation ring of unequal characteristic, with *perfect* residue field of characteristic p. Let

$\mathcal{M}$ be a formal scheme formally locally of finite type over $Spf\, O$. Let X be a p–divisible group over $\mathcal{M}$. We denote by M_X the Lie algebra of the universal extension of X. Its formation is functorial in X and commutes with base change. Let $\mathbf{X}$ be a p–divisible group over κ. We assume given a quasi–isogeny

$$\varrho : \mathbf{X}_{\mathcal{M}_0} \longrightarrow X_{\mathcal{M}_0}.$$

Here $\mathcal{M}_0$ denotes the κ–scheme defined by an ideal of definition of $\mathcal{M}$ containing the uniformizer π. By the rigidity of quasi–isogenies this datum is independent of the choice of such an ideal of definition.

Proposition 5.15 *The quasi–isogeny ϱ induces a canonical and functorial isomorphism of locally free $O_{\breve{\mathcal{M}}^{rig}}$–modules of finite rank, compatible with base change,*

$$\tilde{\varrho} : N(\mathbf{X}) \otimes_{W(\kappa)_{\mathbb{Q}}} \mathcal{O}_{\breve{\mathcal{M}}^{rig}} \xrightarrow{\sim} M_X^{rig}.$$

Here $N(\mathbf{X})$ denotes the isocrystal associated to the p–divisible group $\mathbf{X}$.

Proof: We first treat the case when $\mathcal{M}$ is a π–adic formal scheme, in which case we may assume that $\mathcal{M}_0$ is defined by the image of π. Let $\mathcal{M}'_0$ be defined by the image of p. Then $\mathcal{M}_0 \subset \mathcal{M}'_0$ is a nilpotent immersion. Since O/pO is a κ–algebra we may consider $\mathcal{M}'_0$ as a κ–scheme. By the rigidity of quasi–isogenies the quasi–isogeny ϱ extends in a unique way into a quasi–isogeny of p–divisible groups over $\mathcal{M}'_0$,

$$\varrho' : \mathbf{X}_{\mathcal{M}'_0} \longrightarrow X_{\mathcal{M}'_0}.$$

Since the closed immersion $\mathcal{M}'_0 \subset \mathcal{M}$ has a canonical divided power structure we may apply the theory of Grothendieck–Messing. Therefore ([Me], IV, 2.2.) if $\tilde{\mathbf{X}}$ is any lifting of $\mathbf{X}$ to O and if $N > 0$ is such that $p^N \varrho'$ is an isogeny, there is an induced homomorphism of locally free $O_{\mathcal{M}}$–modules of finite rank,

$$\tilde{\varrho}_N : M_{\tilde{\mathbf{X}}_{\mathcal{M}}} \longrightarrow M_X.$$

Since there exists a morphism from $X_{\mathcal{M}'_0}$ to $\tilde{\mathbf{X}}_{\mathcal{M}'_0}$ such that the compositions with $p^N \varrho'$ are a power of p it follows that $\tilde{\varrho}_N$ induces an isomorphism between the corresponding $\mathcal{O}_{\mathcal{M}^{rig}}$–modules. Furthermore there is a canonical identification (compatibility with base change)

$$M^{rig}_{\tilde{\mathbf{X}}_{\mathcal{M}}} = N(\mathbf{X}) \otimes_{W(\kappa)_{\mathbf{Q}}} \mathcal{O}_{\mathcal{M}^{rig}}.$$

The desired isomorphism is now defined as

$$\tilde{\varrho} = \frac{1}{p^N} \cdot \tilde{\varrho}_N^{rig} : N(\mathbf{X}) \otimes \mathcal{O}_{\mathcal{M}^{rig}} \longrightarrow M_X^{rig}.$$

It has all the properties stated in theorem (5.15).
(In the previous argument when $p = 2$ the application of the result of Grothendieck–Messing is not justified - one also needs the *nilpotence* of the divided power–structure. We leave the modifications of the argument needed in this case to the reader).
We now consider the general case. We may assume $\mathcal{M}$ affine, $\mathcal{M} = Spf\ A$. We recall (cf. (5.5)) the way in which the associated rigid space $\mathcal{M}^{rig}$ was defined. Using the notations introduced there, $\mathcal{M}^{rig}$ is the union of the open subspaces $Spm(B_n \otimes F)$ and $Spf\ B_n$ comes with a morphism to $\mathcal{M}$. Since $Spf\ B_n$ is a π–adic formal scheme we may apply the first part of the construction to the pull-back of X to $Spf\ B_n$. The desired isomorphism $\tilde{\varrho}$ then arises by passing to the limit over the open subspaces $Spm(B_n \otimes F)$.

5.16 Going back to the set–up of (5.13) we see that the quasi–isogenies ϱ appearing in the moduli data induce a canonical isomorphism

$$N(\mathbf{X}) \otimes_{K_0} \mathcal{O}_{\breve{\mathcal{M}}^{rig}} = M_X^{rig}.$$

On the other hand the isocrystal $N(\mathbf{X})$ is identified with $V \otimes K_0$, cf. (3.19), which establishes the desired isomorphism (5.1). This isomorphism is compatible with the actions of B and, in case (PEL), preserves the alternating forms up to a scalar in $\mathbf{Q}_p^\times$.
The surjective homomorphisms $M_\Lambda \to Lie(X_\Lambda)$ induce surjective homomorphisms $M_\Lambda^{rig} \to Lie(X_\Lambda)^{rig}$ which are independent of Λ and will be denoted by

$$M_X^{rig} \longrightarrow Lie(X)^{rig}. \tag{5.2}$$

Recall from chapter 1 the Grassmann variety $\mathcal{F}$ over E parametrizing the B-invariant totally isotropic subspaces of V in the isomorphism class of V_1. The kernel of (5.2) is a point with values in $\breve{\mathcal{M}}^{rig}$ of $\mathcal{F}$, i.e. defines a rigid–analytic morphism $\breve{\pi}^1 : \breve{\mathcal{M}}^{rig} \longrightarrow \mathcal{F} \otimes_E \breve{E}$. We abbreviate $\mathcal{F} \otimes_E \breve{E}$ into $\breve{\mathcal{F}}$. On the other hand recall from (3.52) the morphism $\varkappa$ from $\breve{\mathcal{M}}$ to Δ. We denote by $\breve{\pi}^2$ the resulting morphism of rigid–analytic spaces over $\breve{E}$, $\breve{\pi}^2 : \breve{\mathcal{M}}^{rig} \to \Delta$. Here Δ denotes the discrete rigid–analytic space with Δ as its underlying point set. The product morphism

$$\breve{\pi} = \breve{\pi}^1 \times \breve{\pi}^2 : \breve{\mathcal{M}}^{rig} \longrightarrow \breve{\mathcal{F}}^{rig} \times \Delta$$

is called the *period morphism of the moduli problem.* Recall from chapter 1 the algebraic group J over $\mathbf{Q}_p$ associated to the group G and $b \in G(K_0)$. The group $J(\mathbf{Q}_p)$ acts via

$$h \cdot (X_{\mathcal{L}}, \varrho) = (X_{\mathcal{L}}, \varrho \cdot h^{-1}), \;\; h \in J(\mathbf{Q}_p)$$

on the moduli problem and hence on $\breve{\mathcal{M}}$ (cf. (3.22)). The group $J(\mathbf{Q}_p)$ also acts on $\breve{\mathcal{F}}$, (cf. (1.35)), as well as by translations on Δ, cf. (3.52). We let $J(\mathbf{Q})$ act diagonally on $\breve{\mathcal{F}} \times \Delta$. It follows immediately from the construction above that *$\breve{\pi}$ is $J(\mathbf{Q}_p)$–equivariant.*

Proposition 5.17 *The period morphism $\breve{\pi}$, or equivalently its first component $\breve{\pi}^1$, is étale. In particular $\breve{\mathcal{M}}^{rig}$ is a smooth rigid-analytic space.*

Proof: We are going to use the infinitesimal criterion (5.10). Let Z be as in the statement of this criterion. Then Z is of the form $Z = Spm(R \otimes \breve{E})$, where $R \otimes \breve{E}$ is the Tate algebra associated to a *finite flat* $O_{\breve{E}}$–algebra R such that, denoting by $\mathbf{n} \subset R$ the nilradical, $R/\mathbf{n}$ is a complete discrete valuation ring. Consider the set of R–algebras R' which are finite flat $O_{\breve{E}}$–algebras with $R/\mathbf{n} \simeq R'/\mathbf{n} \cdot R'$ and with $R \otimes_{O_{\breve{E}}} \breve{E} \tilde{\to} R' \otimes_{O_{\breve{E}}} \breve{E}$. Then these form in an obvious way a directed set $\mathcal{S}$ under the inclusion relation. We fix a free $R/\mathbf{n}$-module M_0 of finite rank. There is an obvious functor ("associated rigid module"):

$$\text{"}\varinjlim_{R' \in \mathcal{S}}\text{"} \begin{pmatrix} \text{locally free } R'\text{–modules} \\ M' \text{ of finite rank, together} \\ \text{with an isomorphism} \\ M' \otimes_{R'} R/\mathbf{n} = M_0 \end{pmatrix} \to \begin{pmatrix} \text{locally free } \mathcal{O}_Z\text{–modules} \\ M \text{ of finite rank, together} \\ \text{with an isomorphism} \\ M \otimes_{\mathcal{O}_Z} \mathcal{O}_{Z_0} = M_0^{rig} \end{pmatrix}$$

This functor is exact and an equivalence of categories, compatible with base change $R \to R_1$ where R_1 is another finite flat O–algebra as above.

We now turn to the verification of the *existence* of the dotted arrow in the diagram of (5.10). Replacing R by a larger $R' \in \mathcal{S}$, we may assume that the morphism $Z_0 \to \breve{\mathcal{M}}^{rig}$ is induced by a morphism of formal schemes

$$Spf\ R/\mathbf{a} \longrightarrow \breve{\mathcal{M}},$$

where $\mathbf{a} \subset \mathbf{n}$ is a nilpotent ideal. Proceding inductively we may assume $\mathbf{a}^2 = (0)$. By pullback we obtain an object $(X_{\mathcal{L},0}, \varrho_0)$ of the moduli problem (3.21) over $Spf\ R/\mathbf{a}$ (i.e. over all its truncations). The morphism $Z \to \breve{\mathcal{F}}$ making the solid diagram commutative defines a locally free factor module

$$N(\mathbf{X}) \otimes \mathcal{O}_Z \longrightarrow \mathcal{L}'. \tag{5.3}$$

We equip $\mathbf{a}$ with trivial divided powers. Let M_Λ be the value of the crystal associated to $X_{\Lambda,0}$ on $Spf\ R$. Then M_Λ ($\Lambda \in \mathcal{L}$) is a (polarized) chain and there is a compatible family of natural identifications,

$$N(\mathbf{X}) \otimes \mathcal{O}_Z = M_\Lambda^{rig}.$$

Furthermore, the homomorphism (5.3) is induced by a unique surjective homomorphism onto a locally free module

$$M_\Lambda \longrightarrow \mathcal{L}_\Lambda,$$

at least after replacing R by a larger $R' \in \mathcal{S}$. The transition homomorphisms $M_\Lambda \to M'_\Lambda$ induce homomorphisms $\mathcal{L}_\Lambda \to \mathcal{L}'_\Lambda$ and tensoring with $\otimes_R R/\mathbf{a}$ gives us back

$$M_{X_{\Lambda,0}} \longrightarrow Lie(X_{\Lambda,0}),\ \ \Lambda \in \mathcal{L}.$$

By Grothendieck–Messing, the system above is induced by a unique chain of p–divisible groups $X_{\mathcal{L}} = (X_\Lambda;\ \Lambda \in \mathcal{L})$ over $Spf\ R$ such that the above homomorphism is equal to

$$M_{X_\Lambda} \longrightarrow Lie\, X_\Lambda$$

and restricting to $X_{\mathcal{L},0}$. The quasi–isogeny ϱ_0 lifts automatically. It is obvious that $(X_{\mathcal{L}}, \varrho)$ is an object of $\breve{\mathcal{M}}(Spf\ R)$. The induced rigid–analytic morphism $Z \to \breve{\mathcal{M}}^{rig}$ renders the diagram commutative. The uniqueness assertion is proved in the same way.

Remark 5.18 We recall (cf. (3.29)) the morphisms of formal schemes over $Spf\, O_{\breve{E}}$

$$\begin{array}{ccc} \breve{\mathcal{N}} & \longrightarrow & \hat{\mathbf{M}}^{loc} \\ \downarrow & & \\ \breve{\mathcal{M}} & & \end{array}$$

The associated rigid–analytic space $\breve{\mathcal{N}}^{rig}$ classifies the trivializations

$$\gamma : M_X^{rig} \xrightarrow{\sim} V \otimes \mathcal{O}_{\breve{\mathcal{M}}^{rig}}$$

of the isocrystal of the universal object $(X_{\mathcal{L}}, \varrho)$ over $\breve{\mathcal{M}}$. The construction of the first component of the period morphism above may therefore be interpreted as defining a canonical section to the projection morphism,

$$s : \breve{\mathcal{M}}^{rig} \longrightarrow \breve{\mathcal{N}}^{rig}.$$

Furthermore, $\mathbf{M}^{loc}$ is a projective scheme over $Spec\; O_E$ whose generic fibre may be identified in an obvious way with $\mathcal{F}$. The horizontal morphism in the diagram above induces therefore a morphism $\breve{\mathcal{N}}^{rig} \to \breve{\mathcal{F}}^{rig}$. The first component of the period morphism $\breve{\pi}^1$ is the composition

$$\breve{\mathcal{M}}^{rig} \xrightarrow{s} \breve{\mathcal{N}}^{rig} \longrightarrow \breve{\mathcal{F}}^{rig}.$$

A variant of the above proof that $\breve{\pi}$ is étale (which uses the reference [Ro]) would be to imitate the proof of proposition (3.33). Namely, if the statements from EGA used in that proof could be transposed to the rigid–analytic context, it would follow that the étaleness of π could be checked on diagrams as in (5.10) for $Z = Spm\; K[\varepsilon]$ and $Z_0 = Spm\; K$, for a finite field extension K of the residue field of a point of X. In this case a cofinal system in the set $\mathcal{S}$ appearing in the proof of (5.17) is given by

$$R_i = O_K\left[\frac{\varepsilon}{\pi^i}\right] = O_K[X]/(\pi^i X)^2, \quad i = 0, 1, 2, \ldots \; .$$

Here π denotes a uniformizing element in O_K.

5.19 We give a different definition of the period morphism for formal groups, which is due to G. Faltings.
Let R be a ring such that p is nilpotent in R. Let $R' \to R$ be a PD-extension with nilpotent kernel. Later we will also consider the case where R' is adic and the kernel is topologically nilpotent. We define a category $\mathrm{Ext}_{R' \to R}$ as follows. The objects are quadruples

$$(\mathcal{V}, E, \varrho, X) .$$

Here X is a p-divisible formal group over R, E is a (smooth) formal group over R', $\mathcal{V} \subset E$ is a vector group and $\varrho : E_R \to X$ is a morphism, such that the following sequence is exact:

$$0 \longrightarrow \mathcal{V}_R \longrightarrow E_R \xrightarrow{\varrho} X \longrightarrow 0.$$

We note that $E/\mathcal{V}$ is a formal group over R', which is a lifting of X.
A morphism $(\mathcal{V}, E, \varrho, X) \to (\mathcal{V}', E', \varrho', X')$ is a morphism $v : E \to E'$ of formal groups, such that the following conditions are satisfied.
The map v_R gives rise to a morphism of exact sequences:

$$\begin{array}{ccccccccc} 0 & \longrightarrow & \mathcal{V}_R & \xrightarrow{\iota_R} & E_R & \longrightarrow & X & \longrightarrow & 0 \\ & & v_0 \downarrow & & \downarrow & & \downarrow & & \\ 0 & \longrightarrow & \mathcal{V}'_R & \xrightarrow{\iota'_R} & E'_R & \longrightarrow & X' & \longrightarrow & 0 \end{array}$$

We require that v_0 is a morphism of vector groups. Moreover we require that there exists a lifting of v_0 to a morphism of vector groups $\tilde{v}_0 : \mathcal{V} \to \mathcal{V}'$, such that

$$\iota' \circ \tilde{v}_0 - v|\mathcal{V} : \mathcal{V} \longrightarrow E'$$

is an exponential in the sense of Messing. Here ι' is the given inclusion $\iota' : \mathcal{V} \to E'$.
The conditions on a morphism are easier to explain in Cartier theory ([Z4], esp. (2.3), (2.4)). Let $\mathbf{E}_{R'}$ be the Cartier ring of R'. We denote the Witt ring by $W(R')$ and by $w_n : W(R') \to R'$ the Witt polynomials. Let L be a R'-module. We define a $\mathbf{E}_{R'}$-module $C(L)$ as follows. As an additive group we set

$$C(L) = \prod_{i=0}^{\infty} L\,.$$

An element of this product with component x_i we write formally as

$$\sum_{i=0}^{\infty} \mathbf{V}^i x_i\,.$$

The $\mathbf{E}_{R'}$-module structure is given by the following equations

$$\begin{aligned} \xi\left(\sum_{i=0}^{\infty} \mathbf{V}^i x_i\right) &= \sum_{i=0}^{\infty} \mathbf{V}^i w_n(\xi) x_i\,, \quad \xi \in W(R') \\ \mathbf{V}\left(\sum_{i=0}^{\infty} \mathbf{V}^i x_i\right) &= \sum_{i=0}^{\infty} \mathbf{V}^{i+1} x_i \\ \mathbf{F}\left(\sum_{i=0}^{\infty} \mathbf{V}^i x_i\right) &= \sum_{i=1}^{\infty} \mathbf{V}^{i-1} p x_i \end{aligned}$$

If L is a finite locally free R'-module, this is the Cartier module of the formal group associated to L.

Let $\mathbf{a}$ be the kernel of $R' \to R$. The PD-structure defines an isomorphism of additive groups

$$\begin{aligned} \log: \quad W(\mathbf{a}) &\xrightarrow{\sim} \textstyle\prod_{i=0}^{\infty} \mathbf{a}\,, \\ \xi &\longmapsto \textstyle\prod w_n(\xi)/p^n \end{aligned}$$

where the divided powers give a sense to the expression $w_n(\xi)/p^n$. The map $\mathbf{a} \to \prod_{i=0}^{\infty} \mathbf{a} \simeq W(\mathbf{a})$ that maps a to $(a, 0 \ldots 0)$ is a ring homomorphism, that maps $\mathbf{a}$ to an ideal in $W(R')$.

An object $(\mathcal{V}, E, \varrho, X)$ gives rise to an exact sequence of Cartier modules

$$0 \longrightarrow C(L) \longrightarrow M_E \longrightarrow M \longrightarrow 0\,,$$

where $L = \mathcal{V} + \mathbf{a}\mathrm{Lie}E \subset \mathrm{Lie}E$, M_E is the Cartier module of E, and M that of X.

One has $\mathbf{a}M_E \subset L \subset C(L)$ where the last inclusion is given by $l \longmapsto (l, 0 \ldots 0)$. Then a morphism $(\mathcal{V}, E, \varrho, X) \xrightarrow{v} (\mathcal{V}', E', \varrho', X')$ is simply a morphism of exact sequences

$$\begin{array}{ccccccccc} 0 & \longrightarrow & C(L) & \longrightarrow & M_E & \longrightarrow & M & \longrightarrow & 0 \\ & & v_0 \downarrow & & \downarrow & & \downarrow & & \\ 0 & \longrightarrow & C(L') & \longrightarrow & M_{E'} & \longrightarrow & M' & \longrightarrow & 0 \end{array}$$

such that v_0 is induced by an R'-module homomorphism $L \to L'$. The category $\mathrm{Ext}_{R' \to R}$ lies over the category of p-divisible formal groups on R. By a theorem of Messing the fiber $\mathrm{Ext}_{R' \to R}(X)$ at X has an initial object $E(X)^{\mathrm{univ}}$, i.e. Hom $(E(X)^{\mathrm{univ}}, (\mathcal{V}, E, \varrho))$ consists of a single element.

5.20 Let us consider the following situation. Let k be a perfect field of characteristic p. Let $W = W(k)$ be its Witt vectors. Let $K_0 = W_{\mathbf{Q}}$ be the field of fractions. Let K be a finite field extension of K_0 and $O = O_K$ be its rings of integers. We set $R = O/pO$. Let X be a p-divisible formal group over R and $X_0 = X \otimes_R k$ be its reduction. We denote by $M = M(X)$ respectively $M_0 = M(X_0)$ the Cartier modules of X respectively X_0.
Let $\mathcal{C}_{X_0,K}$ be the category of morphisms of K-vectorspaces

$$M_0 \otimes_W K \longrightarrow L.$$

We consider functors for varying O and X

$$\mathrm{Ext}_{O \to R}(X) \longrightarrow \mathcal{C}_{X_0,K}$$

of the form $E \longmapsto [M_0 \otimes_W K \longrightarrow \mathrm{Lie}\, E \otimes_O K]$. We require that for any extension K'/K there is a commutative diagram

$$\begin{array}{ccc} \mathrm{Ext}_{O \to R}(X) & \longrightarrow & \mathcal{C}_{X_0,K} \\ \text{base change} \downarrow & & \downarrow \otimes_K K' \\ \mathrm{Ext}_{O' \to R'}(X_{R'}) & \longrightarrow & \mathcal{C}_{X_0,K'} \end{array} \quad (5.4)$$

We require that a morphism $(\mathcal{V}, E, \varrho, X) \to (\mathcal{V}', E', \varrho', X')$ gives rise to a commutative diagram

$$\begin{array}{ccc} M_0(X) \otimes_W K & \longrightarrow & \mathrm{Lie}\, E_X \otimes_O K \\ \downarrow & & \downarrow \\ M_0(X') \otimes_W K & \longrightarrow & \mathrm{Lie}\, E_{X'} \otimes_O K. \end{array} \quad (5.5)$$

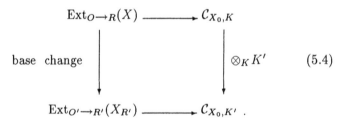

Finally we require that for $K = K_0$ the map

$$M_0(X) \otimes_W K_0 \xrightarrow{\sim} \text{Lie } E(X)^{\text{univ}} \otimes_W K_0 \tag{5.6}$$

is an isomorphism.
We claim that any two families of functors with these properties differ by an automorphism of the functor $X \longmapsto M_0(X) \otimes_W K$.
Indeed for that we verify that $M_0(X) \otimes_W K \to \text{Lie } E(X)^{\text{univ}} \otimes_0 K$ is always an isomorphism. We have $W \subset O$ and hence $k \subset R$. If $X = X_0 \otimes_k R$, we get what we want by base change (5.4). In the general case we have a quasi–isogeny $X_{0,R} \xrightarrow{\alpha} X$, which induces the identity on X_0. We can apply the commutative diagram (5.5) to $p^m\alpha$ and $p^m\alpha^{-1}$ for a large $m \in \mathbf{N}$ to get the assertion. The uniqueness of the families of functors follows from the existence of the universal extension.
We will call a family of functors as above a *period map in Cartier theory.*
Let us replace for a moment the module M_0 in the definition of the category $\mathcal{C}_{X_0,K}$ by Lie $E(X_0)^{\text{univ}}$. Then the crystalline theory [Me] provides us with a period map (family of functors) as above (compare proof of (5.15)). Clearly this crystalline period map agrees with any other period map in Cartier theory via the isomorphism (5.6) by uniqueness.

5.21 For the construction of the period map in Cartier theory, which is independent of the crystalline theory, let us consider a more general situation. Let O be a torsion free p-adic ring. Let $R = O/p$. We will assume that there is a natural number m, such that $x^m = 0$ for any element in the kernel of $R \to R_{\text{red}}$ and that R_{red} is perfect. Let X be a p–divisible formal group over R and let M be its Cartier module. Let M_E be a reduced Cartier module over O and let L be an O-module. Assume we are given an exact sequence of modules over the Cartier ring $\mathbf{E}_O$,

$$0 \longrightarrow C(L) \longrightarrow M_E \xrightarrow{\varrho} M \longrightarrow 0.$$

Then we will construct for large natural numbers l and i homomorphisms of abelian groups $\tau_i : p^l M \to M_E$ such that $p\tau_i = \tau_{i+1}$ and $\varrho\tau_i = p^i$. For the construction we fix a number h, such that $Ker(\mathbf{F}^t : M \to M) = Ker(\mathbf{F}^h : M \to M)$, $t \geq h$. Let M_0 be the Cartier module of X_{red}. Moreover we fix a natural number r such that

$$\mathbf{V}^r M_0 \subset \mathbf{F} M_0.$$

This is possible because the multiplication by $p : X \to X$ is by assumption an isogeny and the ring R_{red} is perfect. Since $p^h M = p^h M_0$, we get $\mathbf{V}^{rn} p^h M \subset \mathbf{F}^n p^h M$ for any natural number n. We have maps for each $i \geq h$:

$$\sigma_i : \mathbf{F}^i M \longrightarrow M_E / p^{i-h} C(L).$$

For the definition we represent $x \in \mathbf{F}^i M$ as $x = \mathbf{F}^i m$ and choose a lifting $\tilde{m} \in M_E$ of m. We define

$$\sigma_i(x) = \mathbf{F}^i \tilde{m}.$$

For a different choice $m', \tilde{m}'$, we have $\mathbf{F}^i(m - m') = 0$ and hence $\mathbf{F}^h(m - m') = 0$. Hence we get $\mathbf{F}^h(\tilde{m} - \tilde{m}') \in C(L)$ and $\mathbf{F}^i(\tilde{m} - \tilde{m}') \in \mathbf{F}^{i-h} C(L) = p^{i-h} C(L)$. Therefore the map is well-defined. Clearly we have $\sigma_i(x) \equiv \sigma_{i+1}(x) \bmod p^{i-h} C(L)$ for $x \in \mathbf{F}^{i+1} M$.
Let us set $\bar{M} = p^h M$. For $n \geq h$ we get maps

$$\mathbf{F}^{nr} \sigma_n \mathbf{V}^{nr} : \bar{M} \longrightarrow M_E / p^{nr+n-h} C(L).$$

Lemma 5.22 *The operator $\mathbf{F}^{nr} \sigma_n \mathbf{V}^{nr}$ is divisible by $p^{(n-h)r}$. For each $n > h$ one has the congruences*

$$\mathbf{F}^{nr} \sigma_n \mathbf{V}^{nr} \equiv p^r \mathbf{F}^{(n-1)r} \sigma_{n-1} \mathbf{V}^{(n-1)r} \bmod p^{nr+n-h-1} C(L)$$

Proof: The first assertion makes sense because M_E has no p-torsion and is an obvious consequence of the second one. To prove the congruence we may on the right hand side replace σ_n by σ_{n-1}. Take $x \in \bar{M}$. We choose $m \in \bar{M}$ such that $\mathbf{V}^{(n-1)r} x = \mathbf{F}^{n-1} m$ and hence $\mathbf{V}^{nr} x = \mathbf{F}^{n-1} \mathbf{V}^r m$. For a lifting $\tilde{m}$ of m we get

$$\begin{aligned} &\mathbf{F}^{nr} \sigma_{n-1} \mathbf{V}^{nr} x \equiv \mathbf{F}^{nr} \mathbf{F}^{n-1} \mathbf{V}^r \tilde{m} \equiv p^r \mathbf{F}^{(n-1)r} \mathbf{F}^{n-1} \tilde{m} \\ &\equiv p^r \mathbf{F}^{(n-1)r} \sigma_{n-1} \mathbf{V}^{(n-1)r} x \bmod p^{nr+n-1-h} C(L). \end{aligned}$$

□

We define

$$\tau_{hr}(x) = \lim_{n \to \infty} \frac{1}{p^{(n-h)r}} \mathbf{F}^{nr} \sigma_n \mathbf{V}^{nr} x.$$

Since $C(L)$ is p-adically complete, the limit exists by the lemma. We have $\varrho\tau_{hr}(x) = p^{hr}x$. Indeed, the difference between two members in the sequence above is in $C(L)$; hence it is enough to check that $\varrho(\mathbf{F}^{hr}\sigma_h\mathbf{V}^{hr}) = p^{hr}$, which is trivial. For $i \geq hr$ we define

$$\tau_i(X) = p^{i-hr}\tau_{hr}(X).$$

For different choices of r and h we would end up with the same τ_i, whenever they are defined.

We note that if the ring R is reduced and perfect, we may choose $h = 0$. Then τ_0 is defined and is a section of ϱ.

Lemma 5.23 *The homomorphisms τ_i have the following properties*

(i) $\tau_i\mathbf{F} = \mathbf{F}\tau_i$

(ii) $\tau_i\mathbf{V}x - \mathbf{V}\tau_i x \in L \subset C(L)$ *for* $x \in M$.

Proof: The verification of (i) is trivial. For the verification of (ii) we note that it is enough to check that

$$\mathbf{F}(\tau_i\mathbf{V} - \mathbf{V}\tau_i) = 0\,,$$

which follows from (i). □

Let us denote by k the ring R_{red}. Let $W = W(k)$ be the Witt ring. The homomorphism $W(O) \to W(k)$ has a section δ, which is defined as follows. For a fixed $w \in W(k)$ we find a sequence $w_n \in W(k)$ such that $\mathbf{F}^n w_n = w$. We define

$$\delta(x) = \lim_{n\to\infty} \mathbf{F}^n \tilde{w}_n\,,$$

where $\tilde{w}_n$ are liftings of w_n and the limit is in the p-adic topology. We denote by δ_R the composite of the maps

$$\delta_R : W(k) \xrightarrow{\delta} W(O) \longrightarrow W(R).$$

The sections δ resp. δ_R commute with the action of the Frobenius.

Lemma 5.24 *For any i such that τ_i is defined, we have*

$$\tau_i(\delta_R(w)x) = \delta(w)\tau_i(x), \quad x \in \bar{M}.$$

Proof: With the notations above we may assume $i = rh$. We fix $n \geq h$ and choose m such that $\mathbf{V}^{nr}x = \mathbf{F}^n m$. This implies $\mathbf{V}^{nr}\delta_R(w)x = \mathbf{F}^n\delta_R(w_{nr+n})m$. Let $\tilde{m} \in M_E$ be a lifting of m. We obtain

$$\frac{1}{p^{(n-h)r}}\mathbf{F}^{nr}\sigma_n\mathbf{V}^{nr}\delta_R(w)x = \frac{1}{p^{(n-h)r}}\mathbf{F}^{nr+n}\delta(w_{nr+n})\tilde{m}$$
$$= \frac{1}{p^{(n-h)r}}\delta(w)\mathbf{F}^{nr+n}\tilde{m} = \frac{1}{p^{(n-h)r}}\delta(w)\mathbf{F}^{nr}\sigma_n\mathbf{V}^{nr}x \bmod p^{nr+n-h}C(L).$$

Passing to the limit we get what we want. □

5.25 Let us localize the exact sequence that we associated to $(\mathcal{V}, E, \varrho, X)$:

$$0 \longrightarrow C(L) \otimes_{\mathbf{Z}} \mathbf{Q} \longrightarrow M_E \otimes_{\mathbf{Z}} \mathbf{Q} \xrightarrow{\varrho} M \otimes_{\mathbf{Z}} \mathbf{Q} \longrightarrow 0\,. \qquad (5.7)$$

We note that the map $\tau = 1/p^i\tau_i$ is a natural $W(k)[F]$-linear splitting of the sequence. Moreover the natural projection $M \rightarrow M_0$ has a kernel annihilated by a power of p and hence induces an isomorphism

$$M \otimes \mathbf{Q} \longrightarrow M_0 \otimes \mathbf{Q}.$$

Hence we may rewrite the sequence as follows

$$0 \longrightarrow C(L) \otimes_{\mathbf{Z}} \mathbf{Q} \longrightarrow M_E \otimes_{\mathbf{Z}} \mathbf{Q} \underset{\varrho}{\overset{\tau}{\leftrightarrows}} M_0 \otimes \mathbf{Q} \longrightarrow 0.$$

We obtain a K_0–linear map

$$\bar{\tau} : M_0 \otimes \mathbf{Q} \longrightarrow (M_E/\mathbf{V}M_E) \otimes \mathbf{Q} = (\text{Lie } E)_{\mathbf{Q}}\,.$$

Proposition 5.26 *The functor that associates to an extension $(\mathcal{V}, E, \varrho, X)$ the map $\bar{\tau} : M(X_0)_{\mathbf{Q}} \rightarrow (\text{Lie } E)_{\mathbf{Q}}$ is a period map in Cartier theory. Let $\mathcal{C}$ be the category of pairs $(M_{0,\mathbf{Q}}, M_{0,\mathbf{Q}} \rightarrow L)$, where the arrow is a K_0-linear map to a K-vectorspace L and $M_{0,\mathbf{Q}}$ is the rational Cartier module of a p-divisible group over k. Then the functor*

$$\begin{array}{rcl} Ext_{\mathcal{O}\rightarrow R} \otimes \mathbf{Q} & \longrightarrow & \mathcal{C} \\ (\mathcal{V}, E, \varrho, X) & \longmapsto & (M(X_0)_{\mathbf{Q}}, \bar{\tau}) \end{array}$$

is fully faithful.

Proof: Let us first prove that the functor is fully faithful. By Cartier theory we know that the functor which associates to an object $(\mathcal{V}, E, \varrho, X)$ of $\text{Ext}_{O \to R} \otimes \mathbf{Q}$ the exact sequence (5.7) is fully faithful. Hence we have to show how to recover this sequence of modules over the Cartier ring $\mathbf{E}_O$ from $M_0(\mathbf{Q})$, $\operatorname{Lie} E \otimes \mathbf{Q}$ and the morphism $\bar{\tau}$.
By (5.19) we have an isomorphism $L_{\mathbf{Q}} \simeq (\operatorname{Lie} E)_{\mathbf{Q}}$. We obtain an isomorphism $C(L) \otimes_{\mathbf{Z}} \mathbf{Q} \simeq C((\operatorname{Lie} E)_{\mathbf{Q}})$. The section τ defines an isomorphism of abelian groups

$$M_E \otimes_{\mathbf{Z}} \mathbf{Q} = C((\operatorname{Lie} E)_{\mathbf{Q}}) \oplus M_{0,\mathbf{Q}}. \tag{5.8}$$

We are done if we describe the action of $\mathbf{E}_O$ on the right hand side of (5.8) in terms of $\bar{\tau}$. Let us first look at the action of the Witt ring $W(O)$. Since $W(R) \otimes \mathbf{Q} = W(k) \otimes \mathbf{Q}$ we get a split exact sequence

$$\begin{array}{ccccccc} 0 \longrightarrow & W(pO) \otimes \mathbf{Q} & \longrightarrow & W(O) \otimes \mathbf{Q} & \overset{\delta}{\leftrightarrows} & W(k) \otimes \mathbf{Q} & \longrightarrow 0 \\ & \simeq \downarrow \log & & & & & \\ & (\prod pO) \otimes \mathbf{Q} & & & & & \end{array}$$

Hence $W(O) \otimes \mathbf{Q} = K_0 \oplus \prod_b K$, where $\prod_b K \subset K^\infty$ are the elements with bounded denominator. Then (5.8) is a decomposition of $K_0[F]$-modules, where K_0 acts on $C(L) \otimes_{\mathbf{Z}} \mathbf{Q}$ by

$$k_0(\sum \mathbf{V}^i l_i) = \sum \mathbf{V}^i \sigma^i(k_0) l_i\,, \quad \sigma = \text{Frobenius}.$$

For $\underline{a} \in \prod_b K$ we have

$$\underline{a}\left(\sum_i \mathbf{V}^i l_i\right) = \sum \mathbf{V}^i p^i a_i l_i$$
$$\underline{a} m = \textstyle\sum_i \mathbf{V}^i \left(a_i \otimes \bar{\tau} \mathbf{F}^i m\right) \in C(L) \otimes \mathbf{Q}.$$

Finally we have to define the action of $\mathbf{V}$ on elements $m \in M_{0,\mathbf{Q}} \subset M_{E,\mathbf{Q}}$. We note that the map $\mathbf{V}\tau - \tau\mathbf{V} : M_{0,\mathbf{Q}} \to (\operatorname{Lie} E)_{\mathbf{Q}}$ given by lemma (5.23) may be identified with $-\bar{\tau}\mathbf{V} : M_{0,\mathbf{Q}} \to (\operatorname{Lie} E)_{\mathbf{Q}}$. Hence $\mathbf{V}$ is given by

$$\mathbf{V}(0, m) = (-\bar{\tau}\mathbf{V} m, \mathbf{V} m).$$

It remains to check that $\bar{\tau}$ is a period map in Cartier theory. We show that the map $M_0(X_0) \otimes_W K_0 \to \operatorname{Lie} E(X)^{\text{univ}} \otimes_W K_0$ is an isomorphism. The other requirements for a period map are obviously fulfilled.
In the case $K = K_0$ the section τ is even a section of the sequence

$$0 \longrightarrow C(L) \longrightarrow M_E \longrightarrow M(X_0) \longrightarrow 0,$$

i.e. without killing torsion. The same argument as above therefore shows that the category of such sequences is equivalent to the category of commutative diagrams

$$\begin{array}{ccc} & & L \\ & & \downarrow \\ M(X_0) & \longrightarrow & \operatorname{Lie} E \\ & \searrow & \downarrow \\ & & M(X_0)/\mathbf{V}M(X_0) \end{array}$$

Hence we get the universal extension if we set $\operatorname{Lie} E = M(X_0)$.

5.27 We now return to the general case and consider the problem of determining the image of the period map associated to a moduli problem of type (EL) or (PEL). By Fontaine [Fo] the image of the first factor of the period morphism $\breve{\pi}^1 : \breve{\mathcal{M}}^{rig} \to \breve{\mathcal{F}}$ lies in the weakly admissible subset.

Proposition 5.28 *Let us assume Fontaine's conjecture ([Fo2], 5.2.5) that a weakly admissible filtered isocrystal $(V, \Phi, \mathcal{F}^\bullet)$ with $\mathcal{F}^2 = (0)$ and $\mathcal{F}^0 = V \otimes_{K_0} K$ comes from a p-divisible group over $\operatorname{Spec} O_K$. Hence to every $x \in \breve{\mathcal{F}}^{wa}(K)$ there is associated via Fontaine's functor a $\mathbf{Q}_p$-vector space $V_p(x)$ with an action of B and a non-degenerate alternating $\mathbf{Q}_p$-linear form up to a constant in $\mathbf{Q}_p^\times$. We make the further assumption that the isomorphism class of $V_p(x)$ is independent of $x \in \breve{\mathcal{F}}^{wa}$. Then the image of the period morphism is of the form*

$$\breve{\mathcal{F}}^{wa} \times \Delta',$$

where Δ' is a union of cosets of the subgroup of finite index in Δ formed by the $\mathbf{Q}_p$-rational cocharacters of the center of G, unless $\breve{\mathcal{M}}^{rig}$ is empty.

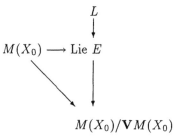

Proof: Let K be a finite extension of $\breve{E}$ and $x \in \breve{\mathcal{F}}^{wa}(K)$. Fontaine's conjecture says in this case that there is a p–divisible group X over $Spec\ O_K$ with rational Dieudonné module of its special fibre equal to $V \otimes K_0$ such that x corresponds to the natural surjection of K–vector spaces,

$$V \otimes K = M_X \otimes_{O_K} K \longrightarrow Lie(X) \otimes_{O_K} K.$$

The p–divisible group is uniquely determined up to isogeny *over* O_K, cf. ([Fo1], IV., 5.2). Consider the rational p-adic Tate module of the generic fibre of X,

$$V_p(X) = V_p(X \otimes_{O_K} K).$$

Then $V_p(X)$ is the image $V_p(x)$ of the filtered isocrystal corresponding to x under the Fontaine functor ([Fo2]). By hypothesis, the isomorphism class of $V_p(x)$ is independent of the point x. Let $x' \in \mathcal{F}^{wa}(K')$ be a point in the image of $\breve{\mathcal{M}}^{rig}(K')$, for a finite extension K' of K. Let $(X'_{\mathcal{L}}, \varrho')$ be an object of $\breve{\mathcal{M}}(Spf\, O_{K'})$ mapping to x'. The p–adic Tate modules of X'_Λ define a polarized multichain of O_B–lattices $\mathcal{L}'$ in $V_p(X')$ whose members are in a natural one-to-one correspondence with the members of $\mathcal{L}$. Let $\mathcal{L}''$ be the image of $\mathcal{L}'$ under a fixed isomorphism

$$V_p(X) \simeq V_p(X').$$

There is a finite extension K'' of K such that $\mathcal{L}''$ is stable under $Gal(\bar{\breve{E}}/K'')$. Hence there is a chain of p–divisible groups X_Λ $(\Lambda \in \mathcal{L})$ over $O_{K''}$ in the isogeny class of $X \otimes_K K''$. Furthermore, the identification $M_{X_\Lambda} \otimes_{O_{K''}} K'' = V \otimes K''$ provides us with a family of quasi–isogenies over the residue field of $O_{K''}$,

$$\varrho_\Lambda : \mathbf{X} \longrightarrow X_\Lambda \otimes_{O_{K''}} \kappa.$$

We claim that $(X_{\mathcal{L}}, \varrho)$ is an object of $\breve{\mathcal{M}}(Spf\, O_{K''})$. We have to check conditions (i) – (v) of (3.21). Consider the determinant condition (iv) on the generic fibre of X_Λ. However, $Lie(X_\Lambda \otimes_{O_{K''}} K'') = Lie(X) \otimes_{O_K} K''$ is a module over the semi–simple algebra $B \otimes K''$ and therefore prescribing the determinants of elements of B comes down to prescribing the dimensions of the corresponding modules over the various simple factors of B. Since $Lie(X) \otimes K$ satisfies the determinant condition, so does the generic fibre

of X_Λ and hence X_Λ itself. In the presence of condition (iv), condition (ii) is equivalent to condition (ii bis) of ((3.23), d). To check this condition we may assume B simple. However, for any two neighbours $\Lambda \subset \Lambda'$ the height of the isogeny $X_\Lambda \to X_{\Lambda'}$ is equal to the order of the cokernel of the induced map on the p-adic Tate modules of the generic fibres. Since the indices of members of $\mathcal{L}$ and corresponding members of $\mathcal{L}''$ are identical the condition (ii) is indeed satisfied. Of the remaining conditions, (iii) and (v) are trivially satisfied and (i) is automatic. Therefore we have found a point $(X_{\mathcal{L}}, \varrho) \in \breve{\mathcal{M}}(Spf\ O_{K''})$ whose image under the first factor of the period morphism is clearly the point x with which we started.
Let Z be the center of G. Then Z is also a central subgroup of J. The assertion follows from the fact that the image of $\breve{\pi}$ is stable under $Z(\mathbf{Q}_p)$ and that the image of $Z(\mathbf{Q}_p)$ under ω_J (3.52) is a subgroup of finite index in Δ. □

Remark 5.29 In (1.19) a cohomology class $cls(x, b) \in H^1(\mathbf{Q}_p, G)$ was associated to $x \in \breve{\mathcal{F}}^{wa}(K)$, if x is admissible. It seems reasonable to expect this class to be constant as x varies through $\breve{\mathcal{F}}^{wa}$. This is true by the formula in (1.20) if G is connected with simply connected derived group. In this sense the last assumption in proposition (5.20) seems reasonable.

5.30 In the unramified case we can say more. Let $(F, B, O_B, V, b, \mu, \mathcal{L})$ be unramified data of type (EL) or (PEL), cf.(3.82). Let $\breve{\mathcal{M}}$ be the corresponding formal scheme over the Witt vectors $Spf\ O_{K_0}$. By the flatness of $\breve{\mathcal{M}}$ (cf.(3.82)) it follows that $\breve{\mathcal{M}}^{rig}$ is non-empty. Let K be a finite extension of K_0 with ring of integers O_K. Let $(X, \varrho) \in \breve{\mathcal{M}}(Spf\ O_K)$ be an object of our moduli problem, cf. (3.82). Then the rational p-adic Tate module $V_p(X)$ of the generic fibre of X is the image of the filtered isocrystal over K associated to X under the Fontaine functor. It contains the p-adic Tate module $T_p(X)$ as a O_B-lattice which in case (PEL) is selfdual with respect to the non-degenerate alternating $\mathbf{Q}_p$-form on $V_p(X)$.

Proposition 5.31 *Let $(B, F, O_B, V, b, \mu, \mathcal{L})$ be unramified and let x be a point of $\mathcal{F}^{wa}(K_0)$. Then x is admissible. Let $V_p(x)$ denote the image under the Fontaine functor. We make the assumption that $V_p(x)$ contains a selfdual O_B-lattice (automatic, if G is connected, cf. (5.33) below). Then there exists $y \in \breve{\mathcal{M}}^{rig}(K_0)$ with $\breve{\pi}^1(y) = x$.*

Proof: We only treat the case (PEL), the case (EL) being similar but simpler. By the theorem of Laffaille [L], proving Fontaine's conjecture in the unramified case, x comes from a p-divisible group X over O_{K_0}. In particular, the first assertion is true. Using similar arguments as in the proof of the previous proposition the second assertion will be proved if we can exhibit a p-divisible group in the isogeny class of X which is equipped with an O_B-action and an O_B-linear principal polarization. For this it suffices to find a selfdual O_F-lattice Λ' in the rational p-adic Tate module $V_p(X)$ which is stable under the action of the Galois group $Gal(\bar{K}_0/K_0)$. Indeed, the stabilizer of Λ' in B is a maximal order O'_B, invariant under $*$, and any element of B which conjugates O'_B into O_B also conjugates Λ' into a Galois stable selfdual O_B-lattice which then yields the desired p-divisible group. The image of the Galois group $Gal(\bar{K}_0/K_0)$ in $\mathrm{GL}(V_p(X))$ is contained in the group of B-linear symplectic similitudes. It is therefore contained in a maximal compact subgroup of this group and therefore stabilizes an O_F-lattice $\Lambda \subset V_p(X)$ with

$$\Lambda \subset \Lambda^* \subset p^{-1} \cdot \Lambda.$$

Indeed, consider a Galois invariant Λ such that $\Lambda \subset \Lambda^*$ and such that the index $[\Lambda^* : \Lambda]$ is minimal. If $p\Lambda^* \not\subseteq \Lambda$ consider the lattice $\Lambda' = \Lambda + p\Lambda^*$. Then $(\Lambda')^* = \Lambda^* \cap p^{-1}\Lambda$. Hence $\Lambda' \subset (\Lambda')^*$ with a smaller index, which is a contradiction. We are going to replace Λ by a Galois invariant O_F-lattice Λ' with $\Lambda \subset \Lambda' = (\Lambda')^* \subset \Lambda^*$. Proceeding inductively we may assume $\Lambda \neq \Lambda^*$ and that there is no Galois invariant O_F-lattice Λ' with $\Lambda \subsetneq \Lambda' \subseteq (\Lambda')^* \subsetneq \Lambda^*$. We want to derive a contradiction.

On the $\mathbf{F}_p$-vector space

$$\mathcal{V} = \Lambda^*/\Lambda$$

we have a non-degenerate alternating bilinear form

$$(\bar{v}, \bar{v}') = p \cdot (v, v') \bmod p, \quad v, v' \in \Lambda^*.$$

The Galois group acts on $\mathcal{V}$ and preserves this form. We remark that $\mathcal{V}$ cannot contain a non-zero Galois-stable totally isotropic subspace. Indeed, the orbit under O_F of the inverse image in Λ^* of such a subspace would yield a Galois invariant O_F-lattice Λ' with $\Lambda \subsetneq \Lambda' \subset (\Lambda')^* \subsetneq \Lambda^*$, contrary to our hypothesis.

Claim: The Galois module $\mathcal{V}$ is trivial.

The Galois representation on $\mathcal{V}$ is defined by a finite flat group scheme over O_{K_0} killed by p (the kernel of the polarization induced on the p–divisible group corresponding to Λ). We may therefore apply Raynaud's theory [Ra2]. It follows (loc.cit, §3) that on each Jordan - Hölder factor of $\mathcal{V}$ the Galois group $Gal(\bar{K}_0/K_0)$ acts through its tame quotient I_t. Furthermore, if $\mathcal{W}$ is a Jordan - Hölder factor of $\mathcal{V}$, then the commutant of I_t in End $\mathcal{W}$ is a finite field $\mathbf{F}$ and $\mathcal{W}$ is an $\mathbf{F}$–vector space of dimension 1. The action of I_t is given by a character

$$I_t \longrightarrow \mu_{q'-1}(K_0) \xrightarrow{\psi} \mathbf{F}^{\times}.$$

Here $q' = p^r$ denotes the cardinality of $\mathbf{F}$. Let χ_0 be the inverse of the *Teichmüller character*, and consider the *fundamental characters*

$$\chi_0, \chi_1 = \chi_0^p, \ldots, \chi_{r-1} = \chi_0^{p^{r-1}}.$$

Any character ψ as above has a unique expression of the form

$$\psi = \chi_0^{n_0} \cdots \chi_{r-1}^{n_{r-1}}, \quad 0 \leq n_i \leq p-1.$$

By Raynaud and since the ramification index of K_0 is one, the character ψ defining the Jordan - Hölder factor $\mathcal{W}$ has as exponents n_i only 0 or 1 $(i = 0, \ldots, r-1)$.

Let now $\mathcal{W}$ be a simple Galois submodule. The bilinear form defines a Galois homomorphism of $\mathcal{W}$ into its contragredient $\mathcal{W}^*$. By the simplicity of $\mathcal{W}$ this homomorphism is either an isomorphism (i.e. the restriction of $(\ , \ )$ to $\mathcal{W}$ is non–degenerate), or zero (i.e. $\mathcal{W}$ is totally isotropic) and the latter case is excluded by our inductive hypothesis. We may write the restriction of $(\ , \ )$ to $\mathcal{W}$ as the trace of an $\mathbf{F}$–bilinear form defined by

$$(aw, w') = \mathrm{Tr}_{\mathbf{F}/\mathbf{F}_p}\, a \cdot (w, w')', \quad a \in \mathbf{F}.$$

The $\mathbf{F}$–valued form satisfies

$$(aw, a'w')' = a \cdot (w, w')' a'^{*}, \quad a, a' \in \mathbf{F},$$

where $*$ is an involution on $\mathbf{F}$. If $\mathcal{W}$ is defined by the character ψ, then $\mathcal{W}^*$ is defined by ψ^{*-1}. Since $\mathcal{W}$ is isomorphic to $\mathcal{W}^*$, we obtain

$$1 = \psi^2 = \chi_0^{2n_0} \cdots \chi_{r-1}^{2n_{r-1}},$$

if $*$ is trivial, and

$$1 = \psi \cdot \psi^* = \chi_0^{n_0+n_s} \cdots \chi_{r-1}^{n_{r-1}+n_{r-1+s}},$$

if $*$ is the non-trivial automorphism of $\mathbf{F} = \mathbf{F}_{p^r} = \mathbf{F}_{p^{2s}}$ over $\mathbf{F}_{p^s}$.
By the uniqueness of the expression in terms of the fundamental characters and since $0 \leq n_i + n_j \leq 2 < p$ we conclude that $\mathcal{W}$ is a trivial Galois module. Replacing $\mathcal{V}$ by the orthogonal complement of $\mathcal{W}$, an obvious induction proves the claim.
We now finish the proof of the proposition. By assumption $V_p(X)$ contains a self-dual O_B-lattice. This implies that there exists a self-dual O_F-lattice $\Lambda' \subset V_p(X)$ with $\Lambda \subset \Lambda'$. By our claim above this lattice is automatically Galois invariant and this contradicts our assumption. □

5.32 We now introduce the tower of rigid-analytic coverings of $\breve{\mathcal{M}}^{rig}$. Here $\breve{\mathcal{M}}$ denotes as before the formal scheme over $Spf\, O_{\breve{E}}$ representing our moduli problem for the data $(F, B, O_B, V, b, \mu, \mathcal{L})$ of type (EL) or (PEL).
Let $x \in \breve{\mathcal{M}}^{rig}(K)$ be represented by $(X_{\mathcal{L}}, \varrho) \in \breve{\mathcal{M}}(Spf\, O_K)$. Then the rational p-adic Tate module $V_p(X)$ is equipped with an action of B and, in case (PEL), with an alternating form up to a constant. Unfortunately we do not know whether the isomorphism class of $V_p(X)$ is independent of $x \in \breve{\mathcal{M}}^{rig}(K)$, cf. remark (5.29). Therefore we fix representatives $V_i', (\ ,\ )_i'$ $(i = 1, \ldots, t)$ of the isomorphism classes of $V_p(X)$, as x varies through $\breve{\mathcal{M}}^{rig}$. Fix i and let $V' = V_i'$. Then V' defines a twisted form of our original group, with

$$G'(\mathbf{Q}_p) = \{g' \in GL_B(V');\ (g'v, g'v')' = c(g')\,(v, v')',\ c(g') \in \mathbf{Q}_p^\times\}.$$

This expression refers to the case (PEL). The group G' is an inner form of G since the B-modules V and V' are isomorphic and since the forms $(\ ,\ )$ and $(\ ,\ )'$ are both compatible with the same involution $*$ on B. Indeed, one easily checks that the B-modules V and V' equipped with their alternating forms become isomorphic over the algebraic closure $\bar{\mathbf{Q}}_p$. We fix a multichain of lattices $\mathcal{L}'$ in V' which under an isomorphism of V' with $V_p(X)(x \in \breve{\mathcal{M}}^{rig})$ maps to the multichain $T_p(X_\Lambda), \Lambda \in \mathcal{L}$, defined by the

p–adic Tate module of $X_{\mathcal{L}}$. Let $K'_{\mathcal{L}'} \subset G'(\mathbf{Q}_p)$ be the corresponding fix group. The conjugacy class of $\mathcal{L}'$ resp. $K'_{\mathcal{L}'}$ is independent of the choice of the isomorphism and of x. We note that the members of $\mathcal{L}$ and $\mathcal{L}'$ are in a natural one-to-one correspondence such that indices of corresponding pairs of lattices are identical, comp. the proof of proposition (5.28).
Let $\breve{\mathcal{M}}_i^{rig} (i = 1, \ldots, t)$ be the subset of points $x \in \breve{\mathcal{M}}^{rig}$ where the rational p–adic Tate module $V_p(X)$ is isomorphic to V_i'. We also use the notation $\breve{\mathcal{M}}_{V'}^{rig}$ or $\breve{\mathcal{M}}_{\mathcal{L}'}^{rig}$ for $\breve{\mathcal{M}}_i^{rig}$ if $V' = V_i'$, and $\mathcal{L}'$ is the multichain of lattices in V' defined above.

Lemma 5.33 *The subset $\breve{\mathcal{M}}_i^{rig}$ is a union of connected components of $\breve{\mathcal{M}}^{rig}, i = 1, \ldots, t$. If G is connected then $V_p(X) \simeq V$ for all $x \in \breve{\mathcal{M}}^{rig}$, i.e. $t = 1$ and $V_1' = V$. In particular this applies to the case (EL).*

Proof: We remark that the p–adic Tate modules $T_p(X_\Lambda)$, $\Lambda \in \mathcal{L}$, for the points $x \in \breve{\mathcal{M}}^{rig}$ piece together to a locally constant $\mathbf{Z}_p$–sheaf for the étale topology on $\breve{\mathcal{M}}^{rig}$ (namely, $T_p(X_\Lambda) \otimes \mathbf{Z}/p^n$ is the generic fibre of the finite flat group scheme of p^n–division of X_Λ which is étale). By Nakayama's lemma $V_p(X)$ is determined by $T_p(X_\Lambda) \otimes \mathbf{Z}/p^n$, $\Lambda \in \mathcal{L}$, for a fixed $n \gg 0$. The first assertion follows since this chain of finite étale group schemes trivializes over a connected finite étale covering of each connected component of $\breve{\mathcal{M}}^{rig}$.
Now assume G connected. Let $x \in \breve{\mathcal{M}}^{rig}$ and let $V', \mathcal{L}'$ denote its rational p–adic Tate module with its natural multichain of lattices. Then $\mathcal{L}'$ is a polarized multichain of O_B–modules on *Spec* $\mathbf{Z}_p$ of type $(\mathcal{L})$, in the sense of chapter 3. Therefore (3.16) there exists an unramified extension L of $\mathbf{Q}_p$ such that there is an isomorphism of polarized multichains,

$$\mathcal{L} \otimes O_L \xrightarrow{\sim} \mathcal{L}' \otimes O_L.$$

Therefore, the difference between V and V' is measured by a 1–cocycle with values in $K_{\mathcal{L}}(O_L)$ where we momentarily denote by $K_{\mathcal{L}}$ the smooth $\mathbf{Z}_p$–form of G defined by $\mathcal{L}$. However, since G is connected, $K_{\mathcal{L}}$ has connected fibres. Therefore by Lang's theorem and a standard approximation argument (comp. [BT], 7.2) we conclude that after enlarging L this 1–cocycle is trivial, and hence $V \simeq V'$.

5.34 We denote by $T_{\mathcal{L}} = \{T_\Lambda;\ \Lambda \in \mathcal{L}\}$ the multichain of locally constant étale $\mathbf{Z}_p$–sheaves defined by the p–adic Tate modules of the universal p–divisible group X_Λ on $\breve{\mathcal{M}}$. Fix i $(i = 1, \ldots, t)$ and let $(V', \mathcal{L}') = (V_i', \mathcal{L}_i')$.

We also have the inner form G' of G. Let $K' \subset K'_{\mathcal{L}'}$ be an open compact subgroup. We consider the finite étale covering $\breve{\mathbf{M}}_{\mathcal{L}',K'}$ of $\breve{\mathcal{M}}^{rig}_{\mathcal{L}'}$ parametrizing the classes modulo K' of trivializations of the local system $T_{\mathcal{L}}$ over $\breve{\mathcal{M}}^{rig}_{\mathcal{L}'}$,

$$\alpha : T_{\mathcal{L}} \xrightarrow{\sim} \mathcal{L}' \pmod{K'}.$$

Then for $K' = K'_{\mathcal{L}'}$ we have obviously $\breve{\mathbf{M}}_{\mathcal{L}',K'_{\mathcal{L}'}} = \breve{\mathcal{M}}^{rig}_{\mathcal{L}'}$. Let $K' \subset K'_{\mathcal{L}'}$ and let $g' \in G'(\mathbf{Q}_p)$ be such that $g'^{-1}K'g' \subset K'_{\mathcal{L}'}$. We are going to construct a canonical isomorphism

$$\iota(g') : \breve{\mathbf{M}}_{\mathcal{L}',K'} \longrightarrow \breve{\mathbf{M}}_{\mathcal{L}',g'^{-1}K'g'}.$$

We content ourselves with giving the construction pointwise. Let us describe the image of a point $x \in \breve{\mathbf{M}}_{\mathcal{L}',K'}(K)$ corresponding to $((X_{\mathcal{L}}, \varrho), \alpha)$. The inverse image under $\alpha \otimes_{\mathbf{Z}_p} \mathbf{Q}_p$ of the multichain $g'\mathcal{L}'$ defines a multichain of p–divisible groups $X'_{\mathcal{L}}$ over O_K with a quasi–isogeny $X_{\mathcal{L}} \to X'_{\mathcal{L}}$. We obtain an object $(X'_{\mathcal{L}}, \varrho')$ of $\breve{\mathcal{M}}$ by taking ϱ' to be the composition of ϱ with the special fibre of this quasi–isogeny. Finally, the trivialization α' of $T_p(X'_{\mathcal{L}})$ is given by the composition

$$T_p(X'_{\mathcal{L}}) \xrightarrow{\alpha} g'\mathcal{L}' \xrightarrow{g'^{-1}} \mathcal{L}'.$$

To perform the same construction starting with a point of $\breve{\mathbf{M}}_{\mathcal{L}',K'}$ with values in the rigid space associated to a p-adic formal scheme $S \to \breve{\mathcal{M}}$ the main point is the construction of $(X'_{\mathcal{L}}, \varrho')$. In general the p-divisible group $(X'_{\mathcal{L}})^{rig}$ over S^{rig} corresponding to the multichain $g'\mathcal{L}'$ extends to a p-divisible group over S only after replacing S by an admissible blowing-up ([BL], thm. 4.1).

The definition of the Hecke correspondences is now done exactly as in the classical case ([De], 2.1.). We therefore obtain for every pair of open compact subgroups $K', K'_1 \subset G'(\mathbf{Q}_p)$ contained in $K'_{\mathcal{L}'}$ and any $g' \in G'(\mathbf{Q}_p)$ with $g'^{-1}K'_1g' \subset K'$ a morphism

$$\pi_{K'_1,K'}(g') : \breve{\mathbf{M}}_{\mathcal{L}',K'_1} \longrightarrow \breve{\mathbf{M}}_{\mathcal{L}',K'}$$

defined by the composition of $\iota(g') : \breve{\mathbf{M}}_{\mathcal{L}',K'_1} \to \breve{\mathbf{M}}_{\mathcal{L}',g'^{-1}K'_1g'}$ and the projection morphism $\breve{\mathbf{M}}_{\mathcal{L}',g'^{-1}K'_1g'} \to \breve{\mathbf{M}}_{\mathcal{L}',K'}$ induced by the inclusion $g'^{-1}K'_1g' \subset K'$. Furthermore, if K'_2 is another open compact subgroup contained in $K'_{\mathcal{L}'}$ and $g'_1 \in G'(\mathbf{Q}_p)$ such that $g_1'^{-1}K'_2g'_1 \subset K'_1$ we have that

$$\pi_{K'_1,K'}(g') \circ \pi_{K'_2,K'_1}(g'_1) = \pi_{K'_1,K'}(g'_1 g').$$

Furthermore if $K' \subset K'_{\mathcal{L}'}$ and $K'_1 \subset K'$ is a normal subgroup the morphisms $\pi_{K'_1,K'_1}(g')$ define an action of K'/K'_1 on $\breve{\mathbf{M}}_{\mathcal{L}',K'_1}$ with

$$\breve{\mathbf{M}}_{\mathcal{L}',K'_1}/(K'/K'_1) \xrightarrow{\sim} \breve{\mathbf{M}}_{\mathcal{L}',K'}.$$

The last formula allows us to define $\breve{\mathbf{M}}_{\mathcal{L}',K'}$ even if K' is not contained in $K'_{\mathcal{L}'}$. Namely, we take a principal congruence subgroup K'_1 which is a normal subgroup of K' and such that $K'_1 \subset K'_{\mathcal{L}'}$. Then we define $\breve{\mathbf{M}}_{\mathcal{L}',K'}$ through the last formula. All we have said above remains valid for arbitrary open compact subgroups $K' \subset G'(\mathbf{Q}_p)$.

The Hecke correspondence on $\breve{\mathbf{M}}_{\mathcal{L}',K'}$ defined by $g' \in G'(\mathbf{Q}_p)$ is given by the diagram

$$\begin{array}{ccc} & \breve{\mathbf{M}}_{\mathcal{L}',K'\cap g'Kg'^{-1}} & \\ \pi \swarrow & & \searrow \pi(g') \\ \breve{\mathbf{M}}_{\mathcal{L}',K'} & \longleftarrow - - - - - - - & \breve{\mathbf{M}}_{\mathcal{L}',K'}. \end{array}$$

We abbreviate $\pi_{K',K}(1)$ into $\pi_{K',K}$.

The tower of rigid–analytic spaces $\{\breve{\mathbf{M}}_{\mathcal{L}',K'};\ K' \subset G'(\mathbf{Q}_p)\}$ with the right action of $G'(\mathbf{Q}_p)$ defined above are our candidates for the conjectured étale coverings of $\breve{\mathcal{F}}^{wa}$, (cf. (1.37)),

$$\breve{\pi}^1_{K'} : \breve{\mathbf{M}}_{\mathcal{L}',K'} \longrightarrow \breve{\mathcal{F}}^{wa}.$$

If $K' \subset K'_{\mathcal{L}'}$, we introduce $\breve{\pi}^2_{K'} : \breve{\mathbf{M}}_{\mathcal{L}',K'} \to \Delta$ as the composition of $\pi_{K',K'_{\mathcal{L}}}$ and $\breve{\pi}^2$. It is clear that this definition extends in the obvious way to all open subgroups $K' \subset G'(\mathbf{Q}_p)$.

5.35 We note that the (left) action of $J(\mathbf{Q}_p)$ on $\breve{\mathcal{M}}^{rig}$ is lifted to each member $\breve{\mathbf{M}}_{\mathcal{L}',K'}$ of the tower.

Lemma 5.36 *Let Z be the center of G. Then we may also identify Z with the center of the inner form G' of G and also with a central subgroup of J. The left action of an element $z \in J(\mathbf{Q}_p)$ with $z \in Z(\mathbf{Q}_p)$ coincides with the right action of $z \in G'(\mathbf{Q}_p)$.*

Proof: We check this on points. Let $x \in \breve{\mathbf{M}}_{\mathcal{L}',K'}(K)$ correspond to a triple $((X_{\mathcal{L}}, \varrho), \alpha)$. The Hecke correspondence associated to $z \in G'(\mathbf{Q}_p)$ sends x to x' corresponding to $((X'_{\mathcal{L}}, \varrho'), \alpha')$ where $X'_{\mathcal{L}}$ is connected to $X_{\mathcal{L}}$ through a quasi-isogeny τ appearing in the following commutative diagram

$$\begin{array}{ccccc} T_p(X_{\mathcal{L}}) & \overset{\alpha}{\underset{\sim}{\longrightarrow}} & \mathcal{L}' & & \\ & & & \subset & \\ \tau\big\downarrow & & & & V' \\ & & & \subset & \\ T_p(X'_{\mathcal{L}}) & \overset{\alpha}{\underset{\sim}{\longrightarrow}} & z\mathcal{L}' & & \end{array}$$

However, z may be considered as an element of $F^\times$ and defines a self-quasi-isogeny of $X_{\mathcal{L}}$. Furthermore, there is an isomorphism of $X'_{\mathcal{L}}$ with $X_{\mathcal{L}}$ making the following diagram commutative

$$\begin{array}{ccc} X_{\mathcal{L}} & = & X_{\mathcal{L}} \\ \tau\downarrow & & \downarrow z \\ X'_{\mathcal{L}} & \simeq & X_{\mathcal{L}} \ , \end{array}$$

as one checks on Tate modules. This isomorphism allows us to identify $X'_{\mathcal{L}}$ with $X_{\mathcal{L}}$. Under this identification ϱ' becomes $z^{-1}\varrho$ and α' becomes α. Hence

$$(X'_{\mathcal{L}}, \varrho', \alpha') \simeq (X_{\mathcal{L}}, z^{-1}\varrho, \alpha) = (X_{\mathcal{L}}, \varrho z^{-1}, \alpha).$$

Here in the last equation z is considered as a self-quasi-isogeny of $\mathbf{X}$. Recalling the definition of the action of $J(\mathbf{Q}_p)$ on the moduli problem the last equation says that the Hecke correspondence $z \in G'(\mathbf{Q}_p)$ acts on $\breve{\mathbf{M}}_{\mathcal{L},K'}$ as $z \in J(\mathbf{Q}_p)$.

Proposition 5.37 *Let $K' \subset G'(\mathbf{Q}_p)$ and consider the first component of the period morphism*

$$\breve{\pi}^1_{K'} : \breve{\mathbf{M}}_{\mathcal{L}',K'} \longrightarrow \breve{\mathcal{F}}^{wa}.$$

Two points $x_1, x_2 \in \breve{\mathbf{M}}_{\mathcal{L}',K'}$ have the same image under $\breve{\pi}^1_{K'}$ if and only if they are mapped into each other under a Hecke correspondence, i.e. if there exists $g' \in G'(\mathbf{Q}_p)$ and $y \in \breve{\mathbf{M}}_{\mathcal{L}',K' \cap g'K'g'^{-1}}$ with

$$x_1 = \pi_{K' \cap g'K'g'^{-1}, K'}(1)(y), \; x_2 = \pi_{K' \cap g'K'g'^{-1}, K'}(g')(y).$$

In other words, the fibre of $\breve{\pi}^1_{K'}$ *through a point* $x \in \breve{\mathbf{M}}_{\mathcal{L}', K'}$ *may be identified with* $G'(\mathbf{Q}_p)/K'$. *Let* $G'(\mathbf{Q}_p)^1$ *be the subgroup of* $G'(\mathbf{Q}_p)$ *where the values of all* $\mathbf{Q}_p$*-rational characters are units. Then the fibre of the period morphism* $\breve{\pi}_{K'} = \breve{\pi}^1_{K'} \times \breve{\pi}^2_{K'}$ *through* $x \in \breve{\mathbf{M}}_{\mathcal{L}', K'}$ *may be identified with* $G'(\mathbf{Q}_p)^1/K'$.

Proof: Let x_i be represented by $((X^i_{\mathcal{L}}, \varrho^i), \alpha^i)$ over $Spf\ O_K$. Since x_1 and x_2 have the same image in $\breve{\mathcal{F}}^{wa}$, the fully faithfulness of the Fontaine functor implies that there is a B–quasi–isogeny $\tau : X^1_{\mathcal{L}} \to X^2_{\mathcal{L}}$ over O_K such that ϱ^2 equals the composition of ϱ^1 with the special fibre of τ. We define $g' \in G'(\mathbf{Q}_p)$ so as to make the following diagram commutative

$$\begin{array}{ccc} T(X^1_{\mathcal{L}}) \otimes \mathbf{Q} & \overset{\alpha^1}{\underset{\sim}{\longrightarrow}} & V' \\ \downarrow \tau & & \downarrow g' \\ T(X^2_{\mathcal{L}}) \otimes \mathbf{Q} & \overset{\alpha^2}{\underset{\sim}{\longrightarrow}} & V'. \end{array}$$

The double coset of g' mod K' is well–determined by α^1 mod K' and α^2 mod K'. It is easy to see that the point y represented by $((X^1_{\mathcal{L}}, \varrho^1), \alpha^1 \text{mod } K' \cap g'K'g'^{-1})$ has the required property. The last assertion follows easily.

5.38 Let $\mathcal{L}_1$ be another multichain of lattices in V and let $\breve{\mathcal{M}}_1$ be the formal scheme over $Spf\ O_{\breve{E}}$ corresponding to the moduli problem where the multichain $\mathcal{L}$ has been replaced by $\mathcal{L}_1$. Assume that $\mathcal{L}_1$ is a refinement of $\mathcal{L}$ and denote by

$$\pi : \breve{\mathcal{M}}_1 \longrightarrow \breve{\mathcal{M}}$$

the obvious morphism. Let V'_i $(i = 1, \ldots, t_1)$ be the set of isomorphism classes of rational p–adic Tate modules of points $x \in \breve{\mathcal{M}}^{rig}_1$. It is obvious that this is a subset of the corresponding set for $\breve{\mathcal{M}}^{rig}$. We now encounter the problem that it may be a *proper* subset (the worst case would occur if $\breve{\mathcal{M}}^{rig}_1$ were empty - which would contradict the flatness conjecture before (3.36)). We note however that if V occurs in one of these sets it occurs in both. Indeed, let $x \in \breve{\mathcal{M}}^{rig}(K)$ be represented by $(X_{\mathcal{L}}, \varrho)$ over $Spf\ O_K$ and let

$$V_p(X) \simeq V$$

be an isomorphism taking the p-adic Tate module chain $T_p(X_{\mathcal{L}})$ into $\mathcal{L}$. The inverse image of $\mathcal{L}_1$ in $V_p(X)$ under this isomorphism is invariant under the Galois group $Gal(\bar{K}/K')$ of a finite extension K' of K and defines in the obvious way an object $(X'_{\mathcal{L}}, \varrho')$ of $\breve{\mathcal{M}}_1^{rig}$ over $Spf\ O_{K'}$.
We now fix V' which occurs in both the set of isomorphism classes corresponding to $\breve{\mathcal{M}}_1$ and to $\breve{\mathcal{M}}$. Again we denote by $\mathcal{L}'_1$ and $\mathcal{L}'$ and G' the corresponding multichains in V' and the inner form of G. Then $\mathcal{L}'_1$ is a refinement of $\mathcal{L}'$. Let $K'_{\mathcal{L}'_1}$ and $K'_{\mathcal{L}'}$ be the corresponding fix groups in $G'(\mathbf{Q}_p)$. We obtain a commutative diagram of rigid–analytic morphisms

$$\begin{array}{ccc} \breve{\mathcal{M}}^{rig}_{1,\mathcal{L}'_1} & \xrightarrow{\iota} & \breve{\mathbf{M}}_{\mathcal{L}',K'_{\mathcal{L}'_1}} \\ \pi^{rig}_{\mathcal{L}'_1}\Big\downarrow & & \Big\downarrow\pi_{K'_{\mathcal{L}'_1},K'_{\mathcal{L}'}} \\ \breve{\mathcal{M}}^{rig}_{\mathcal{L}'} & = & \breve{\mathbf{M}}_{\mathcal{L}',K'_{\mathcal{L}'}}. \end{array}$$

Since π is proper, so is $\pi^{rig}_{\mathcal{L}'_1}$ and therefore also ι. On the other hand, if $x \in \breve{\mathcal{M}}^{rig}_{\mathcal{L}'}(K)$ is a point represented by $(X_{\mathcal{L}}, \varrho)$ over $Spf\ O_K$, to give a point $x_1 \in \breve{\mathcal{M}}^{rig}_{1,\mathcal{L}'_1}(K)$ represented by $(X_{\mathcal{L}_1}, \varrho)$ over $Spf\ O_K$ mapping to x is equivalent to giving a refinement of type $\mathcal{L}'_1$ of the multichain $T_p(X_{\mathcal{L}})$ of $V_p(X)$, which is $Gal(\bar{K}/K)$-invariant. It follows that ι induces a bijection on points. Similarly one shows using the infinitesimal criterion (5.11) that $\pi^{rig}_{\mathcal{L}'_1}$ is étale. Hence ι is finite, étale and bijective and therefore an isomorphism. It follows that the towers $\{\breve{\mathbf{M}}_{\mathcal{L}'_1,K'}; K' \subset G'(\mathbf{Q}_p)\}$ and $\{\breve{\mathbf{M}}_{\mathcal{L}',K'}; K' \subset G'(\mathbf{Q}_p)\}$ defined by $\breve{\mathcal{M}}^{rig}_{1,\mathcal{L}'_1}$ and $\breve{\mathcal{M}}^{rig}_{\mathcal{L}'}$ are canonically identified.

5.39 Finally we mention that the tower $\{\breve{\mathbf{M}}_{\mathcal{L}',K'}; K' \subset G'(\mathbf{Q}_p)\}$ is independent of the choice of $\mathcal{L}'$. More precisely, let $h' \in G'(\mathbf{Q}_p)$ and let $\mathcal{L}'_1 = h'\mathcal{L}'$ be another candidate for the polarized multichain in V'. We define as follows an isomorphism of towers

$$\begin{array}{rcl} \varphi : \breve{\mathbf{M}}_{\mathcal{L}',K'} & \longrightarrow & \breve{\mathbf{M}}_{\mathcal{L}'_1,h'K'h'^{-1}} \\ (X_{\mathcal{L}}, \varrho, \alpha) & \longmapsto & (X_{\mathcal{L}}, \varrho, h'\alpha) \end{array}$$

One verifies easily that this isomorphism is $\mathrm{Int}(h')$–equivariant with respect to the action of $G'(\mathbf{Q}_p)$ by Hecke operators on both sides,

$$\varphi \circ \pi(g') = \pi_1(h'g'h'^{-1}) \circ \varphi.$$

Here $\pi(g')$ resp. $\pi_1(g')$ refers to the action of $G'(\mathbf{Q}_p)$ on the tower defined by $\mathcal{L}'$ resp. by $\mathcal{L}'_1$. Furthermore, using that $h'K'_{\mathcal{L}'}$ is uniquely defined by $\mathcal{L}'$ and $\mathcal{L}'_1$, one sees that this isomorphism depends only on $\mathcal{L}'$ and $\mathcal{L}'_1$.

5.40 We now want to compare the rigid–analytic Hecke correspondences with their formal counterparts, cf. chapter 4. We start with the data for our moduli problem $(F, B, O_B, V, b, \mu, \mathcal{L})$. We fix $V' = V_i$ $(i = 1, \ldots, t)$ and introduce again the multichain $\mathcal{L}'$, the inner form G' of G and the subgroup $K'_{\mathcal{L}'} \subset G'(\mathbf{Q}_p)$. For an element $g' \in G'(\mathbf{Q}_p)$ we have the rigid–analytic Hecke correspondence

$$T(g')_{\mathcal{L}'} = \breve{\mathbf{M}}_{\mathcal{L}', K'_{\mathcal{L}'} \cap g' K'_{\mathcal{L}'} g'^{-1}} \subset \breve{\mathbf{M}}_{\mathcal{L}', K'_{\mathcal{L}'}} \times \breve{\mathbf{M}}_{\mathcal{L}', K'_{\mathcal{L}'}}.$$

We fix an indexing by $\mathbf{Z}$ of the members of $\mathcal{L}$ and therefore also of $\mathcal{L}'$. The element $g' \in G'(\mathbf{Q}_p)$ defines the function $t = t(g') = (t_{ij}(g')) = (t_{ij}(g'\mathcal{L}', \mathcal{L}'))$, cf. (4.12) and (4.18). (As a matter of fact, g' defines such a function for each simple factor of B, cf. (4.41)). We stress that we are applying the considerations of chapter 4 to V' and $\mathcal{L}'$, not to V and $\mathcal{L}$. Consider the formal Hecke correspondence associated to $t = t(g')$, cf. (4.43) resp. (4.57),

$$\mathcal{C}(t) = \mathcal{C}orr\ (t) \subset \breve{\mathcal{M}} \times \breve{\mathcal{M}}.$$

Let $\mathcal{C}(t)^{rig}_{\mathcal{L}'}$ be the pullback of $\mathcal{C}(t)^{rig}$ under the morphism

$$\breve{\mathcal{M}}^{rig}_{\mathcal{L}'} \times \breve{\mathcal{M}}^{rig}_{\mathcal{L}'} \longrightarrow \breve{\mathcal{M}}^{rig} \times \breve{\mathcal{M}}^{rig}.$$

For any point $((X_{\mathcal{L}}, \varrho), (X'_{\mathcal{L}}, \varrho')) \in \mathcal{C}(t)(Spf\ O_K)$ defining a K–rational point of $\mathcal{C}(t)^{rig}_{\mathcal{L}'}$ the quasi–isogeny $\varrho'\varrho^{-1}$ on the special fibre lifts uniquely to a quasi–isogeny from $X_{\mathcal{L}}$ to $X'_{\mathcal{L}}$ over $Spec\ O_K$. Consider the multichains of their p–adic Tate modules $T_p(X_{\mathcal{L}})$ and $T_p(X'_{\mathcal{L}})$ in their common rational p–adic Tate module $V_p(X) = V_p(X')$. It is obvious that they are in relative position t. Therefore there exists an isomorphism α of $V_p(X)$ with V' which sends $T_p(X_{\mathcal{L}})$ to $g'\mathcal{L}'$ and $T_p(X'_{\mathcal{L}})$ to $\mathcal{L}'$. This construction defines a rigid–analytic morphism

$$\iota : \mathcal{C}(t)^{rig}_{\mathcal{L}'} \longrightarrow T(g'^{-1})_{\mathcal{L}'} ,$$

compatible with the projections to $\breve{\mathcal{M}}^{rig}_{\mathcal{L}'} = \breve{\mathbf{M}}_{\mathcal{L}',K'_{\mathcal{L}'}}$.

Proposition 5.41 *The above morphism is an isomorphism.*

Proof: Since $\mathcal{C}(t)$ is a locally closed formal subscheme in $\breve{\mathcal{M}} \times \breve{\mathcal{M}}$, it follows (5.3), (iv) that $\mathcal{C}(t)^{rig}_{\mathcal{L}'}$ is an admissible open in a Zariski–closed subspace $\overline{\mathcal{C}(t)}^{rig}_{\mathcal{L}'}$ of $\breve{\mathcal{M}}^{rig}_{\mathcal{L}'} \times \breve{\mathcal{M}}^{rig}_{\mathcal{L}'}$. It is obviously contained in $T(g'^{-1})_{\mathcal{L}'}$,

$$\iota : \mathcal{C}(t)^{rig}_{\mathcal{L}'} \subset \overline{\mathcal{C}(t)}^{rig}_{\mathcal{L}'} \subset T(g'^{-1})_{\mathcal{L}'}.$$

However, since the relative position of $(X_{\mathcal{L}}, \varrho)$ and $(X'_{\mathcal{L}}, \varrho')$ over $Spf\ O_K$ can be checked on their p–adic Tate modules, it follows that ι is bijective on points. Since $T(g'^{-1})_{\mathcal{L}'}$ is reduced, ι is an isomorphism.

Remark 5.42 As a consequence of this proposition, $\mathcal{C}(t)^{rig}_{\mathcal{L}'}$ is Zariski–closed in $\breve{\mathcal{M}}^{rig}_{\mathcal{L}'} \times \breve{\mathcal{M}}^{rig}_{\mathcal{L}'}$. This is in contrast with the fact pointed out in (4.44) that $\mathcal{C}(t)$ very often is not Zariski–closed in $\breve{\mathcal{M}} \times \breve{\mathcal{M}}$.
Consider quite generally a locally closed formal subscheme of a formal scheme formally locally of finite type over a complete discrete valuation ring O,

$$i : \mathcal{Y} \to \mathcal{X}.$$

Assume that $i^{rig} : \mathcal{Y}^{rig} \to \mathcal{X}^{rig}$ is a closed embedding. Then, as R. Huber pointed out to us, i is a closed embedding, provided that $\mathcal{Y}$ is O–flat.
The upshot is that the formal Hecke correspondences are formal models of the rigid–analytic Hecke correspondences which are *not flat over* $O_{\breve{E}}$.

5.43 Recall that we defined a Weil descent datum on $\breve{\mathcal{M}}$ over $Spf\ O_E$, cf. (3.48). We are now going to construct a Weil descent datum α on $\breve{\mathcal{F}} \times \Delta$ over $Spec\ E$, compatible with the period morphism. (Here we transpose our terminology from (3.44) - (3.47) which referred to the descent from $Spf\ O_{\breve{E}}$ to $Spf\ O_E$ to the descent problem from $\breve{E}$ to E.)
We place ourselves in the general context of (1.35) and (1.36). Let $\tau = \sigma^f$ be the Frobenius in $Gal(\breve{E}/E)$. Let R be a $\breve{E}$–algebra. We then define the isomorphism α as follows,

$$\begin{array}{rcl} \alpha : \breve{\mathcal{F}} \times \Delta(R) & \longrightarrow & (\breve{\mathcal{F}} \times \Delta)^{\tau}(R) = \breve{\mathcal{F}}(R \otimes_{K_0,\tau^{-1}} K_0) \times \Delta \\ (\mathcal{F}^{\bullet}_{\mu}, \omega) & \longmapsto & (b\sigma)^f(\mathcal{F}^{\bullet}_{\mu}, \omega) = ((b\sigma)^f \mathcal{F}^{\bullet}_{\mu}, \omega + f \operatorname{ord}(b)). \end{array}$$

Here $\operatorname{ord}(b)$ is defined by the identity

$$\langle \operatorname{ord}(b), \chi \rangle = \operatorname{ord}_p \chi(b), \quad \chi \in X^*_{\mathbf{Q}_p}(G).$$

Since under the isomorphism of isocrystals

$$(b\sigma)^f : V \otimes K_0 \longrightarrow (V \otimes K_0)_{[\tau]}$$

the filtration $\mathcal{F}^{\bullet}_{\mu}$ is carried into $(b\sigma)^f(\mathcal{F}^{\bullet}_{\mu})$ it is obvious that α respects the admissible open subset $\breve{\mathcal{F}}^{wa} \times \Delta$.

5.44 Let $\nu = \nu_b$ be the slope homomorphism associated to $b \in G(K_0)$. Let $s > 0$ be a sufficiently divisible integer such that $s\nu$ factors through $\mathbf{G}_m$. Let $\bar{\gamma}_s = s\nu(p) \in G(K_0)$. Obviously $\operatorname{ord} \bar{\gamma}_s = s[\nu] \in \Delta$ where $[\nu]$ is defined by the identity

$$\langle [\nu], \chi \rangle = \chi \circ \nu \in \mathbf{Q}, \; \chi \in X^*_{\mathbf{Q}_p}(G).$$

Under either of the following two conditions the quotients of $\breve{\mathcal{F}} \times \Delta$ resp. $\breve{\mathcal{F}}^{wa} \times \Delta$ by $\bar{\Gamma}_s = \bar{\gamma}_s^{\mathbf{Z}}$ exist in the category of rigid–analytic spaces.

(i) b is basic.
Indeed, in this case the action of $\bar{\Gamma}_s$ on the first factor is trivial.

(ii) $[\nu] \neq 0$.
Indeed, in this case the action of $\bar{\Gamma}_s$ is obviously properly discontinuous for the Zariski topology on $\breve{\mathcal{F}} \times \Delta$.

Since $\bar{\gamma}_s$ commutes with $b\sigma$ the descent datum α induces a Weil descent datum on the quotient,

$$\alpha : (\breve{\mathcal{F}} \times \Delta)/\bar{\Gamma}_s \longrightarrow (\breve{\mathcal{F}} \times \Delta/\bar{\Gamma}_s)^{\tau}.$$

Proposition 5.45 *We assume condition* (i) *or* (ii) *above satisfied. The Weil descent datum on* $(\breve{\mathcal{F}} \times \Delta)/\bar{\Gamma}_s$ *is effective and defines a projective scheme* $\mathcal{F}_s$ *over* E. *There is an admissible open rigid–analytic subset* $\mathcal{F}_s^{wa}$ *in* $\mathcal{F}_s$ *with an action of* $J(\mathbf{Q}_p)$ *defined over* E *such that*

$$\mathcal{F}_s^{wa}(\mathbf{C}) = (\check{\mathcal{F}}^{wa}(\mathbf{C}) \times \Delta)/\bar{\Gamma}_s.$$

If s' is divisible by s, then the action of $\bar{\Gamma}_s/\bar{\Gamma}_{s'}$ on $\mathcal{F}_{s'}$ is defined over E and

$$\mathcal{F}_s = \mathcal{F}_{s'}/(\bar{\Gamma}_s/\bar{\Gamma}_{s'}),\ \mathcal{F}_s^{wa} = \mathcal{F}_{s'}^{wa}/(\bar{\Gamma}_s/\bar{\Gamma}_{s'}).$$

Proof: We first assume that b is a decent element satisfying a decency equation for s, cf. (1.8). Then $\bar{\gamma}_s \in G(\mathbf{Q}_{p^s})$ and $b \in G(\mathbf{Q}_{p^s})$. Furthermore, the equation $(b\sigma)^s = \bar{\gamma}_s \cdot \sigma^s$ shows that $\alpha^s = \alpha^{\tau^{s-1}} \ldots \alpha^\tau \alpha$ coincides with the Weil descent datum for the obvious form $((\mathcal{F} \otimes_E E_s) \times \Delta)/\bar{\Gamma}_s$ over the unramified extension E_s of E of degree s. In particular, α^s is effective. Since we are dealing with projective schemes it follows (cf. (3.47)) that α itself is also effective. In order to construct the admissible open subset $\mathcal{F}_s^{wa}$ of $\mathcal{F}_s$ it suffices by proposition (1.34) to define a closed subset

$$\mathcal{H} \subset \mathcal{F}_s \times (T \otimes_{\mathbf{Q}_p} E)$$

such that

$$\mathcal{F}_s^{wa}(\mathbf{C}) = (\check{\mathcal{F}}^{wa}(\mathbf{C}) \times \Delta)/\bar{\Gamma}_s = \mathcal{F}_s \setminus \bigcup_{t \in T(\mathbf{Q}_p)} \mathcal{H}_t.$$

We return to the proof of (1.36). Let V be a faithful $\mathbf{Q}_p$–rational representation. For T we take the same scheme as in the proof of (1.36) except that we apply the considerations there to s' where $\mathbf{Q}_{p^{s'}}.E = E_s$. Therefore T is a $\mathbf{Q}_p$–form of the $\mathbf{Q}_{p^{s'}}$–scheme T' parametrizing the subspaces V' of $V \otimes \mathbf{Q}_{p^{s'}}$ compatible with the isotypical decomposition. Instead of giving $\mathcal{H}$ we give

$$\mathcal{H}' \subset (\mathcal{F}_s \otimes_E E_s) \times (T' \otimes_{\mathbf{Q}_{p^{s'}}} E_s)$$

invariant under the descent datum. Let R be a E_s–algebra. We put

$$\mathcal{H}'(R) = \{((\mathcal{F}^\bullet, \omega), V') \in (\mathcal{F}(R) \times \Delta/\bar{\Gamma}_s) \times T'(R); \sum i \operatorname{rk}(gr^i_{\mathcal{F} \cap V'}(V')) > \sum \lambda \operatorname{rk}(V'_\lambda)\}.$$

This proves the second statement of the proposition in this case. The last statement is obvious and allows us to reduce the case of an arbitrary $s > 0$ to the case where s is sufficiently divisible.

If now $\bar{b}$ is a decent σ–conjugacy class and $b \in \bar{b}$ is an arbitrary element, write $b = gb'\sigma(g)^{-1}$ with $b' \in \bar{b}$ descent and $g \in G(K_0)$. Then $\mathcal{F}^\bullet \mapsto g^{-1}(\mathcal{F}^\bullet)$ defines an isomorphism between the quotients of $\tilde{\mathcal{F}}'$ by $s\nu_b(p)$ and by $s\nu_{b'}(p)$ compatible with the respective Weil descent data. Therefore the result already proved for b' implies the corresponding result for b. The assertion for a non–decent σ–conjugacy class is treated as in the proof of (ii) of proposition (1.36) by embedding the group G in a larger one, G_1, in which it becomes decent. □

5.46 We claim that the period morphism

$$\breve{\pi} : \breve{\mathcal{M}}^{rig} \longrightarrow \breve{\mathcal{F}} \times \Delta$$

is compatible with the Weil descent data. Here we must be careful that we are working with covariant Dieudonné theory. To define the Weil descent datum on the period domain we use the pair $(w(p)^{-1}b, w^{-1}\mu)$ where

$$w : \mathbf{G}_m \longrightarrow G, \ \ t \longmapsto t \cdot \mathrm{id}_V$$

is the obvious central cocharacter. We remark that the spaces $\breve{\mathcal{F}}^{wa}$ obtained from (b, μ) and $(w(p)^{-1}b, w^{-1}\mu)$ are canonically isomorphic, via $\mathcal{F}^\bullet \mapsto (\mathcal{F} \otimes \mathbf{1}(-1))^\bullet$. The compatibility of the Weil descent data now follows immediately from the fact that the Frobenius morphism of the p–divisible group $\mathbf{X}$ relative to the residue field κ of E induces on $V \otimes K_0 = M(\mathbf{X}) \otimes \mathbf{Q}$ the inverse of the f–th power of the Verschiebung,

$$(Frob_E^{-1})^* = \mathbf{V}^{-f} = (w(p^{-1}) \cdot b \cdot \sigma)^f .$$

We also remark that if we exclude the uninteresting case that $\mathbf{X}$ is étale, the condition (ii) of (5.44) relative to $(w(p)^{-1}b, w^{-1}\mu)$ is satisfied. Applying proposition (5.45) we obtain for varying s a projective system of morphisms of rigid spaces defined over E,

$$\pi : \mathcal{M}_s \longrightarrow \mathcal{F}_s^{wa} .$$

5.47 It follows immediately from the definition (3.48) that the Weil descent datum α on $\breve{\mathcal{M}}$ lifts to the universal p–divisible group on $\breve{\mathcal{M}}$. We therefore also obtain a Weil descent datum on the towers $\{\breve{\mathbf{M}}_{\mathcal{L}',K'};\ K' \subset G'(\mathbf{Q}_p)\}$, compatible with the period morphisms $\breve{\pi}_{K'}$, $K' \subset G'(\mathbf{Q}_p)$ for the various forms V_i', $i = 1, \dots, t$. After passage from $\breve{\mathcal{M}}$ to $\breve{\mathcal{M}}_s$, the finite étale cover $\breve{\mathbf{M}}_{\mathcal{L}',K'}$ descends to a finite étale cover $\mathbf{M}_{\mathcal{L}',K',s}$ of the descended rigid space $\mathcal{M}^{rig}_{\mathcal{L}',s}$ of $\breve{\mathcal{M}}^{rig}_{\mathcal{L}',s}$. For varying s these form a projective system.

5.48 We now give some examples for the period morphism. We start with the Drinfeld example (3.58). We will compare the period map (5.16) in the Drinfeld case to the map given in [Dr2]. This compatibility result was communicated to us by G. Faltings. We keep the notation of (1.44): $D, F, \varepsilon : F \to \bar{\mathbf{Q}}_p \subset \mathbf{C}_p, E = \varepsilon(F), \pi \in F, \Pi \in D, \tau$. The invariant of D is $1/d$.
In (3.54) we constructed a crystal $(\widetilde{\mathbf{M}}, \widetilde{\mathbf{V}}, \widetilde{\mathbf{F}})$ of a special formal O_D-module $\mathbf{X}$ over $\bar{\mathbf{F}}_p$. Let $\mathbf{M} \subset \widetilde{\mathbf{M}}$ be the associated $\tau - W_F(\bar{\mathbf{F}}_p)$–crystal with operators $\mathbf{V}$ and $\mathbf{F}$. We identify the O_F-module of invariants $\mathbf{M}_0^{\Pi\mathbf{V}^{-1}}$ with O_F^d (3.73). By (1.46) the period space Ω_F^d in this example consists of the points of $\mathbf{P}_F^{d-1}$ which do not lie on any F-rational hyperplane.
By (5.16) we have a map of rigid analytic spaces over $\breve{E}$:

$$\breve{\pi}^1 : \breve{\mathcal{M}}^{rig} \longrightarrow \Omega_F^d \otimes_{F,\varepsilon} \breve{E}\,.$$

Let $K \subset \mathbf{C}_p$ be a finite extension of K_0, which contains $\varepsilon(F)$. Then $\breve{\pi}^1$ is defined on the K-valued points as follows. A point of $\breve{\mathcal{M}}^{rig}$ is given by a special formal O_D-module X over O_K, and a quasi–isogeny $\varrho : \mathbf{X}_R \to X_R$ over the $\bar{\mathbf{F}}_p$-algebra $R = O_K/pO_K$. We use ϱ to identify the isocrystals of $\mathbf{X}_R$ and X_R. Hence the value of the isocrystal of X_R at the PD-extension $Spec\,R \to Spec\,O_K$ is identified with $\widetilde{\mathbf{M}}_{\mathbf{Q}} \otimes_{K_0} K$. The Hodge filtration associated to X gives a map

$$\widetilde{\mathbf{M}}_{\mathbf{Q}} \otimes_{K_0} K \twoheadrightarrow \mathrm{Lie}\; X \otimes_{O_K} K\,.$$

By the decompositions in (3.56) and (3.58) we find a point of $\mathbf{P}_F^d(K)$,

$$F^d = \mathbf{M}_{\mathbf{Q},0}^{\Pi\mathbf{V}^{-1}} \twoheadrightarrow \mathrm{Lie}^0 X \otimes_{O_K} K\,.$$

It is weakly admissible by Fontaine and therefore in $\Omega_F^d(K)$. This defines the image of (X, ϱ) under $\breve{\pi}^1$.

5.49 On the other hand we have Drinfeld's period map, which is defined by the isomorphism

$$\breve{\mathcal{M}} \longrightarrow \hat{\Xi}_F^d \times_{Spf\, O_F} Spf\, O_{\breve{E}}\,.$$

By (3.71) we have a canonical isomorphism of rigid analytic spaces $\Xi_F^d \simeq \Omega_F^d \times \mathbf{Z}$. Hence we get the Drinfeld period map

$$\breve{\pi}_{\mathrm{Dr}} : \breve{\mathcal{M}}^{rig} \longrightarrow \Omega_F^d \otimes_F \breve{E}\,.$$

On the K-valued points this map is explicitly described as follows. Let $\widetilde{M}$ be the Cartier module of the p-divisible group X over O_K. Let $M \subset \widetilde{M}$ be the Cartier module relative to O_F and $N = \bigoplus_{i=0}^{d-1} M/U$ together with the operator φ the module defined by (3.75). Let T be any artinian quotient of O_K and $X_T, \widetilde{M}_T, M_T, N_T$, etc. the objects obtained by base change. Then by Drinfeld [Dr2] (compare 3.65) for any surjection $T \to \bar{T}$ we have an isomorphism

$$\eta_T = N_T^{\varphi} \xrightarrow{\sim} \eta_{\bar{T}} = N_{\bar{T}}^{\varphi}\,.$$

Considering the degree zero part with respect to the $\mathbf{Z}/d$-grading we get a map

$$\begin{aligned} \eta_{\bar{\mathbf{F}}_p,0} \subset \varprojlim_T N_{T,0} = N_0 \quad &\longrightarrow \quad M_0/\mathbf{V}M_0 \\ (x_0,\ldots,x_d) \quad &\longrightarrow \quad x_0 \bmod \mathbf{V}M_0. \end{aligned} \tag{5.9}$$

The quasi-isogeny $\mathbf{X} \to X_{\bar{\mathbf{F}}_p}$ provides us with an isomorphism

$$F^d = \mathbf{M}_{0,\mathbf{Q}}^{\Pi V^{-1}} \xrightarrow{\sim} \eta_{\bar{\mathbf{F}}_p,0} \otimes_{O_F} F\,.$$

Combining this with the map (5.9) we get the definition of the Drinfeld period of (X, α):

$$F^d \cong \eta_{\bar{\mathbf{F}}_p,0} \otimes_{O_F} F \longrightarrow M_0/\mathbf{V}M_0 \otimes_{O_K} K = \mathrm{Lie}^0 X \otimes_{O_K} K\,. \tag{5.10}$$

To show that this map coincides with the period given by $\breve{\pi}^1$, it is enough to verify that the map deduced from (5.10)

$$M_{\bar{\mathbf{F}}_p,0} \otimes \mathbf{Q} = \eta_{\bar{\mathbf{F}}_p,0} \otimes_{O_F,\varepsilon} W_F(\bar{\mathbf{F}}_p)_{\mathbf{Q}} \longrightarrow \mathrm{Lie}^0 X \otimes_{O_K} K$$

coincides with the period map given by proposition (5.26).
Let T be any quotient of O_K. Consider the maps

$$\begin{array}{rcl} M_T & \xrightarrow{\iota} & N_T \xrightarrow{\lambda} M_T \\ m & \longmapsto & (m, 0 \ldots 0) \\ (m_0, \ldots, m_{d-1}) & \longmapsto & \Pi^{d-1} m_0 + \Pi^{d-2} \mathbf{V} m_1 + \cdots + \mathbf{V}^{d-1} m_{d-1} \end{array}$$

We have $\iota \circ \lambda = \Pi^{d-1}$ and $\lambda \circ \iota = \Pi^{d-1}$. Hence $\iota \otimes \mathbf{Q}$ is an isomorphism

$$M_T \otimes \mathbf{Q} \simeq N_T \otimes \mathbf{Q}\,, \tag{5.11}$$

which we use to identify these modules. It is easily checked from the defining relation $\lambda \mathbf{L} = \mathbf{F}$ for the operator $\mathbf{L}$, that φ induces on the left hand side of (5.11) the operator $\Pi \mathbf{V}^{-1}$. Hence from (5.11) we get an inclusion $\eta_{\bar{\mathbf{F}}_p,0} \subset M_0 \otimes \mathbf{Q}$. Its composition with the canonical map $M_0 \otimes \mathbf{Q} \to \mathrm{Lie}^0 X \otimes \mathbf{Q}$ is the Drinfeld period. The coincidence of the period maps follows if we show that this composition $\eta_{\bar{\mathbf{F}}_p,0} \to \mathrm{Lie}^0 X \otimes \mathbf{Q}$ coincides with the map $\bar{\tau}$ of proposition (5.26). More precisely $\bar{\tau}$ is the map associated to the extension of Cartier modules

$$0 \longrightarrow C(p\mathrm{Lie}X) \longrightarrow \widetilde{M}_X \longrightarrow \widetilde{M}_{X_R} \longrightarrow 0\,,$$

where $R = O_K/pO_K$.

Let $\bar{m} \in \eta_{\bar{\mathbf{F}}_p,0} \subset M_{\bar{\mathbf{F}}_p} \otimes \mathbf{Q}$. The inclusion $\eta_{\bar{\mathbf{F}}_p,0} \subset M \otimes \mathbf{Q}$ provides us with a lifting $m \in M \otimes \mathbf{Q}$, which satisfies $\Pi m = \mathbf{V} m$. The claimed coincidence would follow from the equation

$$\tau \bar{m} = m\,. \tag{5.12}$$

Here the notation τ has the meaning of (5.25). Since M is torsion free as an abelian group we may assume that $m \in M \subset M \otimes \mathbf{Q}$.

We set $n = [F : \mathbf{Q}_p]$. Then the equation $\mathbf{V}\bar{m} = \Pi \bar{m}$ implies $\widetilde{\mathbf{V}}^{nd}\bar{m} = \varepsilon p \bar{m} \subset \widetilde{\mathbf{F}} \widetilde{M}_{\bar{\mathbf{F}}_p}$ since $\Pi^{d \cdot e(F/\mathbf{Q}_p)} = \varepsilon p$, with $\varepsilon \in O_F^{\times}$. Since the same equation holds for m we conclude

$$\sigma_r \widetilde{\mathbf{V}}^{rnd} \bar{m} = \widetilde{\mathbf{V}}^{rnd} m \bmod p^{r-h} C(p\mathrm{Lie}X)$$

and

$$\widetilde{\mathbf{F}}^{rnd} \sigma_r \widetilde{\mathbf{V}}^{rnd} \bar{m} = p^{nrd} m \bmod p^{nr+n-h} C(p\mathrm{Lie}X)\,,$$

where h is defined in (5.21). Hence we obtain

$$\tau_{hnd}\bar{m} = \lim_{r\longrightarrow\infty} \frac{1}{p^{(r-h)nd}} p^{nrd} m = p^{hnd} m\,.$$

This proves (5.12) and finishes the proof that the period map used by us agrees with Drinfeld's.

5.50 We now consider the moduli problem of formal O_F–modules of dimension 1 and height d, cf. (3.78). In this case the period morphism was constructed by Gross and Hopkins [HG1,2] in a different way which is, however, essentially equivalent to our definition. It is a morphism

$$\breve{\pi} : \breve{\mathcal{M}} \longrightarrow \mathbf{P}^{d-1}_{\breve{E}} \times \mathbf{Z}.$$

We noted in (1.48) that all points in the target space are weakly admissible. On the other hand, Laffaille [L] has proved Fontaine's conjecture in the one–dimensional case, i.e. for formal O–modules of dimension 1. (In [L] he only considers the case $O = \mathbf{Z}_p$, but the proof seems to work in the general case.)

The moduli problem is of type (EL) hence there is no twisting, i.e. $G' = G = Res_{F/\mathbf{Q}_p} GL_d$. Let $K \subset GL_d(F)$ be an open and compact subgroup and consider the period morphism which is equivariant with respect to the action of the multiplicative group of the central division algebra with invariant $= 1(\text{mod } d)$ over F,

$$\breve{\pi}_K : \breve{\mathbf{M}}_K \longrightarrow \mathbf{P}^{d-1}_{\breve{E}} \times \mathbf{Z}.$$

It follows that $\breve{\pi}_K$ is a surjective étale morphism, with fibres of the form

$$\breve{\pi}_K^{-1}(\breve{\pi}_K(x)) = GL_d(F)^1/K.$$

Here $GL_d(F)^1 = \{g \in GL_d(F);\ \det(g) \in O_F^\times\}$.

5.51 We now consider the period morphism for the example considered in (3.80) resp. (1.50). We continue to use the same notations. In this case we had a decomposition into a disjoint sum

$$\breve{\mathcal{M}} = \coprod_{GL_n(\mathbf{Q}_p)/GL_n(\mathbf{Z}_p)\times GL_n(\mathbf{Q}_p)/GL_n(\mathbf{Z}_p)} \breve{\mathcal{M}}^{(g_1,g_2)}.$$

We abbreviate $(\check{\mathcal{M}}^{(1,1)})^{rig}$ into $\check{\mathbf{M}}^0$ and consider the restriction $\bar{\pi}$ of the first component of the period morphism to $\check{\mathbf{M}}^0$,

$$\bar{\pi} : \check{\mathbf{M}}^0 = (Spf\ W(\bar{\mathbf{F}}_p)[[T_{11}, \ldots, T_{nn}]])^{rig} \longrightarrow Grass_n(V \otimes K_0).$$

The source of this morphism is identified with the open unit disc of dimension n^2. The period morphism in this case is due to Dwork (comp. [Ka1]). The weakly admissible subset was determined in (1.50). It is the big cell formed by the subspaces transversal to $V_+ \subset V$. Let

$$\tau_{ij},\quad i,j = 1, \ldots, n$$

be the canonical coordinates on the big cell $\mathbf{A}^{n^2}$, i.e. τ_{ij} is the function corresponding to the element in $\mathrm{Hom}\,(V_-, V_+)$ which sends e_i into e_{n+j} and all other basis elements to zero. Recall from the proof of (3.81) the rigid–analytic functions on $\check{\mathbf{M}}^0$,

$$q_{ij},\quad i,j = 1, \ldots, n.$$

Proposition 5.52 *The period morphism $\bar{\pi}$ is given by the following formula,*

$$\bar{\pi}^*(\tau_{ij}) = \log q_{ij},\quad i,j = 1, \ldots, n.$$

Proof: Let

$$\nabla : M_X^{rig} \longrightarrow M_X^{rig} \otimes \Omega^1_{\check{\mathbf{M}}^0/K_0}$$

be the Gauss–Manin connection on the Lie algebra of the universal extension of the universal p–divisible group. Its sheaf of locally constant sections is the constant sheaf $V \otimes K_0$. Let $\mathcal{F}$ be the universal Hodge filtration,

$$0 \to \mathcal{F} \to M_X^{rig} \to Lie\ (X)^{rig} \to 0.$$

Since $\mathcal{F}$ is transversal to $V_+ \otimes K_0$ at every point, $\mathcal{F}$ projects isomorphically to $V_- \otimes \mathcal{O}_{\check{\mathbf{M}}^0}$ (where the latter is considered as the quotient by $V_+ \otimes \mathcal{O}_{\check{\mathbf{M}}^0}$). Let ω_i be the global section of $\mathcal{F}$ projecting to $e_i \otimes 1 \in V_- \otimes \mathcal{O}_{\check{\mathbf{M}}^0}$, $i = 1, \ldots, n$. Then obviously,

$$\omega_i = e_i \otimes 1 + \sum_j e_{n+j} \otimes (\tau_{ij} \circ \bar{\pi}).$$

Hence

$$\nabla(\omega_i) = \sum_j e_{n+j} \otimes d(\tau_{ij} \circ \breve{\pi}), \quad i = 1, \ldots, n.$$

On the other hand, by Katz [Ka3], 4.3.1.,

$$\nabla(\omega_i) = \sum_j e_{n+j} \otimes d \log q_{ij}, \quad i = 1, \ldots, n.$$

The result follows since both functions in (5.52) vanish at the origin. For the right hand side this is obvious. For the left hand side it follows from the fact that $\mathcal{F}_o = V_- \otimes K_0$ since the origin corresponds to the canonical lifting $\hat{\mathbf{G}}_m^n \times (\mathbf{Q}_p/\mathbf{Z}_p)^n$ of $\mathbf{X}$ to $W(\bar{\mathbf{F}}_p)$.

5.53 In the examples (5.50) and (5.51) the period morphism is not quasi–compact. In fact, in these cases the fibres of $\breve{\pi}$ are infinite. We now give an example which seems to show that the phenomenon that $\breve{\pi}$ may not be quasi–compact, lies deeper.
We take up the example of type (EL) with rational data introduced in (1.49)(i), for integers r and s such that $r+s$ is prime to d. The corresponding variety $\mathcal{F}$ is defined over $E = \mathbf{Q}_p$ and is an unramified twist of a Grassmannian,

$$\mathcal{F} \times_{Spec\, \mathbf{Q}_p} Spec\, K_0 = Grass_{d-r}(W).$$

Here W denotes a d–dimensional K_0–vector space. There is only one choice of a lattice chain $\mathcal{L}$ possible. We denote by $\breve{\mathcal{M}}$ the corresponding formal scheme over $Spf\, W(\bar{\mathbf{F}}_p)$. We have a decomposition of $\breve{\mathcal{M}}$ according to the height of the quasi–isogeny ϱ,

$$\breve{\mathcal{M}} = \coprod_{n \in \mathbf{Z}} \breve{\mathcal{M}}^{(n)}.$$

We consider the first component of the period morphism restricted to one of these summands,

$$\pi^{(n)} : \breve{\mathcal{M}}^{(n)rig} \longrightarrow Grass_{d-r}(W).$$

In this case there is no twisting, i.e. $G = G'$, and $K_{\mathcal{L}}$ is the unique maximal compact subgroup of $G(\mathbf{Q}_p)$. Furthermore, $K_{\mathcal{L}}$ is the stabilizer of $\breve{\mathcal{M}}^{(n)rig}$ in the group of Hecke correspondences. Proposition (5.37) implies that $\pi^{(n)}$ is injective on points. On the other hand, we noted in (1.49)(i) that all points of $Grass_{d-r}(W)$ are weakly admissible. A slight variant of proposition (5.28) shows that $\pi^{(n)}$ is surjective, provided that $\breve{\mathcal{M}}^{(n)rig}$ is non-empty and that Fontaine's conjecture holds true. Therefore under this hypothesis, if $\pi^{(n)}$ were quasi-compact it would follow that $\pi^{(n)}$ is an isomorphism. This is absurd since the rigid-analytic finite étale coverings $\breve{\mathbf{M}}_{\mathcal{L},K'}$ of $\breve{\mathcal{M}}^{rig}$ would define non-trivial algebraic étale coverings of the Grassmannian, which is impossible. We therefore see (assuming the above hypothetical statements) that the period morphism is not quasi-compact in a very serious sense in this case.

5.54 We conclude this chapter with some more details on the conjecture in (1.37). We return to the general setting of (1.37). We therefore fix an algebraic group G over $\mathbf{Q}_p$ and a conjugacy class of cocharacters μ and denote by $\mathcal{F}$ the corresponding partial flag variety over the Shimura field E. Let $b \in G(K_0)$ and let J be the corresponding algebraic group over $\mathbf{Q}_p$. We had constructed a 1-cocycle with values in G which measured the difference between two fibre functors on $\mathcal{REP}(G)$, cf.(1.19). The class of this cocycle should only depend on the conjugacy class of μ, cf. (1.20). Let G' be the corresponding inner form of G (defined by the image 1-cocycle with values in G_{ad}). The conjecture in (1.37) would define a $G'(\mathbf{Q}_p)$-equivariant tower of rigid-analytic étale coverings of $(\mathcal{F}_b^{wa})'$, where $(\mathcal{F}_b^{wa})'$ maps to $\mathcal{F}_b^{wa}$ by a bijective étale morphism,

$$\{\breve{\mathbf{M}}_{\mathcal{L},K'};\ K' \subset G'(\mathbf{Q}_p)\},$$

where each member is equipped with a lifting of the action of $J(\mathbf{Q}_p)$ on $\mathcal{F}_b^{wa}$.

We also mention the following speculation. Assume that b is basic and denote by $\check{G} = J$ the corresponding inner form of G. We also assume that the 1-cocycle measuring the difference between the two fibre functors is trivial, so that $G' = G$. Let $\{\breve{\mathbf{M}}_K; K \subset G(\mathbf{Q}_p)\}$ be the corresponding tower. We identifiy G_{K_0} with $\check{G}_{K_0}$. Let $\check{b} \in \check{G}(K_0)$ correspond to b^{-1} under this identification and let the conjugacy class of cocharacters $\check{\mu}$ of $\check{G}$ correspond to μ^{-1} under this identification. The corresponding partial flag variety $\check{\mathcal{F}}$ is the *opposite* of $\mathcal{F}$. Since the pair $(\check{b}, \check{\mu})$ satisfies the same hypotheses relative to $\check{G}$ as (b, μ) relative to G and since the inner form of $\check{G}$ defined by $\check{b}$ is equal to G, we obtain a tower $\{\check{\check{\mathbf{M}}}_{\check{K}}; \check{K} \subset \check{G}(\mathbf{Q}_p)\}$ of étale coverings of $(\check{\mathcal{F}}_{\check{b}}^{wa})'$ on which $\check{G}(\mathbf{Q}_p)$ acts as a group of the Hecke correspondences and with a lifting of the action of $G(\mathbf{Q}_p)$ on $\check{\mathcal{F}}_{\check{b}}^{wa}$. The conjecture is that there exists a rigid–analytic space X with an action of $G(\mathbf{Q}_p) \times \check{G}(\mathbf{Q}_p)$ such that

$$\check{\mathbf{M}}_K = X/K \ (K \subset G(\mathbf{Q}_p)), \check{\check{\mathbf{M}}}_{\check{K}} = X/\check{K} \ (\check{K} \subset \check{G}(\mathbf{Q}_p)),$$

compatible with the actions of $G(\mathbf{Q}_p)$ and $\check{G}(\mathbf{Q}_p)$ on both sides. This is the natural generalization of a modification of a conjecture of B. Gross. That Gross's original formulation had to be modified was pointed out by Drinfeld. Gross was considering the case where (G, b, μ) is the Drinfeld example, cf.(1.46) and (3.54). In this case $(\check{G}, \check{b}, \check{\mu})$ leads to the moduli problem of formal O_F–modules of dimension $d-1$ and height d (dual in some sense to the moduli problem of formal O_F–modules of dimension 1 and height d considered in (5.50)).

6. The p-adic uniformization of Shimura varieties

In this chapter we establish non-archimedean uniformization theorems for Shimura varieties. We define a global moduli problem and prove a non-archimedean uniformization theorem for its formal completion along an isogeny class. In the case of a *basic* isogeny class this theorem can be strengthened considerably. Finally we establish a *p-adic* uniformization theorem under very special hypotheses.

6.1 In this chapter we use different notations. Let B be a finite dimensional semisimple algebra over $\mathbf{Q}$. Let $*$ be a positive involution on B. Then the center F of B is a product of CM fields and totally real number fields. Let V be a $\mathbf{Q}$-vector space with an alternating bilinear form (,) with values in $\mathbf{Q}$. Assume that V is equipped with a B-module structure, such that

$$(bv, w) = (v, b^* w), \ v, w \in V, \ b \in B.$$

Let $G \subset GL_B(V)$ be the closed algebraic subgroup over $\mathbf{Q}$, such that

$$G(\mathbf{Q}) = \{g \in GL_B(V) \mid (gv, gw) = c(g)(v, w), c(g) \in \mathbf{Q}\}.$$

We set $\mathcal{S} = \mathrm{Res}_{\mathbf{C}/\mathbf{R}} \mathbf{G}_{m,\mathbf{C}}$. Let $h : \mathcal{S} \to G_{\mathbf{R}}$ be a morphism that defines on $V_{\mathbf{R}}$ a Hodge structure of type (1,0), (0,1), such that $(v, h(\sqrt{-1})w)$ is a symmetric positive bilinear form on $V \otimes \mathbf{R}$, i.e. the Riemann period relations are fullfilled. We get a decomposition

$$V_{\mathbf{C}} = V_{0,\mathbf{C}} \oplus V_{1,\mathbf{C}} \tag{6.1}$$

where $\mathcal{S}$ acts on $V_{0,\mathbf{C}}$ via the character $\bar{z}$ and on $V_{1,\mathbf{C}}$ via z. These data define by Deligne [De] a Shimura variety over the Shimura field

$$E = \mathbf{Q}[\{Tr_{\mathbf{C}}(b \mid V_{0,\mathbf{C}})\}_{b \in B}]$$

In fact we denote here by h, what is $\bar{h}_0^{-1}$ in the notation of [De] 4.9., where $\bar{h}_0$ denotes the composition of h_0 with the complex conjugation $\mathcal{S} \longrightarrow \mathcal{S}$. But we retain the same reciprocity law as in [De], i.e. the reciprocity law obtained from $\bar{h}^{-1}$ by loc. cit. 3.9.1.
Making suitable hypotheses for the prime number p that ensure good reduction, Kottwitz [Ko3] defines a moduli scheme of abelian varieties that is defined over $O_E \otimes \mathbf{Z}_{(p)}$. The general fibre of the moduli scheme contains the Shimura variety mentioned above as a connected component. The other connected components are Shimura varieties too, but they may correspond to different groups.
We will consider the moduli problem of Kottwitz but for less restrictive conditions on p.

6.2 We consider an order O_B of B, such that $O_B \otimes \mathbf{Z}_p$ is a maximal order of $B \otimes \mathbf{Q}_p$. We assume that $O_B \otimes \mathbf{Z}_p$ is invariant under the involution. The existence of such an order is a condition on p. Assume that we are given a selfdual multichain $\mathcal{L}$ of $O_B \otimes \mathbf{Z}_p$-lattices in $V \otimes \mathbf{Q}_p$ with respect to the antisymmetric form (,).
Let $C^p \subset G(\mathbf{A}_f^p)$ be an open compact subgroup. We fix embeddings $\bar{\mathbf{Q}} \longrightarrow \mathbf{C}$ and $\bar{\mathbf{Q}} \xrightarrow{\nu} \bar{\mathbf{Q}}_p$. The ν-adic completion of the Shimura field E will be denoted by $E_\nu \subset \bar{\mathbf{Q}}_p$.
Using the isomorphism $\mathcal{S}_{\mathbf{C}} \simeq \mathbf{C}^* \times \mathbf{C}^*, z \longmapsto z \times \bar{z}$, we define μ to be the composition

$$\mathbf{C}^* \xrightarrow{(z,1)} \mathbf{C}^* \times \mathbf{C}^* \xrightarrow{h_{\mathbf{C}}} G_{\mathbf{C}}\,.$$

Let $K \subset \bar{\mathbf{Q}}_p$ be a finite extension of $\mathbf{Q}_p$, such that (6.1) is defined over K,

$$V_K = V_0 \oplus V_1.$$

Let $\mathbf{V}$ be the $\mathbf{Z}_p$-scheme that represents the functor $R \longmapsto O_B \otimes_{\mathbf{Z}} R$ on the category of $\mathbf{Z}_p$-algebras R. Then as in (3.23), a) we may view

$$\det_{\mathrm{K}}(b\,;\,V_0), \qquad b \in B$$

as a morphism of schemes $\mathbf{V}_{O_K} \to \mathbf{A}^1_{O_K}$. One verifies that this morphism is defined over O_{E_ν}.

6.3 We will work with a category AV, which is defined as follows: The objects of AV are pairs (A, ι), where A is an abelian scheme over some base S and ι is a homomorphism

$$\iota : O_B \otimes \mathbf{Z}_{(p)} \longrightarrow \mathrm{End} A \otimes \mathbf{Z}_{(p)}.$$

The homomorphisms between two objects (A_1, ι_1) and (A_2, ι_2) are homomorphisms between abelian varieties that respect the O_B-actions, tensored with $\mathbf{Z}_{(p)}$

$$\mathrm{Hom}_{AV}(A_1, A_2) = \mathrm{Hom}_{O_B}(A_1, A_2) \otimes \mathbf{Z}_{(p)}.$$

We call AV the category of *abelian O_B-varieties up to isogeny of order prime to p*. It is a fibred category over the category of schemes. An isogeny in AV is a quasi–isogeny of abelian schemes relative to some base scheme S, which is a morphism in AV. We note that the p-divisible group of an object of AV is well-defined, i.e. functorial. An isogeny $A_1 \to A_2$ defines an isogeny of the corresponding p-divisible groups $X_1 \to X_2$ in the usual sense. We define the kernel of an isogeny $A_1 \to A_2$ over a base scheme S to be the kernel K of $X_1 \to X_2$. Then K is a finite locally free group scheme whose order is locally a power of p. Let $B \otimes \mathbf{Q}_p = \prod_{i=1}^r B_i$ be a decomposition into simple algebras such that

$$O_B \otimes \mathbf{Z}_p = \prod O_{B_i}\,.$$

Then K decomposes into a product of finite locally free group schemes

$$K = \prod K_i\,,$$

where O_B acts on K_i via the projection to O_{B_i}. Assume that the order of the group schemes K_i is constant on S. Then we denote by $h_K(i), i = 1, \ldots, r$ the height of K_i, i.e. $p^{h_K(i)}$ is the order of the finite locally free scheme K_i. We call the function $h(i) = h_K(i)$ the *height of the isogeny* $A_1 \to A_2$.

6.4 Let A be an object of AV with O_B-action by ι. For an element $a \in B^\times$ that normalizes $O_B \otimes \mathbf{Z}_{(p)}$ we define as in the local case a new action

$$\iota^a(b) = \iota(a^{-1}ba), \qquad b \in O_B .$$

We use the notation A^a for the object (A, ι^a). The multiplication by $\iota(a)$ is a quasi-isogeny in AV

$$a : A^a \longrightarrow A.$$

We have a canonical identification $(A^{a_2})^{a_1} = A^{a_1 a_2}$. If (A, ι) is an object of AV, we define an action of O_B on the dual abelian variety $\hat{A}$ by the formula:

$$\hat{\iota}(b) = \iota(b^*)^\wedge .$$

Hence $\hat{A}$ becomes an object of AV. For an element $a \in B^\times$ that normalizes $O_B \otimes \mathbf{Z}_{(p)}$ we have a natural identification

$$(A^a)^\wedge = (\hat{A})^{(a^*)^{-1}} .$$

A *polarization* of an object A of AV is a quasi-isogeny $\lambda : A \to \hat{A}$ in AV, such that $n\lambda$ for a suitable natural number n is induced by an ample line bundle on A. We call the polarization *principal* if λ is an isomorphism in AV. The condition that λ respects the O_B-action is equivalent to the condition that the Rosati involution induced by λ is the given involution $*$.

Definition 6.5 *Let $\mathcal{L}$ be a multichain of $O_B \otimes \mathbf{Z}_p$-lattices in $V \otimes \mathbf{Q}_p$. A $\mathcal{L}$-set A of abelian varieties is a functor from the category $\mathcal{L}$ to the category AV*

$$\Lambda \longmapsto A_\Lambda, \quad \varrho_{\Lambda',\Lambda} : A_\Lambda \longrightarrow A_{\Lambda'} ,$$

such that the following conditions hold.

(i) $\varrho_{\Lambda',\Lambda}$ *is an isogeny of height $h(i)$ equal to* $\log_p |\Lambda'_i/\Lambda_i|$.

(ii) For any $a \in B^\times \cap (O_B \otimes \mathbf{Z}_{(p)})$ that normalizes $O_B \otimes \mathbf{Z}_{(p)}$ there are periodicity isomorphisms θ_a, such that the following diagrams are commutative

$$\begin{array}{ccc} \theta_a : A^a_\Lambda & \longrightarrow & A_{a\Lambda} \\ & {}_a\searrow \quad \swarrow_{\varrho_{\Lambda,a\Lambda}} & \\ & A_\Lambda & . \end{array}$$

It follows easily that the isomorphisms θ_a are uniquely determined and functorial in Λ. For any two lattices $\Lambda', \Lambda \in \mathcal{L}$ there are quasi–isogenies

$$\varrho_{\Lambda',\Lambda} : A_\Lambda \longrightarrow A_{\Lambda'} ,$$

such that for any $\Lambda'' \subset \Lambda \cap \Lambda', \Lambda'' \in \mathcal{L}$ we have

$$\varrho_{\Lambda,\Lambda'} \circ \varrho_{\Lambda',\Lambda''} = \varrho_{\Lambda,\Lambda''} .$$

With this definition there are isomorphisms θ_a for any $a \in B^\times$ that normalizes $O_B \otimes \mathbf{Z}_{(p)}$, such that the diagram under (ii) is commutative.

Let us now assume that $\mathcal{L}$ is a polarized multichain. To any $\mathcal{L}$-set of abelian varieties a dual $\mathcal{L}$-set $\tilde{A}$ is defined as follows:

$$\begin{aligned} \tilde{A}_\Lambda &= (A_{\Lambda^*})^\wedge \\ \tilde{\varrho}_{\Lambda',\Lambda} &= (\varrho_{\Lambda^*,\Lambda'^*})^\wedge \\ \tilde{\theta}_a &= ((\theta_{(a^*)^{-1}})^{-1})^\wedge \end{aligned}$$

Definition 6.6 *A polarization of an $\mathcal{L}$-set of abelian varieties $A = \{A_\Lambda\}$ is a quasi–isogeny of $\mathcal{L}$-sets, i.e. a rational multiple of an isogeny of $\mathcal{L}$-sets*

$$\lambda : A \longrightarrow \tilde{A} ,$$

such that the morphism

$$A_\Lambda \xrightarrow{\lambda_\Lambda} \tilde{A}_\Lambda = (A_{\Lambda^*})^\wedge \xrightarrow{\varrho^\wedge_{\Lambda^*,\Lambda}} (A_\Lambda)^\wedge$$

is a polarization of the abelian variety A_Λ for each Λ. The polarization is called principal if λ_Λ is an isomorphism in the category AV for each Λ.

A polarization satisfies the symmetry condition

$$\lambda^\wedge_\Lambda = \lambda_{\Lambda^*} .$$

Indeed this follows from the commutative diagram

$$\begin{array}{ccc} A_\Lambda & \xrightarrow{\lambda_\Lambda} & (A_{\Lambda^*})^\wedge \\ {\scriptstyle \varrho_{\Lambda^*,\Lambda}}\Big\downarrow & & \Big\downarrow{\scriptstyle (\varrho_{\Lambda^*,\Lambda})^\wedge} \\ A_{\Lambda^*} & \xrightarrow[\lambda_{\Lambda^*}]{} & (A_\Lambda)^\wedge \end{array}$$

and the symmetry of the polarization $(\varrho_{\Lambda^*,\Lambda})^\wedge \circ \lambda_\Lambda$.

Definition 6.7 *A* $\mathbf{Q}$*-homogeneous polarization is a set* $\bar{\lambda}$ *of quasi–isogenies of* $\mathcal{L}$*-sets of abelian varieties*

$$A \longrightarrow \tilde{A}$$

such that locally two elements of $\bar{\lambda}$ *differ by a factor from* $\mathbf{Q}^\times$ *and such that there exists* $\lambda \in \bar{\lambda}$*, such that* λ *is a polarization. If* λ *may be chosen to be principal the* $\mathbf{Q}$*-homogeneous polarization is called principal.*

6.8 Assume we are given an $\mathcal{L}$-set A of abelian varieties over an algebraically closed field. For $l \neq p$ the rational Tate module $V_l(A_\Lambda) = T_l(A_\Lambda) \otimes \mathbf{Q}$ is well-defined by the isomorphism class of A_Λ in AV. Moreover the quasi–isogenies $\varrho_{\Lambda',\Lambda}$ define isomorphisms $V_l(A_\Lambda) \simeq V_l(A_{\Lambda'})$. Hence the rational Tate module $V_l(A)$ of the $\mathcal{L}$-set A is well defined. We will also denote it by $H_1(A, \mathbf{Q}_l)$. As usual we can form the restricted product

$$H_1(A, \mathbf{A}_f^p) = \left(\prod_{l \neq p} T_l(A_\Lambda) \right) \otimes \mathbf{Q}\,.$$

Again this is well-defined independently of Λ. If A is given over any base S, we view $H_1(A, \mathbf{A}_f^p)$ as a $\mathbf{A}_f^p$-sheaf for the étale topology. If A is polarized there is the Riemann form

$$H_1(A, \mathbf{A}_f^p) \times H_1(A, \mathbf{A}_f^p) \longrightarrow \mathbf{A}_f^p(1)\,.$$

We are now ready to define a moduli problem of abelian varieties associated to the set of data

$$(B, O_B, V, (\ ,\ ), \mu, \mathcal{L}, C^p, \nu)$$

over the ring of integers O_{E_ν} of the completion of the Shimura field at the place ν.

Definition 6.9 *A point of the functor $\mathcal{A}_{C^p}$ with values in an O_{E_ν}-scheme S is given by the following set of data up to isomorphism.*

1. *An $\mathcal{L}$-set of abelian varieties $A = \{A_\Lambda\}$.*

2. *A $\mathbf{Q}$-homogeneous principal polarization $\bar{\lambda}$ of the $\mathcal{L}$-set A.*

3. *A C^p-level structure*

$$\bar{\eta} : H_1(A, \mathbf{A}_f^p) \simeq V \otimes \mathbf{A}_f^p \bmod C^p$$

that respects the bilinear forms on both sides up to a constant in $(\mathbf{A}_f^p)^\times$ (Kottwitz [Ko3] § 5).

We require an identity of polynomial functions for each Λ

$$\text{(i)} \qquad \det_{\mathcal{O}_S}(b\,;\, Lie A_\Lambda) = \det_K(b\,;\, V_0), \qquad b \in O_B$$

The representability of the functor $\mathcal{A}_{C^p}$ by a quasiprojective scheme over O_{E_ν} follows from Mumford's theorem in a standard way (Kottwitz [Ko3] §5). For varying C^p the schemes $\mathcal{A}_{C^p}$ form a projective system $\mathcal{A}$. The transition maps $\mathcal{A}_{C_1^p} \to \mathcal{A}_{C_2^p}$ for $C_1^p \subset C_2^p$ are finite. They are étale if C_2^p is small enough, i.e. is contained in a principal congruence subgroup of level $N \geq 3, (N,p) = 1$. We define a right action of $G(\mathbf{A}_f^p)$ on $\mathcal{A}$. An element $g \in G(\mathbf{A}_f^p)$ acts on $\mathcal{A}$ by morphisms

$$g : \mathcal{A}_{C^p} \longrightarrow \mathcal{A}_{g^{-1}C^p g}$$

which are defined by $(A, \bar{\lambda}, \bar{\eta})g = (A, \bar{\lambda}, g^{-1}\bar{\eta})$ on the S-valued points.

6.10 Assume that p is locally nilpotent on S. Then we denote by M_Λ the Lie algebra of the universal extension of A_Λ. We have shown ((3.23), c)) that by the condition (i) we have locally on S an isomorphism

$$M_\Lambda \simeq \Lambda \otimes \mathcal{O}_S$$

of $O_B \otimes \mathcal{O}_S$-modules, and that for any two neighbours $\Lambda \subset \Lambda'$ of $\mathcal{L}$ we have locally on S

$$M_{\Lambda'}/M_\Lambda \simeq \Lambda'/\Lambda \otimes \mathcal{O}_S\,.$$

Assume that S is the spectrum of an algebraically closed field of characteristic 0. Then for each Λ the p-adic Tate module $T_p(A_\Lambda)$ is well-defined. Again the condition (i) implies that there is an isomorphism of O_B-modules

$$T_p(A_\Lambda) \simeq \Lambda\,.$$

Indeed, for this it is enough to show that there is an isomorphism of B-modules

$$V_p(A_\Lambda) \simeq V \otimes \mathbf{Q}_p\,.$$

To show this we may assume that the O_{E_ν}-algebra $\mathcal{O}_S$ is equal to the O_{E_ν}-algebra $\mathbf{C}$. For $b \in O_B$ we get an identity in $\bar{\mathbf{Q}}$ in the sense of the chosen embeddings

$$\begin{aligned}
&\det\nolimits_{\mathbf{Q}_p}(b\,;\,V_p(A_\Lambda)) = \det\nolimits_{\mathbf{C}}(b\,;\,H_1(A,\mathbf{C}))\\
&= \det\nolimits_{\mathbf{C}}(b\,;\,\mathrm{Lie}A_\Lambda) + \overline{\det\nolimits_{\mathbf{C}}(b\,;\,\mathrm{Lie}A_\Lambda)}\\
&= \det\nolimits_{\mathbf{C}}(b\,;\,V_{0,\mathbf{C}}) + \det\nolimits_{\mathbf{C}}(b\,;\,V_{1,\mathbf{C}}) = \det\nolimits_{\mathbf{Q}}(b\,;\,V),
\end{aligned}$$

which implies the assertion. It is also a consequence of the existence of a C^p-level structure.

Consider again the decomposition of B into simple algebras over $\mathbf{Q}_p$, cf. (6.3). It induces a decomposition of the Tate module

$$T_p(A_\Lambda) = \prod_{i=1}^{r} T_p(A)_{\Lambda_i}.$$

Let $\Lambda \subset \Lambda'$ be neighbours in our multichain. The periodicity isomorphisms θ_a imply that $T_p(A)_{\Lambda'_i}/T_p(A)_{\Lambda_i}$ is a vector space over the residue class field κ_i of O_{B_i}. By the condition on the heights of definition (6.5) (i) the dimension of this vector space is equal to the dimension of the κ_i-vectorspace Λ'_i/Λ_i. Hence we get an isomorphism of O_{B_i}-modules

$$T_p(A)_{\Lambda'_i}/T_p(A)_{\Lambda_i} \simeq \Lambda'_i/\Lambda_i\,,$$

and hence an isomorphism of O_B-modules

$$T_p(A_{\Lambda'})/T_p(A_\Lambda) \simeq \Lambda'/\Lambda\,.$$

Therefore the $\mathcal{L}$-sets of our moduli problem satisfy stronger conditions than those of definition (6.5).

Definition 6.11 *An $\mathcal{L}$-set of abelian varieties A over a base scheme S is called a $\mathcal{L}$-multichain of abelian varieties if for any lattice Λ of $\mathcal{L}$ and for any pair of neighbours $\Lambda \subset \Lambda'$ of $\mathcal{L}$ the following conditions hold*

(i) *For any geometric point η of S of characteristic different from p there are isomorphisms of O_B-modules*

$$T_p(A_{\Lambda,\eta}) \simeq \Lambda, \quad T_p(A_{\Lambda',\eta})/T_p(A_{\Lambda,\eta}) \simeq \Lambda'/\Lambda$$

(ii) *For any geometric point η of S of characteristic p there are isomorphisms of*
$O_B \otimes_{\mathbf{Z}_p} \mathcal{O}_\eta$-modules

$$M_{\Lambda,\eta} \simeq \Lambda \otimes_{\mathbf{Z}_p} \mathcal{O}_\eta, \quad M_{\Lambda',\eta}/M_{\Lambda,\eta} \simeq \Lambda'/\Lambda \otimes_{\mathbf{Z}_p} \mathcal{O}_\eta\,,$$

where $M_{\Lambda,\eta}$ denotes the Lie algebra of the universal extension of $A_{\Lambda,\eta}$.

We may replace the word $\mathcal{L}$-set by $\mathcal{L}$-multichain in the definition (6.9) without changing the functor $\mathcal{A}_{C^p}$.

6.12 Let us denote by κ the residue class field of O_{E_ν} and by $\bar{\kappa}$ its algebraic closure. We consider a point of $\mathcal{A}_{C^p}(\bar{\kappa})$, which is given by the data $(A_0, \bar{\lambda}_0, \bar{\eta}_0)$. We will give a description of the set of points $(A', \bar{\lambda}', \bar{\eta}')$ which are isogenous to $(A_0, \bar{\lambda}_0, \bar{\eta}_0)$, i.e. such that there exists a quasi–isogeny $A_0 \longrightarrow A'$ that respects the $\mathbf{Q}$-homogeneous polarizations. Let us fix a principal polarization $\lambda_0 \in \bar{\lambda}_0$, and an isomorphism $\eta_0 \in \bar{\eta}_0$.
Let K_0 be the field of fractions of $W(\bar{\kappa})$. Let N be the isocrystal associated to A_0. The principal polarization induces an antisymmetric polarization form on N. There is an isomorphism of $B \otimes K_0$-modules

$$N \simeq V \otimes K_0 \qquad (6.2)$$

that respects the antisymmetric forms on both sides. Indeed, the polarized multichain of abelian varieties $A_0 = \{A_{0\Lambda}\}$ induces a polarized multichain of $O_B \otimes_{\mathbf{Z}_p} W(\bar{\kappa})$-lattices in N. Since the scheme $Spec\ W(\bar{\kappa})$ does not have nontrivial etale coverings, it follows from theorem (3.16) that this multichain is isomorphic to the standard multichain $\mathcal{L} \otimes_{\mathbf{Z}_p} W(\bar{\kappa})$. Hence we get a fortiori the isomorphism of $B \otimes K_0$–modules above.
In the discussion that follows the isomorphism above will be fixed. The Frobenius operator $\mathbf{F}$ on the left hand side may be written as $b \otimes \sigma$, for a uniquely defined $b \in G(K_0)$. Moreover we have by construction $c(b) = p$ (compare (3.20)). Let $\breve{\mathcal{M}}$ be the formal scheme over $O_{\breve{E}_\nu}$ associated to the data $(F, B, O_B, V, b, \mu, \mathcal{L})$ and $\mathcal{M} = \varprojlim \mathcal{M}_s$ be the formal proscheme over O_{E_ν} constucted from $\breve{\mathcal{M}}$.

6.13 We are going to define a morphism of functors over $Nilp\ O_{E_\nu}$

$$\mathcal{M} \times G(\mathbf{A}_f^p) \longrightarrow \mathcal{A}_{C^p}$$

such that the image of the $\bar{\kappa}$-valued points are the points of $\mathcal{A}_{C^p}(\bar{\kappa})$ that are isogenous to $(A_0, \bar{\lambda}_0, \bar{\eta}_0)$.
The definition is based on the following construction. Let A' an object of the category AV of abelian O_B-varieties up to isogeny of order prime to p over a base scheme S. Let X' be the p-divisible group of A'. Then to any quasi–isogeny $\xi : X' \to X''$ of p-divisible groups over S with an $O_B \otimes \mathbf{Z}_p$-action there is an object A'' of AV whose p-divisible group is X'' and a quasi–isogeny $A' \to A''$ that induces $X' \to X''$. The arrow $\xi_a : A' \to A''$ in AV is uniquely determined. We will use the notation $\xi_* A'$ for A''. If A' carries a polarization λ' we will use the notation $\xi_* \lambda'$ for the polarization $(\xi_a^{-1})^\wedge \lambda' \xi_a^{-1}$. If S is a $\mathbf{Z}_{(p)}$-scheme and A' has a rigidification $\eta : V^p(A') \xrightarrow{\sim} V \otimes \mathbf{A}_f^p$, we denote the rigidification $\eta \circ V^p(\xi_a^{-1})$ by $\xi_* \eta$. Clearly this construction is functorial in ξ,

$$(\xi_2 \xi_1)_* A' = \xi_{2*}(\xi_{1*} A') .$$

6.14 Let $\mathbf{X}$ be the polarized p-divisible group over $\bar{\kappa}$ used to define the functor $\breve{\mathcal{M}}$. Its polarized isocrystal is identified with $(V \otimes K_0, b \otimes \sigma) \simeq (N, \mathbf{F})$. There is an object $\mathbf{A} \in AV$ which is isogenous in AV to $A_{0,\Lambda}$ and whose p–divisible group is isomorphic to $\mathbf{X}$. It is well-defined up to isogeny of order prime to p and it inherits a $\mathbf{Q}$-homogeneous polarization $\bar{\lambda}_0$ and a rigidification $\bar{\eta}_0$ from the quasiiaogeny $\mathbf{A} \to A_{0,\Lambda}$. We fix a lifting $\tilde{\mathbf{X}}$ of $\mathbf{X}$ to the Witt ring $W(\bar{\kappa}) = W$ and denote by $\tilde{\mathbf{A}}$ the corresponding lifting of $\mathbf{A}$.

Let $(X_\Lambda, \varrho_\Lambda)$ be a point of $\breve{\mathcal{M}}(S)$. Then each ϱ_Λ lifts to a quasi–isogeny $\tilde{\varrho}_\Lambda : \tilde{\mathbf{X}}_S \to X_\Lambda$. Applying the construction given above we obtain a polarized $\mathcal{L}$-multichain of abelian varieties $(\tilde{\varrho}_\Lambda)_* \tilde{\mathbf{A}}_S = A_\Lambda$. The rigidification of $V^p(\tilde{\mathbf{A}})$ obtained from $\bar{\eta}_0$ carries over to a rigidification of $V^p(A_\Lambda)$ denoted by $(\tilde{\varrho}_\Lambda)_* \bar{\eta}_0$.

Hence we have constructed a morphism of functors on $\mathrm{Nilp}_{O_{\breve{E}_\nu}}$

$$\begin{array}{rcl} \Theta : \breve{\mathcal{M}} \times G(\mathbf{A}_f^p)/C^p & \longrightarrow & \mathcal{A}_{C^p} \times \mathit{Spec}\ O_{\breve{E}_\nu} \\ (X_\Lambda, \varrho_\Lambda) \times g & \longmapsto & (A = (\tilde{\varrho}_\Lambda)_* \tilde{\mathbf{A}}_S,\ g^{-1}(\tilde{\varrho}_\Lambda)_* \bar{\eta}_0) \end{array} \tag{6.3}$$

The morphism Θ is equivariant with respect to the right $G(\mathbf{A}_f^p)$–action on the projective systems on both sides of (6.3).

6.15 Let us denote by $I(\mathbf{Q})$ the group of quasi–isogenies in $\mathrm{End}^0_{AV}(\mathbf{A})$ that respect the given homogeneous polarization. Here we view I as an algebraic group over $\mathbf{Q}$. By (6.12) the polarized isocrystal of $\mathbf{A}$ is identified with $(V \otimes K_0, b\sigma)$. Let us denote as before by $J(\mathbf{Q}_p)$ the group of automorphisms of the isocrystal $(V \otimes K_0, b\sigma)$ that respect the polarization form up to a factor in $\mathbf{Q}_p^\times$. We obtain a homomorphism

$$\alpha_p : I(\mathbf{Q}_p) \hookrightarrow J(\mathbf{Q}_p).$$

On the other hand we have a homomorphism

$$\alpha^p : I(\mathbf{Q}) \longrightarrow G(\mathbf{A}_f^p)$$

which is defined by the following commutative diagram

$$\begin{array}{ccc} V^p(\mathbf{A}) & \xrightarrow{V^p(\xi)} & V^p(\mathbf{A}) \\ \eta_0 \Big\downarrow & & \Big\downarrow \eta_0 \\ V \otimes \mathbf{A}_f^p & \xrightarrow[\alpha^p(\xi)]{} & V \otimes \mathbf{A}_f^p \end{array}$$

for $\xi \in I(\mathbf{Q})$ and some choice $\eta_0 \in \bar{\eta}_0$ which is fixed.
We define a left action of the group $I(\mathbf{Q})$ on $\breve{\mathcal{M}} \times G(\mathbf{A}_f^p)$

$$(X_\Lambda, \varrho_\Lambda) \times g \longmapsto (X_\Lambda, \varrho_\Lambda \circ \alpha_p(\xi^{-1})_S) \times \alpha^p(\xi)g.$$

We claim that two points in the same orbit of $I(\mathbf{Q})$ have the same image by Θ. Indeed, since the quasi–isogeny $\alpha_p(\xi^{-1}) : \mathbf{X} \to \mathbf{X}$ is induced by the quasi–isogeny of abelian varieties $\xi^{-1} : \mathbf{A} \to \mathbf{A}$ we have $(\xi^{-1})_*\mathbf{A} = \mathbf{A}$. It is clear that this equality respects the polarization. It follows that there is a canonical isomorphism

$$(\varrho_\Lambda \circ \alpha_p(\xi^{-1})_S)_*\mathbf{A} = (\varrho_\Lambda)_*\mathbf{A}\,.$$

We need to verify that the induced rigidifications on both sides are the same if we multiply the left hand side by $g^{-1}\alpha_p(\xi^{-1})$ and the right hand side by g^{-1}. This follows from the equality

$$g^{-1}\alpha_p(\xi^{-1})\eta_0 V_p(\xi\varrho_{\Lambda,a}^{-1}) = g^{-1}\eta_0 V_p(\varrho_{\Lambda,a}^{-1})$$

where $\varrho_{\Lambda,a} : \mathbf{A}_{\bar{S}} \to A_{\Lambda,\bar{S}} = (\varrho_\Lambda)_*\mathbf{A}_{\bar{S}}$ denotes the isogeny that induces ϱ_Λ on the p-divisible groups.
Hence the map Θ factors through a map

$$I(\mathbf{Q})\backslash\breve{\mathcal{M}} \times G(\mathbf{A}_f^p)/C^p \longrightarrow \mathcal{A}_{C^p} \times \operatorname{Spec} O_{\breve{E}_\nu}\,. \tag{6.4}$$

We call this map the *uniformization morphism.* It depends on our choices $(A_0, \bar{\lambda}_0, \bar{\eta}_0)$ and the isomorphism (6.2).
Source and target of the uniformization morphism have an effective Weil descent datum relative to $O_{\breve{E}_\nu}/O_{E_\nu}$. We will prove that (6.4) is compatible with these descent data.
First we verify that Θ is continuous in the following sense.

Proposition 6.16 *Let $S \in Nilp_{O_{\breve{E}_\nu}}$ be an affine scheme. Then there exists a natural number $t \in \mathbf{N}$, such that $\Theta(S) : \breve{\mathcal{M}}(S) \times G(\mathbf{A}_f^p)/C^p \to \mathcal{A}_{C^p}(S)$ factors through $\mathcal{M}_t(S)$.*

Consider the quasi–isogeny γ_s cf. (3.41). It is defined for sufficiently divisible s. We have the relation

$$(\gamma_{s\cdot t}) = (\gamma_s)^t\,.$$

We postpone the proof of the proposition above and first show several lemmas.

Let us consider the natural injection

$$\mathrm{End}^0 \mathbf{A} \otimes_{\mathbf{Q}} \mathbf{Q}_p \hookrightarrow \mathrm{End}^0 \mathbf{X}.$$

Lemma 6.17 *The endomorphism γ_s of $\mathbf{X}$ lies in the center of the algebra $End^0(\mathbf{A}) \otimes \mathbf{Q}_p$. It satisfies the equation*

$$\gamma_s \cdot \gamma_s^* = p^s .$$

Proof: The homogeneously polarized abelian variety $\mathbf{A}$ is obtained from some $\bar{\mathbf{A}}$ over $\mathbf{F}_q$. Replacing $\mathbf{F}_q$ by an extension we may assume that

$$\mathrm{End}^0 \mathbf{A} = \mathrm{End}^0 \bar{\mathbf{A}} .$$

Let $\bar{\mathbf{X}}$ be the p-divisible group of $\bar{\mathbf{A}}$. Since the slope decomposition is defined over any perfect field we have $\gamma_s \in \mathrm{End}\, \bar{\mathbf{X}}$. We conclude $\gamma_s \in \mathrm{End}^0 \bar{\mathbf{A}} \otimes \mathbf{Q}_p$, since by a theorem of Tate $\mathrm{End}^0 \bar{\mathbf{X}} = (\mathrm{End}^0 \bar{\mathbf{A}}) \otimes \mathbf{Q}_p$. The remaining statements of the lemma are obvious. □

We remark that γ_s is even in the maximal order of the center of $(\mathrm{End}^0 \mathbf{A}) \otimes \mathbf{Q}_p$. Indeed this follows because there is a crystal isogenous to that of $\bar{\mathbf{X}}$ that splits into a direct sum of isotypic components and therefore is invariant by γ_s.

Let us denote by Z the center of $\mathrm{End}^0 \mathbf{A}$. Let $\tilde{\gamma}_s$ be the image of γ_s by the natural inclusion

$$Z \otimes \mathbf{Q}_p \longrightarrow Z \otimes \mathbf{A}_f .$$

Lemma 6.18 *For any congruence subgroup $C_Z \subset (Z \otimes \mathbf{A}_f)^\times$ there is an element $z \in Z$ and a natural number r, such that $z \equiv \tilde{\gamma}_s^r \bmod C_Z$ and $z \cdot z^* = p^{rs}$.*

Proof: The image of $\tilde{\gamma}_s$ in the finite group $Z^* \backslash (Z \otimes \mathbf{A}_f)^* / C_Z$ has some finite order r. Hence there is an element $z \in Z$, such that $z \equiv \tilde{\gamma}_s^r \bmod C_Z$. To find in addition a z that satisfies $z \cdot z^* = p^{rs}$ we may assume that the group C_Z is invariant under conjugation. Let $\tilde{p}$ be the image of p by the map $Z \otimes \mathbf{Q}_p \to Z \otimes \mathbf{A}_f$. We find that

$$(z \cdot z^*)^{-1}(\gamma_s \gamma_s^*)^r = (zz^*)^{-1}\tilde{p}^{rs} \in C_Z\,.$$

Since some power of $\tilde{p}p^{-1}$ is contained in C_Z we get that for a possibly bigger r we have $(zz^*)^{-1}p^{rs} \in C_Z$. The last element is a unit in Z. The topology induced by the congruence groups on the group of units is the profinite topology. For small C_Z our element is therefore in the group of units of the maximal real subfield of Z and is there a square

$$(zz^*)^{-1}p^{rs} = u^2, \quad u = u^*\,.$$

The desired element is $z \cdot u$. □

Let us denote the rigidification of $V^p(\mathbf{A})$ obtained from the rigidification η_0 of $V^p(A_0)$ by the same letter. We do the same with $\bar{\lambda}_0$.

Lemma 6.19 *There is an isomorphism of homogeneously polarized and rigidified objects of AV for a suitable natural number r*

$$(\gamma_s^r)_*(\mathbf{A}, \bar{\lambda}_0, \bar{\eta}_0) \cong (\mathbf{A}, \bar{\lambda}_0, \bar{\eta}_0)\,.$$

Proof: The element z of the last lemma induces an isogeny $z : \mathbf{A} \to \mathbf{A}$ that respects $\bar{\lambda}_0$ and $\bar{\eta}_0$. Hence we get an isomorphism $z_*(\mathbf{A}, \bar{\lambda}_0, \bar{\eta}_0) \cong (\mathbf{A}, \bar{\lambda}_0, \bar{\eta}_0)$, where the last z denotes the quasi-isogeny $z : \mathbf{X} \to \mathbf{X}$. If we prove that $(\gamma_s^r z^{-1})_*(\mathbf{A}, \bar{\lambda}_0, \bar{\eta}_0) \simeq (\mathbf{A}, \bar{\lambda}_0, \bar{\eta}_0)$ we are done. For this it is enough to show that $\gamma_s^r z^{-1} : \mathbf{X} \to \mathbf{X}$ is an isomorphism. But this is certainly an isomorphism if $(\gamma_s^r z^{-1})$ is in a sufficiently small open compact subgroup of $(Z \otimes \mathbf{Q}_p)^\times$. Indeed we have a continuous faithful representation on the $\mathbf{Q}_p$-vectorspace N. For the small open subgroup we may take the inverse image of the subgroup of $\mathrm{Aut}_{\mathbf{Q}_p} N$ that fixes the lattice of N given by the crystal $\mathbf{M}$ of $\mathbf{X}$. Hence we are done if we choose $C_{Z,p}$ small enough. □

We can strengthen the last lemma as follows. Let $k \geq 1$. Let $\tilde{\mathbf{A}}_k$ be the restriction of $\tilde{\mathbf{A}}$ to $O_{\breve{E}_\nu}/p^k O_{\breve{E}_\nu}$.

Corollary 6.20 *For a suitable number r we have an isomorphism*

$$(\gamma_s^r)_*(\tilde{\mathbf{A}}_k, \bar{\lambda}_0, \bar{\eta}_0) \simeq (\tilde{\mathbf{A}}_k, \bar{\lambda}_0, \bar{\eta}_0)$$

Proof: The proof we have just given works, if we can show that $\gamma_s^r z^{-1}$ induces an isomorphism of the p-divisible group $\tilde{\mathbf{X}}_k$ of $\tilde{\mathbf{A}}_k$. For this it is enough to show that $\gamma_s^r z^{-1}$ lifts to a homomorphism. By a theorem of Grothendieck and Messing this is the case if $\gamma_s^r z^{-1}$ respects the Hodge filtration on $\mathbf{M} \otimes_{O_{\breve{E}_\nu}} O_{\breve{E}_\nu}/p^k$. But a sufficiently small compact open subgroup of $(Z \otimes \mathbf{Q}_p)^\times$ acts identically on the last space. This proves the corollary.

Proof of proposition (6.16): Let $(X_\Lambda, \varrho_\Lambda)$ be a point of $\breve{\mathcal{M}}$ over a $O_{\breve{E}}/p^k O_{\breve{E}}$-scheme S for some natural number k. Let $\tilde{\varrho}_\Lambda : \tilde{\mathbf{X}}_{k,S} \to X_\Lambda$ be the quasi-isogeny that lifts ϱ_Λ. Then the proposition asserts that for suitable r

$$(\tilde{\varrho}_\Lambda \gamma_s^r)_*(\tilde{\mathbf{A}}_k, \bar{\lambda}_0, \bar{\eta}_0) = \tilde{\varrho}_{\Lambda *}(\tilde{\mathbf{A}}_k, \bar{\lambda}_0, \bar{\eta}_0).$$

But this is obvious by the last corollary. □

Theorem 6.21 *The morphism (6.3) defines a morphism over $Spf\, O_{E_\nu}$*

$$\Theta : \mathcal{M} \times G(\mathbf{A}_f^p)/C^p \longrightarrow \hat{\mathcal{A}}_{C^p},$$

where the left hand side is the pro-formal scheme over O_{E_ν} given by (3.51) and the right hand side is the p-adic completion of the scheme $\mathcal{A}_{C^p}$. The action of $I(\mathbf{Q})$ on $\breve{\mathcal{M}} \times G(\mathbf{A}_f^p)/C^p$ given by (6.15) descends to an action of $I(\mathbf{Q})$ on the source of the above morphism and Θ is invariant with respect to this action.

Proof: We need to verify that the action of the group $I(\mathbf{Q})$ and Θ commute with the descent data given on both source and target of the arrow above and that $I(\mathbf{Q})$ acts on the schemes $\mathcal{M}_t$. The assertions for $I(\mathbf{Q})$ hold because they hold for the action of $J(\mathbf{Q}_p)$ by definition (3.22). To show the assertion about Θ let us first recall what the descent data on $\hat{\mathcal{A}}_{C^p}$ relative to $O_{\breve{E}_\nu}/O_{E_\nu}$ is. Let us denote by $\tau : Spec\, O_{\breve{E}_\nu} \to Spec\, O_{\breve{E}_\nu}$ the morphism induced by the Frobenius automorphism of $\breve{E}_\nu$ relative to E_ν.
Let S be a $O_{\breve{E}_\nu}$-scheme and denote by $\varphi : S \to Spec\, O_{\breve{E}_\nu}$ the structure morphism. Let $S_{[\tau]}$ be the same scheme as S but with structure morphism $\tau\varphi$. Then the natural descent datum on $\mathcal{A}_{C^p} \times Spec\, O_{\breve{E}_\nu}$ is given by the map

$$\mathcal{A}_{C^p}(S) \longrightarrow \mathcal{A}_{C^p}(S_{[\tau]})$$

that views a polarized multichain A_Λ on S as a polarized multichain on $S_{[\tau]}$.

It is enough to show that the restriction of the map Θ to $\breve{\mathcal{M}} \times 1$ is compatible with the descent data,

$$\breve{\mathcal{M}}(S) \longrightarrow \hat{\mathcal{A}}_{C^p}(S).$$

Let us first assume that S is a scheme over $O_{\breve{E}}/pO_{\breve{E}}$. Let $(X_\Lambda, \varrho_\Lambda)$ be a point of $\mathcal{M}(S)$ and let $\varrho_{\Lambda*}\varphi^*(\mathbf{A}, \bar{\eta}_0)$ be its image by the above map. The descent datum on $\breve{\mathcal{M}}$ associates to $(X_\Lambda, \varrho_\Lambda)$ the point $(X_\Lambda, \varrho_\Lambda\varphi^*(\mathrm{Frob}_E^{-1}))$. Hence we have to show that the following points of $\hat{\mathcal{A}}_{C^p}(S_{[\tau]})$ are the same:

$$(\varrho_\Lambda\varphi^*(\mathrm{Frob}_E^{-1}))_*\varphi^*\tau^*(\mathbf{A}, \bar{\eta}_0) \cong \varrho_{\Lambda*}\varphi^*(\mathbf{A}, \bar{\eta}_0).$$

This reduces to the obvious fact that there is an isomorphism

$$(\mathrm{Frob}_E^{-1})_*\tau^*(\mathbf{A}, \bar{\eta}_0) \cong (\mathbf{A}, \bar{\eta}_0)\,.$$

In the general case let us denote by $(\bar{A}_\Lambda, \eta)$ be the polarized rigidified multichain associated to the reduction $(\bar{X}_\Lambda, \varrho_\Lambda)$ of $(X_\Lambda, \varrho_\Lambda)$ by the above map. Then the image of $(X_\Lambda, \varrho_\Lambda)$ is just given by the lifting A_Λ of $\bar{A}_\Lambda$ that is determined by the lifting X_Λ of the p-divisible group $\bar{X}_\Lambda$ of $\bar{A}_\Lambda$. Hence if we know that the images of $(\bar{X}_\Lambda, \varrho_\Lambda)$ and $(\bar{X}_\Lambda, \varrho_\Lambda\bar{\varphi}^*(\mathrm{Frob}_E^{-1}))$ are isomorphic we know the same for the images of the points $(X_\Lambda, \varrho_\Lambda)$ and $(X_\Lambda, \varrho_\Lambda\varphi^*(\mathrm{Frob}_E^{-1}))$. Hence the proof of the theorem 6.21 is finished. □

6.22 Let X be a scheme. We consider a family of closed subschemes $\mathcal{T} = \{T_i\}_{i\in \boldsymbol{I}}$, such that each T_i meets only finitely many members of $\mathcal{T}$. Then we define the completion $X/\mathcal{T}$ of X along $\mathcal{T}$ as follows. For any point $x \in \bigcup_{i\in \boldsymbol{I}} T_i$ we define

$$\mathcal{T}(x) = \left(\bigcup_{x\in T_i} T_i\right) \setminus \left(\bigcup_{x\notin T_j} T_j\right)$$

This is a locally closed subset of X. The underlying topological space of $X/\mathcal{T}$ is to be $\bigcup_{i\in \boldsymbol{I}} T_i$ with the inductive limit topology, i.e. a subset $\mathcal{U} \subset \bigcup_{i\in \boldsymbol{I}} T_i$ is open, iff $T_i \cap \mathcal{U}$ is open for each i. The formal open subscheme of $X/\mathcal{T}$ with the underlying topological space $\mathcal{T}(x)$ is defined to be the completion of X along the locally closed subset $\mathcal{T}(x)$.

Theorem 6.23 *Let $\boldsymbol{I}$ be the set of $I(\mathbf{Q})$-orbits of irreducible components of $\breve{\mathcal{M}} \times G(\mathbf{A}_f^p)/C^p$. Let us denote by T_i the image of any irreducible component in the $I(\mathbf{Q})$-orbit $i \in \boldsymbol{I}$ by the uniformization morphism (6.4). Then each T_i is a projective subscheme of $\mathcal{A}_{C^p} \times \mathit{Spf}\ O_{\breve{E}_\nu}$, that meets only finitely many members of the family $\{T_j\}_{j \in \boldsymbol{I}} = \mathcal{T}$. The morphism Θ induces a $G(\mathbf{A}_f^p)$-equivariant isomorphism of sheaves for the étale topology,*

$$\Theta : I(\mathbf{Q})\backslash\breve{\mathcal{M}} \times G(\mathbf{A}_f^p)/C^p \longrightarrow (\mathcal{A}_{C^p} \times \mathit{Spf}\ O_{\breve{E}_\nu})/\mathcal{T}\,.$$

We note that a geometric point of $\mathcal{A}_{C^p} \times Spf\ O_{\breve{E}_\nu}$ is in the union of the sets T_i, iff the underlying $\mathbf{Q}$-homogeneously polarized O_B-varieties A_Λ are isogenous to $(A_0, \bar{\lambda}_0)$.

Proof: We remark that the map is formally étale. This is easily verified, because quasi–isogenies of p-divisible groups extend to infinitesimal neighbourhoods in the category $\mathrm{Nilp}_{O_{\breve{E}_\nu}}$.
Since the map is compatible with the action of $G(\mathbf{A}_f^p)$ it is enough to prove the assertion for sufficiently small open compact subgroups C^p. We claim that for small C^p the sheaf for the étale topology $I(\mathbf{Q})\backslash\breve{\mathcal{M}} \times G(\mathbf{A}_f^p)/C^p$ is a formal algebraic space. We note that

$$I(\mathbf{Q}) \hookrightarrow J(\mathbf{Q}_p) \times G(\mathbf{A}_f^p)$$

is a discrete subgroup. Indeed, consider an open subgroup $U \subseteq J(\mathbf{Q}_p) \times G(\mathbf{A}_f^p)$. For U sufficiently small the $\mathbf{Z}_\ell$-lattice $\eta_0(T_\ell(\mathbf{A}))$ for $\ell \neq p$ is fixed by U and the quasi–isogenies of $\mathbf{X}$ defined by the elements in the image of the projection $U \to J(\mathbf{Q}_p)$ are isomorphisms. Hence an element of $I(\mathbf{Q}) \cap U$ induces an isomorphism of the polarized abelian variety $(\mathbf{A}, \lambda_0)$, where $\lambda_0 \in \bar{\lambda}_0$ a polarization. Hence $U \cap I(\mathbf{Q})$ is finite. We write

$$I(\mathbf{Q})\backslash\breve{\mathcal{M}} \times G(\mathbf{A}_f^p)/C^p = \bigsqcup_{\Gamma} \Gamma\backslash\breve{\mathcal{M}}\,,$$

where Γ runs through a countable set of subgroups of $J(\mathbf{Q}_p)$ of the form $(J(\mathbf{Q}_p) \times (gC^pg^{-1})) \cap I(\mathbf{Q}) \subset J(\mathbf{Q}_p)$. This is a discrete subgroup by what we have said.
We claim that for small groups C^p the groups Γ are torsion free. Indeed, let $h \in \Gamma \subset I(\mathbf{Q})$ be an element of finite order m. Then we find a polarized abelian variety A in the isogeny class of A_0, such that h induces an

automorphism of A fixing a polarization and the n-division points for some $n \geq 3$ prime to p, so that by Serre's lemma $h = 1$.
Indeed, let $M_0 \subset N$ be the Cartier module of some abelian variety in the isogeny class A_0. We set $M = M_0 + hM_0 + \cdots + h^{m-1}M_0$. We assume that C^p fixes a lattice $V_{\mathbf{Z}^p} \subset V \otimes \mathbf{A}_f^p$ and acts trivially on $V_{\mathbf{Z}^p}/nV_{\mathbf{Z}^p}$. Then we may take for A the abelian variety with Tate module $V^p(A) = \eta_0^{-1}(gV_{\mathbf{Z}^p})$ and Cartier module M.
It follows from (2.37) that $\Gamma\backslash\breve{\mathcal{M}}$ is a formal algebraic space. Next we remark that the map Θ is injective on geometric points. This is easy to check from the definition of $I(\mathbf{Q})$. Since the formal algebraic space $I(\mathbf{Q})\backslash\breve{\mathcal{M}} \times G(\mathbf{A}_f^p)/C^p$ is formally locally of finite type over $Spf\, O_{\breve{E}_\nu}$ and has projective components, the subsets T_i have the properties stated in the theorem.
The morphism Θ is quasifinite. Indeed, since the geometric fibres are finite this follows because the formal algebraic spaces involved are locally formally of finite type over Spf $O_{\breve{E}_\nu}$. Hence by Knudsen [Kn] II 6.7 it follows that $I(\mathbf{Q})\backslash\breve{\mathcal{M}} \times G(\mathbf{A}_f^p)/C^p$ is a formal scheme.
The reader checks easily that a morphism of finite type of locally noetherian schemes that is unramified, proper and radical is a closed immersion. Let

$$\Theta : \mathcal{X} \longrightarrow \mathcal{Y}$$

be the morphism of the theorem. It is formally étale, radical, surjective and the map $\Theta_{red} : \mathcal{X}_{red} \to \mathcal{Y}_{red}$ is proper. We claim that any morphism Θ of locally noetherian formal schemes with these properties is an isomorphism. Clearly we may assume that $\mathcal{X}$ and $\mathcal{Y}$ are affine. We set $\mathcal{X} = \mathit{Spf}\,(A, I)$ and $\mathcal{X}_n = \mathit{Spec}\, A/I^n$. Let us first consider the case where $\mathcal{Y}$ is a reduced scheme. Then $\mathcal{X}_n \to \mathcal{Y}$ is an unramified, proper and radical morphism of schemes and hence a closed immersion. Since $\mathcal{Y}$ is reduced and the last map is surjective it is an isomorphism. In general this shows that the map Θ is adic. Hence it is enough to prove the assertion in the case where $\mathcal{X}$ and $\mathcal{Y}$ are schemes. In this case Θ is a closed immersion that is étale and surjective and hence an isomorphism. The theorem is proved. □

We note that for any affine scheme $S \in \mathrm{Nilp}_{O_{\breve{E}_\nu}}$ there is an integer s such that the canonical map

$$(I(\mathbf{Q})\backslash\breve{\mathcal{M}} \times G(\mathbf{A}_f^p)/C^p)(S) \longrightarrow (I(\mathbf{Q})\backslash\mathcal{M}_s \times G(\mathbf{A}_f^p)/C^p)(S)$$

is bijective. This follows exactly in the same way as proposition (6.16).

The theorem (6.23) shows that the family $\{T_i\}$ of projective subvarieties of $\mathcal{A}_{C_p} \times \mathcal{S}pec\, \bar{\kappa}$ is invariant by the action of $\mathrm{Gal}(\breve{E}_\nu/E_\nu)$. Hence the formal scheme $\mathcal{A}_{C^p}/\mathcal{T}$ is defined over the local Shimura field O_{E_ν}.

Theorem 6.24 *There is a $G(\mathbf{A}_f^p)$-equivariant isomorphism of formal schemes over O_{E_ν}*

$$\Theta : I(\mathbf{Q})\backslash\mathcal{M} \times G(\mathbf{A}_f^p)/C^p \longrightarrow \mathcal{A}_{C^p}/\mathcal{T}\,.$$

Proof: This is a combination of the theorems (6.23) and (6.21). □

6.25 The theorem takes a much simpler form if the element $b \in G(K_0)$ given by (6.12) is basic. By definition basic means that the corresponding slope morphism $\nu : \mathbf{D} \to G$ which is defined over K_0, is central. We use this definition even in the case that G is not connected.
Let $F \otimes \mathbf{Q}_p = \prod_{\mathbf{p}/p} F_{\mathbf{p}}$ be the decomposition as a product of local fields. We get a corresponding decomposition of the isocrystal $(N, \boldsymbol{F}) \cong (V \otimes K_0, b\sigma)$

$$N = \bigoplus_{\mathbf{p}} N_{\mathbf{p}}\,.$$

The assumption that b is basic implies that each of the $N_{\mathbf{p}}$ is an isotypic isocrystal. Indeed, assume that $s\nu$ factors through $\mathbf{G}_m$. Then $s\nu(p) \in G(K_0)$ acts on the isotypic component $N_{r/s}$ of slope r/s of N by multiplication by p^r. By definition $s\nu(p)$ is in the center of $G(K_0)$. Going to the algebraic closure (compare 1.39) we see that the center of $G(K_0)$ is

$$\{f \in F \otimes K_0\,;\, ff^* \in K_0\}\,.$$

Hence $s\nu(p) \in F \otimes K_0$. On the other hand $s\nu(p)$ commutes with $b \in G(K_0)$ and with $b\sigma$. We deduce that $s\nu(p)$ commutes with σ and therefore $s\nu(p) \in F \otimes \mathbf{Q}_p$. This implies that $N_{\mathbf{p}}$ is isotypic.

6.26 Let X be a p-divisible group over a scheme S where p is locally nilpotent with an action $\iota : O_F \to \mathrm{End}X$. Then the set of points $s \in S$ such that the isocrystal of $X_{\overline{\kappa(s)}}$ with its F-action is isomorphic to the isocrystal $(N, \boldsymbol{F}) \bigotimes_{K_0} W(\overline{\kappa(s)})_{\mathbf{Q}}$ is closed. This is a variant of Grothendieck's theorem on the specialization of the Newton polygon. The reader may verify this using (4.30) and the relation between Hodge and Newton polygons explained in Katz [Ka2]. More generally we have the following result (cf. [RR]).

Theorem 6.27 *Assume again that $(N, \mathbf{F})$ is given by a basic element $b \in G(K_0)$. Let X over S as above be a p-divisible group equipped with an O_B-action and a polarization that induces the given involution. Then the set of points $s \in S$, such that there exists an isomorphism of the B-isocrystal of $X_{\overline{\kappa(s)}}$ and the B-isocrystal $(N, \mathbf{F}) \bigotimes_{K_0} W(\overline{\kappa(s)})_{\mathbf{Q}})$ that respects the polarization forms up to a constant is closed.*

Before we state the uniformization theorem for basic isogeny classes we need a lemma.

Lemma 6.28 *Let (A, λ) be a polarized abelian variety over a finite field $\mathbf{F}_q$. Let $K \subset \mathrm{End}^0 A$ be a commutative subalgebra such that the Rosati involution induces an automorphism of K. Let N be the rational Cartier module of A. Consider the decomposition $K \otimes \mathbf{Q}_p = \prod_{\mathbf{p}|p} K_{\mathbf{p}}$ into local fields. We assume that in the corresponding decomposition of $N = \bigoplus_{\mathbf{p}|p} N_{\mathbf{p}}$ each of the Cartier modules $N_{\mathbf{p}}$ is isoclinic. Then some power of the Frobenius morphism $\mathrm{Fr} : A \to A$ over $\mathbf{F}_q$ is contained in K.*

Proof: Since $N_{\mathbf{p}}$ is isoclinic there is a $W(\mathbf{F}_q)$-lattices $M_{\mathbf{p}} \subset N_{\mathbf{p}}$ stable under Frobenius and Verschiebung such that $\mathbf{F}^{s_{\mathbf{p}}} M_{\mathbf{p}} = p^{r_{\mathbf{p}}} M_{\mathbf{p}}$. We may assume that $M_{\mathbf{p}}$ is fixed by $O_{K_{\mathbf{p}}}$. We may assume that the Cartier-module of A is $\oplus M_{\mathbf{p}}$ and that $O_K \subset \mathrm{End} A$, by changing A in its isogeny class. Let us assume that all $s_{\mathbf{p}}$ are equal to s and that $q = p^s$, which we may do without loss of generality. Hence we have $\mathrm{Fr}\ M_{\mathbf{p}} = p^{r_{\mathbf{p}}} M_{\mathbf{p}}$. Let us denote by $\mathrm{ord}_{\mathbf{p}}$ an order function on $K_{\mathbf{p}}$ normalized by the condition $\mathrm{ord}_{\mathbf{p}} p = 1$. We are looking for an element $u \in K$ that is a unit at all non archimedian places not lying over p and that satisfies the equations

$$\mathrm{ord}_{\mathbf{p}} u = r_{\mathbf{p}} \qquad \bar{u} u = q\,, \tag{6.5}$$

where $\bar{u}$ is the Rosati involution applied to u. Since we may enlarge the field $\mathbf{F}_q$ it is enough to find a solution if we replace the integers $r_{\mathbf{p}}$ by a multiple $m r_{\mathbf{p}}$ and q by q^m. Hence we find an element u with the required properties except for the second identity above. By the presence of the polarization we have $r_{\mathbf{p}} + r_{\bar{\mathbf{p}}} = s$. It follows that for $u' = q \cdot u/\bar{u}$ we get the equations

$$\mathrm{ord}_{\mathbf{p}} u' = 2 r_{\mathbf{p}} \qquad u' \bar{u'} = q^2\,.$$

Hence we may assume the existence of an $u \in K$ with (6.5). It follows that $\varepsilon = u^{-1}\mathrm{Fr}$ is an automorphism of A that fixes the polarization. Hence by the lemma of Serre we conclude that some power of ε is 1. □

Corollary 6.29 *Let A be an abelian variety over $\bar{\mathbf{F}}_p$ with a polarization λ and an embedding $\iota : K \to \mathrm{End}^0 A$ that satisfies the assumptions of the previous lemma over $\bar{\mathbf{F}}_p$. Assume we have a second variety (A', λ', ι') satisfying the same assumptions. Then*

$$\begin{aligned} Hom_K^0(A, A') \otimes \mathbf{Q}_\ell &= Hom_K(V_\ell(A), V_\ell(A')) \quad \text{for } \ell \neq p \\ Hom_K^0(A, A') \otimes \mathbf{Q}_p &= Hom_K((N, \mathbf{F}), (N', \mathbf{F})) \end{aligned}$$

□

Theorem 6.30 *Let $(A_0, \bar{\lambda}_0, \bar{\eta}_0)$ be a point of $\mathcal{A}_{C^p}(\bar{\kappa})$. Assume that the isocrystal*
$(V \otimes K_0, b\sigma)$ associated to it via (6.12) is basic. Let us denote by $\mathcal{M}$ the pro-formal scheme associated to the data $(B \otimes \mathbf{Q}_p, V \otimes \mathbf{Q}_p, b, \mu, \mathcal{L})$. Let us denote by $Z \subset \mathcal{A}_{C^p}$ the closed set given by theorem (6.27). Then there is an open and closed subset $Z' \subset Z$ such that the uniformization morphism

$$\Theta : I(\mathbf{Q})\backslash\mathcal{M} \times G(\mathbf{A}_f^p)/C^p \longrightarrow \mathcal{A}_{C^p/Z'}$$

given by $(A_0, \bar{\lambda}_0, \bar{\eta}_0)$ is an isomorphism over $\mathrm{Spf}\ O_{E_\nu}$. The source of this morphism is a finite disjoint sum of formal schemes of the form $\Gamma\backslash\mathcal{M}$ where $\Gamma \subset J(\mathbf{Q}_p)$ is a discrete subgroup which is co-compact modulo center.
The group I is an inner form of G. If G satisfies the Hasse principle the sets Z' and Z coincide.

Proof: Let us first verify that I is an inner form of G. Let $L = \mathrm{End}_B^0 V$ and $L_a = \mathrm{End}_B^0 A_0$. Let us denote the involutions on L (respectively L_a) induced by $(\ ,\ )$ (respectively by $\bar{\lambda}$) by $*$. By the Corollary (6.29) we have $L_a \otimes \mathbf{Q}_\ell \simeq \mathrm{End}_B(V_\ell(A_0))$. The existence of the symplectic similitude implies, that $(L_a \otimes \mathbf{Q}_\ell, *)$ and $(L \otimes \mathbf{Q}_\ell, *)$ are isomorphic as $F \otimes \mathbf{Q}_\ell$-algebras with involution $*$. Hence there exists such an isomorphism $\varphi : L \otimes \bar{\mathbf{Q}} \to L_a \otimes \bar{\mathbf{Q}}$ over $\bar{\mathbf{Q}}$. It defines an isomorphism between the groups $G = \{\ell \in L^\times; \ell\ell^* \in \mathbf{Q}\}$ and $I = \{\ell \in L_a^\times; \ell\ell^* \in \mathbf{Q}\}$ over $\bar{\mathbf{Q}}$. Clearly it is enough to show that φ is unique up to an inner automorphism by an element of $G(\bar{\mathbf{Q}})$. Indeed, let $\varrho : L \otimes \bar{\mathbf{Q}} \to L \otimes \bar{\mathbf{Q}}$ be an automorphism of $F \otimes \bar{\mathbf{Q}}$-algebras that respects the involution $*$. Then by Skolem–Noether there is an $a \in L \otimes \bar{\mathbf{Q}}$, such that $\varrho(\ell) = a\ell a^{-1}$. Since ϱ respects the involution, we have that a^*a lies in the center $F \otimes \bar{\mathbf{Q}}$. Hence $a^*a = f^2$ for some $f \in F \otimes \bar{\mathbf{Q}}$. In the case of an involution of the first kind we may replace a by $f^{-1}a \in G(\bar{\mathbf{Q}})$. Indeed in this case ϱ is the inner automorphism by $f^{-1}a$. The case of an involution of the second kind is similar.

We now verify the remaining statements. Clearly it is enough to verify these over $\mathcal{S}pf\ O_{\breve{E}_\nu}$. By definition of Θ we are also allowed to replace $\mathcal{M}$ by $\breve{\mathcal{M}}$. The homogenously polarized abelian varieties representing points of $Z(\bar{\kappa})$ are divided into finitely many isogeny classes $(\mathbf{A}_1, \bar{\lambda}_1), \ldots, (\mathbf{A}_m, \bar{\lambda}_m)$. Indeed, since the $(\mathbf{A}_i, \bar{\lambda}_i)$ together with the subalgebra $F \subset \mathrm{End}^0 \mathbf{A}_i$ satisfy the assumptions of the lemma above, they are all isogenous as abelian varieties with a B-action. Hence we may assume $\mathbf{A}_1 = \ldots = \mathbf{A}_m$. We have to show that only finitely many homogeneous polarizations on $\mathbf{A}_1$ up to isogeny come from points of $Z(\bar{\kappa})$. Let us fix a polarization λ of $\mathbf{A}_1$ induced from $Z(\bar{\kappa})$. Any other polarization of this type is of the form $\lambda\alpha$, where $\alpha \in \mathrm{End}_B \mathbf{A}_1, \alpha = \alpha^*$ and α is totally positive. The two homogeneous polarizations $\bar{\lambda}$ and $\overline{\lambda\alpha}$ are isogenous, iff there exists $\beta \in \mathrm{End}_B^0 \mathbf{A}_1$ such that $\alpha = g\beta\beta^*$ for some $g \in \mathbf{Q}^\times$. We consider the solution of this equation as a torsor for the algebraic group I defined over $\mathbf{Q}$. Two homogeneous polarizations $\overline{\lambda\alpha_1}$ and $\overline{\lambda\alpha_2}$ are isogenous, iff the corresponding torsors are isomorphic. The existence of the rigidifications η implies that the torsors are locally isomorphic for any prime $\ell \neq p$. The same is true for $\ell = p$ by the definition of Z and at the infinite place since α is totally positive. The isomorphism classes of the torsors are therefore in the kernel of the map

$$H^1(\mathbf{Q}, I) \longrightarrow \prod_w H^1(\mathbf{Q}_w, I).$$

This is known to be finite.

To each $(\mathbf{A}_i, \lambda_i)$ we choose a point of $Z(\bar{\kappa})$ belonging to that isogeny class. It defines a uniformization morphism Θ_i. We consider the disjoint union of these morphisms:

$$\Theta : I(\mathbf{Q}) \backslash \sqcup \breve{\mathcal{M}} \times G(\mathbf{A}_f^p)/C^p \xrightarrow{\sqcup \Theta_i} \mathcal{A}_{C^p/Z} \times \mathcal{S}pf\ O_{\breve{E}_\nu}. \tag{6.6}$$

It is enough to show that (6.6) is an isomorphism. We already know that Θ induces an isomorphism with $\mathcal{A}_{C^p}/T \times \mathcal{S}pf\ O_{\breve{E}_\nu}$ for some family of closed

subspaces $\mathcal{T} = \{T_i\}$. If we show that Θ is surjective it follows that $\mathcal{T}$ consists of the irreducible components of Z and we are done.
By construction Θ is surjective on the $\bar{\kappa}$-valued points. Let us denote the morphism Θ by $\mathcal{X} \to \mathcal{Y}$. Consider a point $y \in \mathcal{Y}$. We prove the surjectivity by induction on the transcendence degree of $\kappa(y)/\bar{\kappa}$. Take a regular point $x \in \overline{\{y\}}$ of codimension 1. It is enough to find an extension L of $\kappa(y)$ such that a L-valued point centered at y is in the image of $\mathcal{X}(L) \to \mathcal{Y}(L)$. A suitable extension of $\hat{O}_{\overline{\{y\}},x}$ is a power series ring in one variable $P[\![T]\!]$ over an algebraically closed field P. Let L be the quotient field of $P[\![T]\!]$. Let us denote by $(A, \bar{\lambda}, \bar{\eta})$ the given point of the moduli functor over $P[\![T]\!]$. It is enough to show that $(A_\Lambda, \bar{\lambda})_L$ is isogenous to $(\mathbf{A}_i, \bar{\lambda}_i)_L$ for some i. By induction assumption this is true, if we replace the index L by P. A theorem of Katz ([Ka2] 2.71) applied to A_Λ tells us that the crystal of A_Λ is isogenous to a constant crystal and hence to the crystal of $(A_i)_{P[\![T]\!]}$ for some $A_i \in \mathbf{A}_i$. Therefore the isogeny $\alpha_P : (A_i, \bar{\lambda}_i)_P \longrightarrow (A_\Lambda, \bar{\lambda})_P$ that exists by induction assumption extends to an isogeny of the crystals of $(A_i)_{P[\![T]\!]}$ and A_Λ. The Hodge filtration which is given on the values at $P[\![T]\!]$ of these crystals is respected if we multiply α_P by p. Therefore we may apply the result of Grothendieck-Messing [Me] and lift α_P step by step to an isogeny $(A_i, \bar{\lambda}_i)_{P[\![T]\!]/(T^n)} \to (A_\Lambda, \bar{\lambda})_{P[\![T]\!]/(T^n)}$. Finally by Grothendieck's existence theorem we get an isogeny $(A_i, \bar{\lambda}_i)_{P[\![T]\!]} \to (A_\Lambda, \bar{\lambda})$. Hence the proof of the surjectivity is finished.
The previous lemma implies that $I(\mathbf{A}_f^p) = G(\mathbf{A}_f^p)$ and $I(\mathbf{Q}_p) = J(\mathbf{Q}_p)$. The assertion that $\Gamma \subset J(\mathbf{Q}_p)$ is discrete and cocompact modulo center follows from the compactness modulo center of $I(\mathbf{R})$, and the finiteness assertion from the finiteness of $I(\mathbf{Q}) \setminus I(\mathbf{A}_f^p)/C^p$. □

6.31 We are going to formulate uniformization theorems on the level of rigid analytic spaces. Let us denote by X the general fibre of the scheme $\mathcal{A}_{C^p}$. The isogeny class $\mathcal{I}$ given by $(A_0, \bar{\lambda}_0)$ defines a family of closed subvarieties $\mathcal{T}$ of the special fiber. We have a morphism of rigid analytic spaces:

$$(\mathcal{A}_{C^p}/\mathcal{T})^{rig} \longrightarrow X^{rig}.$$

We call $(\mathcal{A}_{C^p}/\mathcal{T})^{rig}$ the tubular neighbourhood of $\mathcal{I}$ in X^{rig}. We will denote it by $X^{rig}(\mathcal{I})$. If $\mathcal{I}$ is basic, $\mathcal{T}$ is a closed subscheme of the special fibre and the above morphism identifies $X^{rig}(\mathcal{I})$ with an admissible open subset of X^{rig}, the tube over $\mathcal{T}$, cf. (5.7).

6.32 Let us define a moduli problem over E_ν, that is represented by a union of connected components of X. Let $C_p \subset G(\mathbf{Q}_p)$ be the subgroup that fixes the polarized chain $\mathcal{L}$, i.e.

$$C_p = \{g \in G(\mathbf{Q}_p);\ g\Lambda = \Lambda, \quad \Lambda \in \mathcal{L}\}.$$

Let us denote by $C \subset G(\mathbf{A}_f)$ the subgroup $C = C_p C^p$. We define a functor Sh_C on the category of E_ν-schemes S as follows. A point of $Sh_C(S)$ consists of the following data:

1. An abelian scheme A over S up to isogeny and an injection of algebras

$$\iota : B \to End^0(A)$$

2. A $\mathbf{Q}$-homogeneous polarization $\bar{\lambda}$ on A that induces the given involution $*$ on B.

3. A C-level structure

$$\bar{\eta} : H_1(A, \mathbf{A}_f) \to V \otimes \mathbf{A}_f$$

   that respects the bilinear forms on both sides up to a factor in $\mathbf{A}_f^\times$.

We require an identity of polynomial functions

$$\det_{\mathcal{O}_S}(b\,;\ \mathrm{Lie}A) = \det_K(b\,;\ V_0), \qquad b \in O_B.$$

Consider the universal polarized multichain $\tilde{A}_\Lambda$ of abelian varieties over X. The Tate modules $T_p(\tilde{A}_\Lambda)$ form in each geometric point of X a polarized multichain of O_B-modules. The set of points where this multichain is isomorphic to the multichain $\mathcal{L}$ form a union of connected components $X_\mathcal{L}$ of X.

Lemma 6.33 *The scheme $X_\mathcal{L}$ represents the functor Sh_C.*

Proof: Indeed, assume we are given a point $(A, \iota, \bar{\lambda}, \bar{\eta})$ of Sh_C. The inverse image of the polarized multichain $\mathcal{L}$ by $\bar{\eta}_p$ defines a polarized $\mathcal{L}$-set of abelian varieties A_Λ and hence a point of X that is clearly in $X_\mathcal{L}$.

6.34 Let us assume we are given a point $(A, \iota, \bar{\lambda}, \bar{\eta})$ of Sh_C, such that A has good reduction. Then the reduction of $\{A_\Lambda\}$ defines a point of the special fibre of $\mathcal{A}_{C^p}$. The isogeny class $\mathcal{I}$ of this point in the special fibre is a union of projective varieties defined over $\bar{\kappa}$, the algebraic closure of the residue class field of E_ν. Hence it is the isogeny class of some $(A_0, \bar{\lambda}_0)$ for which we defined the uniformization morphism, cf. (6.13). Let $Sh_C^{rig}(\mathcal{I})$ be the inverse image of $X^{rig}(\mathcal{I})$ over the connected component Sh_C of X. Then $Sh_C^{rig}(\mathcal{I})$ is non–empty since $(A, \iota, \bar{\lambda}, \bar{\eta})$ lies in the image of the morphism into Sh_C^{rig}. The theorem (6.24) applies to the isogeny class $\mathcal{I}$.
For the rigid version of the uniformization theorem we work with the projective limit of rigid analytic spaces $\mathcal{M}^{rig} = \varprojlim \mathcal{M}_s^{rig}$ over E_ν. Let us denote by $\mathcal{M}_\mathcal{L}^{rig}$ the union of connected components of $\mathcal{M}^{rig}$, where the Tate modules $T_\mathcal{L}$ (compare 5.32) of the universal p-divisible groups form a polarized chain of O_B-modules which is isomorphic to $\mathcal{L}$. From theorem (6.24) we get an isomorphism of rigid analytic spaces,

$$I(\mathbf{Q})\backslash \mathcal{M}_\mathcal{L}^{rig} \times G(\mathbf{A}_f^p)/C^p \longrightarrow Sh_C^{rig}(\mathcal{I}).$$

We note that by (5.33) the spaces $\mathcal{M}^{rig}$ and $\mathcal{M}_\mathcal{L}^{rig}$ coincide if the group G is connected.

6.35 Let $\breve{C}_p \subset G(\mathbf{Q}_p)$ be an open compact subgroup contained in C_p. Let $\breve{\mathbf{M}}_{\mathcal{L},\breve{C}_p}$ be the space parametrizing trivializations of the local system $T_\mathcal{L}$ on $\breve{\mathcal{M}}_\mathcal{L}^{rig}$,

$$\alpha : T_\mathcal{L} \longrightarrow \mathcal{L} \bmod \breve{C}_p.$$

Then $\breve{\mathbf{M}}_{\mathcal{L},\breve{C}_p}$ is a finite étale covering of $\breve{\mathcal{M}}_\mathcal{L}^{rig}$. We denote by $\mathbf{M}_{\mathcal{L},\breve{C}_p}$ the corresponding pro–rigid space defined over E_ν which maps to $\mathcal{M}_\mathcal{L}^{rig}$.
Let $\breve{C} \subset C$ be a subgroup which contains a principal congruence subgroup. Clearly the functor $Sh_{\breve{C}}$ makes sense. We define $Sh_{\breve{C}}^{rig}(\mathcal{I})$ as the pullback of $Sh_C^{rig}(\mathcal{I})$ by the morphism $Sh_{\breve{C}}^{rig} \to Sh_C^{rig}$. We allow us to call this the *tubular neighbourhood of* $\mathcal{I}$.

Theorem 6.36 *Let $\breve{C} = \breve{C}_p\breve{C}^p \subset G(\mathbf{A}_f)$ be an open and closed subgroup that contains a principal congruence subgroup, and such that $\breve{C}_p \subset C_p$. Consider a point of $Sh_{\breve{C}}$ such that the corresponding abelian variety has good reduction. The reduction defines an isogeny class $\mathcal{I}$ of points in the* *special fibre of* $Sh_{C_p\check{C}^p}$. *Then we have an isomorphism of rigid analytic spaces over* E_ν,

$$I(\mathbf{Q})\backslash \mathbf{M}_{\mathcal{L},\check{C}_p} \times G(\mathbf{A}_f^p)/\check{C}^p \to Sh_{\check{C}}^{rig}(\mathcal{I}).$$

This isomorphism is for variable $\check{C}$ *equivariant with respect to the action as correspondences by* $G(\mathbf{A}_f)$ *on both sides. On the left hand side the action of* $G(\mathbf{Q}_p)$ *is the one defined in (5.34), whereas the action of* $G(\mathbf{A}_f^p)$ *is the obvious one. On the right hand side the action of* $G(\mathbf{A}_f)$ *is through Hecke correspondences. In the case of a basic isogeny class* $\mathcal{I}$ *the space on the right is an admissible open subset of* $Sh_{\check{C}}^{rig}$.

This theorem follows because the etale covering $Sh_{\check{C}} \to Sh_{C_p\check{C}^p}$ may be described as the classifying space of the trivializations of the polarized chain of Tate modules $T_p(\tilde{A}_\Lambda) \to \mathcal{L}$ on $Sh_{C_p\check{C}^p}$. By the morphism in (6.34) the polarized chains $T_{\mathcal{L}}$ and $T_p(\tilde{A}_\Lambda)$ are identified. □

6.37 In the end of this chapter we give examples of Shimura varieties, which are moduli schemes for abelian varieties with a given PEL–structure and admit a p–adic uniformization by products of Drinfeld's formal schemes $\hat{\Omega}_F^d$. Let us start by defining the datum (G, h) that gives rise to the Shimura variety. We use slightly different notations.
Let D be a central division algebra of degree d^2 over a number field K. Let $*$ be a positive involution of the second kind on D. Then K is a CM–field and the invariants by $*$ on K form the maximal totally real subfield F.
Let V be a left D–module and ψ an alternating $\mathbf{Q}$–bilinear form on V, which satisfies the equations

$$\psi(\ell v, w) = \psi(v, \ell^* w), \quad \ell \in D, \quad v, w \in V.$$

Let G be the reductive group over $\mathbf{Q}$ defined by

$$G(\mathbf{Q}) = \{g \in GL_D(V); \psi(gv, gw) = c(g)\psi(v, w), \ c(g) \in \mathbf{Q}\}.$$

Then $G_{\mathbf{R}}$ is a product of unitary groups as follows. Let us fix a CM–type $\Phi \subset \mathrm{Hom}\,(K, \mathbf{C})$ of K. We choose isomorphisms of $\mathbf{C}$–algebras

$$D \otimes_{K,\varepsilon} \mathbf{C} \simeq M_d(\mathbf{C}), \tag{6.7}$$

such that the tensor product of $*$ with the complex conjugation takes the form $X \mapsto {}^t\bar{X}$ on the right hand side. The decomposition

$$D_{\mathbf{R}} = D \otimes \mathbf{R} \simeq \prod_{\varepsilon \in \Phi} D \otimes_{K,\varepsilon} \mathbf{C}$$

induces an orthogonal decomposition with respect to ψ

$$V_{\mathbf{R}} = V \otimes \mathbf{R} = \prod_{\varepsilon \in \Phi} V \otimes_{K,\varepsilon} \mathbf{C}.$$

Since by (6.7) the algebra $M_d(\mathbf{C})$ acts on each factor we may write

$$V \otimes_{K,\varepsilon} \mathbf{C} \simeq \mathbf{C}^d \otimes_{\mathbf{C}} W_\varepsilon. \tag{6.8}$$

Here W_ε is a $\mathbf{C}$–vector space and the action of $M_d(\mathbf{C})$ on the right hand side is via the first factor. We define an antihermitian form h_ε on W_ε by the equation

$$\psi(Z_1 \otimes W_1,\ Z_2 \otimes W_2) = Tr_{\mathbf{C}/\mathbf{R}}({}^t\bar{Z}_1 Z_2 h_\varepsilon(W_1, W_2)), \qquad \begin{array}{l} W_1, W_2 \in W_\varepsilon \\ Z_1, Z_2 \in \mathbf{C}^d. \end{array}$$

Choosing a suitable isomorphism $W_\varepsilon \simeq \mathbf{C}^m$ we may write h_ε in normal form $h_\varepsilon(W_1, W_2) = {}^t\bar{W}_1 H_\varepsilon W_2$, where

$$H_\varepsilon = \operatorname{diag}(-\sqrt{-1}, \ldots, -\sqrt{-1}; \sqrt{-1}, \ldots, \sqrt{-1}).$$

We denote by r_ε (resp. $r_{\bar{\varepsilon}}$) the number of places, where $-\sqrt{-1}$ (resp. $\sqrt{-1}$) appears in H_ε. Let $J_\varepsilon : V \otimes_{K,\varepsilon} \mathbf{C} \to V \otimes_{K,\varepsilon} \mathbf{C}$ be the endomorphism given by the matrix $\mathrm{id}_{\mathbf{C}^d} \otimes -H_\varepsilon$. It has the property that the $\mathbf{R}$–bilinear form $\psi_\varepsilon(x, J_\varepsilon y)$ in $x, y \in V \otimes_{K,\varepsilon} \mathbf{C}$ is symmetric and positive definite. The endomorphism $J = \oplus J_\varrho$, $J^2 = -1$, defines a complex structure of $V_{\mathbf{R}}$. For this complex structure we have

$$Tr_{\mathbf{C}}(\ell; V_{\mathbf{R}}) = \sum_{\alpha : K \to \mathbf{C}} r_\alpha \alpha(Tr^0 \ell), \quad \ell \in D, \tag{6.9}$$

where $Tr^0\ell$ denotes the reduced trace of D over K.
Let

$$GU(r_\varrho, r_{\bar{\varrho}}) = \{A \in M_m(\mathbf{C});\, {}^t\bar{A} H_\varrho A = c(A) H_\varrho,\ c(A) \in \mathbf{R}^\times\}$$

be the group of unitary similitudes. We have an injection

$$G_{\mathbf{R}} \longrightarrow \prod_{\varrho \in \Phi} GU(r_\varrho, r_{\bar{\varrho}}), \quad r_\varrho + r_{\bar{\varrho}} = m = \frac{1}{d} \dim_K V$$

such that $G_{\mathbf{R}}$ is a normal subgroup with a torus cokernel. We define a homomorphism $h : Res_{\mathbf{C}/\mathbf{R}} \mathbf{G}_{m,\mathbf{C}} \to G_{\mathbf{R}}$ by the condition that $h(r)$ for $r \in \mathbf{R}^\times$ acts on $V_{\mathbf{R}}$ by multiplication with r and $h(\sqrt{-1})$ acts as J. The pair (G, h) gives rise to a Shimura variety Sh, which is defined over the number field $E \subset \mathbf{C}$ generated over $\mathbf{Q}$ by the numbers (6.9).

6.38 We will obtain examples for p–adic uniformization with Drinfeld's Ω only in the cases where $r_\varrho \in \{0, 1, m, m-1\}$ for all $\varrho : K \to \mathbf{C}$ and $m = d$. As before we fix embeddings $\bar{\mathbf{Q}} \to \mathbf{C}$, $\nu : \bar{\mathbf{Q}} \to \mathbf{C}_p$. Then ν defines a p–adic place of the Shimura field E. We define a model of Sh over the ring of integers O_{E_ν}, if ν satisfies the following conditions.
Let $\mathbf{p}_1, \ldots, \mathbf{p}_t$ be the prime ideals of O_F lying over p. We assume that K/F is unramified at these primes. Let us assume that for some s, $1 \leq s \leq t$,

$$\mathbf{p}_i = \mathbf{q}_i \bar{\mathbf{q}}_i \quad \text{for } i = 1, \ldots, s,$$

where $\mathbf{q}_i \neq \bar{\mathbf{q}}_i$ are prime ideals of O_K, and

$$\mathbf{p}_i O_K = \mathbf{q}_i \quad \text{for } i = s+1, \ldots, t$$

are prime ideals.
Let us assume that $D_{\mathbf{q}_i}$ is a matrix algebra over $K_{\mathbf{q}_i}$ for $i = s+1, \ldots, t$. In fact, this is implied by the existence of the involution of the second kind $*$. Moreover we make the assumption that there is a maximal order $O_{D,p} \subset D \otimes \mathbf{Q}_p$, which is invariant by the involution $*$. By the methods of the appendix to chapter 3 one checks the following.

Lemma 6.39 *There exists up to isomorphism a unique symplectic $O_{D,p}$-module (T_p, ψ_p), such that*

(i) T_p is a free $\mathbf{Z}_p$-module of rank $\dim_{\mathbf{Q}_p} V$.

(ii) The pairing $\psi_p : T_p \times T_p \to \mathbf{Z}_p$ is perfect and satisfies

$$\psi_p(\ell v, w) = \psi_p(v, \ell^* w) \text{ for } \ell \in O_{D,p}, v, w \in T_p.$$

We assume that there exists an isomorphism of symplectic $D \otimes \mathbf{Q}_p$–modules

$$(V \otimes_{\mathbf{Q}} \mathbf{Q}_p, \psi) \simeq (T_p, \psi_p) \otimes_{\mathbf{Z}_p} \mathbf{Q}_p. \tag{6.10}$$

Let us denote by $\Lambda \subset V \otimes_{\mathbf{Q}} \mathbf{Q}_p$ the image of T_p by this isomorphism. We have the decomposition

$$\Lambda = \bigoplus \Lambda_{\mathbf{p}_i}.$$

The multiples of the $\Lambda_{\mathbf{p}_i}$ generate a multichain $\mathcal{L}$ (see definition 3.13).
Let us denote by $\mu : \mathbf{G}_{m,\mathbf{C}} \to G_{\mathbf{C}}$ the cocharacter such that $h_{\mathbf{C}} = \mu \times \bar{\mu}$, cf. (6.2). We have defined a moduli problem (6.9) which is for small C^p representable by a quasi–projective scheme $\mathcal{A}_{C^p}$ over O_{E_ν} attached to the data $(D, O_D, V, \psi, \mu, \mathcal{L}, C^p, \nu)$ where $C^p \subset G(\mathbf{A}_f^p)$ is a congruence subgroup. We note that in our case the whole $\mathcal{L}$–set of abelian varieties is determined by the single abelian variety up to isogeny of order prime to p, A_Λ, which in the following will be simply denoted by A.
Let $(A, \bar{\lambda}, \bar{\eta})$, $\bar{\eta} : H_1(A, \mathbf{A}_f^p) \xrightarrow{\sim} V \otimes_{\mathbf{Q}} \mathbf{A}_f^p \bmod C^p$ be a point of the moduli problem over $\mathbf{C}$.
Since $\bar{\lambda}$ contains by definition a polarization λ of order prime to p, it induces a perfect pairing on the Tate–module

$$E^\lambda : T_p(A) \times T_p(A) \longrightarrow \mathbf{Z}_p.$$

Hence by lemma (6.40) there is an isomorphism

$$(T_p(A), E^\lambda) \simeq (T_p, \psi_p).$$

We conclude that locally at any finite place w of $\mathbf{Q}$ there is a similitude between the symplectic D–modules $(H_1(A, \mathbf{Q}_w), E^\lambda)$ and $(V, \psi) \otimes \mathbf{Q}_w$. By the condition ((6.9), i) such a similitude exists also at the infinite place.
If our group satisfied the Hasse principle this would imply the existence of a symplectic similitude between the D–modules $(H_1(A, \mathbf{Q}), E^\lambda)$ and (V, ψ). This would show that $\mathcal{A}_{C^p} \otimes_{O_{E_\nu}} E_\nu$ is isomorphic to $Sh_C \otimes_E E_\nu$, where the subgroup $C \subset G(\mathbf{A}_f)$ is of the form $C = C_p \cdot C^p$. The subgroup $C_p \subset G(\mathbf{Q}_p)$ is given by $C_p = G(\mathbf{Q}_p) \cap \operatorname{End}_{\mathbf{Z}_p} \Lambda$.
Unfortunately the group G does not satisfy the Hasse principle. Therefore we obtain for any symplectic B–module (V_i, ψ_i), which is locally isomorphic to (V, ψ) a Shimura variety Sh_i on which $G(\mathbf{A}_f)$ acts. If $(V, \psi) = (V_1, \psi_1), \ldots, (V_h, \psi_h)$ are all classes of such modules we have

$$\mathcal{A}_{C^p} \otimes_{O_{E_\nu}} E_\nu \simeq \coprod_{i=1}^{h} Sh_{i,C}.$$

6.40 We consider the p–adic uniformization of $\mathcal{A}_{C^p}$ under a whole series of assumptions which we explain now. We assume that the invariants of the division algebra D are as follows at the primes over p

$$\begin{array}{ll} \mathrm{inv}_{\mathbf{q}_i} D = 1/d & \text{for } i = 1, \ldots, r \\ \mathrm{inv}_{\mathbf{q}_i} D = 0 & \text{for } i = s+1, \ldots, t. \end{array}$$

Here $r \leq s$ is a fixed integer. The other invariants are arbitrary but satisfy

$$\mathrm{inv}_{\mathbf{q}_i} D = -\mathrm{inv}_{\bar{\mathbf{q}}_i} D, \quad i = 1, \ldots, t.$$

We will also assume that V is a free D–module of rank 1, i.e. $m = d$. The next assumption is the existence of a CM–type Φ of K, which has the following properties and which will be fixed.

Let us first note that the chosen embeddings $\nu : \bar{\mathbf{Q}} \to \bar{\mathbf{Q}}_p$ and $\bar{\mathbf{Q}} \to \mathbf{C}$ allow us to identify the following sets

$$\mathrm{Hom}\,(K, \mathbf{C}) \simeq \mathrm{Hom}\,(K, \bar{\mathbf{Q}}_p). \tag{6.11}$$

We require that $\Phi \cap \mathrm{Hom}\,(K_{\mathbf{p}_i}, \bar{\mathbf{Q}}_p) = \mathrm{Hom}\,(K_{\mathbf{q}_i}, \bar{\mathbf{Q}}_p)$ for $i = 1, \ldots, s$. If $\mathbf{p}_i \cdot O_K = \mathbf{q}_i$ is a prime ideal, we consider the maximal subfields $K^t_{\mathbf{q}_i} \subset K_{\mathbf{q}_i}$ and $F^t_{\mathbf{p}_i} \subset F_{\mathbf{p}_i}$, which are unramified over $\mathbf{Q}_p$. Since by assumption $K^t_{\mathbf{q}_i}/F^t_{\mathbf{p}_i}$ is a quadratic extension, it makes sense to consider a CM–type $\Phi^t_{\mathbf{q}_i} \subset \mathrm{Hom}\,(K^t_{\mathbf{q}_i}, \bar{\mathbf{Q}}_p)$ relative to $K^t_{\mathbf{q}_i}/F^t_{\mathbf{p}_i}$. We require that $\Phi \cap \mathrm{Hom}\,(K_{\mathbf{q}_i}, \bar{\mathbf{Q}}_p)$ is the inverse image of a CM–type $\Phi^t_{\mathbf{q}_i}$ for the extension $K^t_{\mathbf{q}_i}/F^t_{\mathbf{p}_i}$ by the map

$$\mathrm{Hom}\,(K_{\mathbf{q}_i}, \bar{\mathbf{Q}}_p) \longrightarrow \mathrm{Hom}\,(K^t_{\mathbf{q}_i}, \bar{\mathbf{Q}}_p).$$

For any i, such that $1 \leq i \leq r$ we choose arbitrarily an embedding $\alpha_i : K_{\mathbf{q}_i} \to \bar{\mathbf{Q}}_p$.

The condition we put on the numbers r_α is as follows.

$$r_\alpha = \begin{cases} 1, & \text{if } \alpha \in \{\alpha_1, \ldots, \alpha_r\} \\ d-1, & \text{if } \alpha \in \{\bar{\alpha}_1, \ldots, \bar{\alpha}_r\} \\ 0, & \text{if } \alpha \in \Phi \setminus \{\alpha_1, \ldots, \alpha_r\} \\ d, & \text{if } \alpha \in \Phi \setminus \{\bar{\alpha}_1, \ldots, \bar{\alpha}_r\} \end{cases}$$

Whether this condition is fullfilled depends of course on the place ν, which was used for the identification (6.11).
For each prime ideal $\mathbf{q}$ of K over p, we define

$$E_{\nu,\mathbf{q}} = \mathbf{Q}_p[\sum_{\alpha: K_\mathbf{q} \rightarrow \bar{\mathbf{Q}}_p} r_\alpha \alpha(Tr^0_{D_\mathbf{q}}(\ell));\ \ell \in D_q].$$

We find easily

$$\begin{array}{ll} E_{\nu,\mathbf{q}_i} = E_{\nu,\bar{\mathbf{q}}_i} = \alpha_i(K_{\mathbf{q}_i}), & \text{for } i = 1, \ldots, r \\ E_{\nu,\mathbf{q}_i} = E_{\nu,\bar{\mathbf{q}}_i} = \mathbf{Q}_p, & \text{for } i = r+1, \ldots, s \\ Gal(\bar{\mathbf{Q}}_p/E_{\nu,\mathbf{q}_i}) = \{\tau \in Gal(\bar{\mathbf{Q}}_p/\mathbf{Q}_p); \tau\Phi^t_{\mathbf{q}_i} = \Phi^t_{\mathbf{q}_i}\}, & \text{for } i = s+1, \ldots, t. \end{array}$$

The localization E_ν of the Shimura field is the composite of the fields $E_{\nu,\mathbf{q}_i}$ for $i = 1, \ldots, t$.

Let $(A_0, \bar{\lambda}_0, \bar{\eta}_0) \in \mathcal{A}_{C^p}(\bar{\kappa}_\nu)$ be a point over the residue class field of $\breve{E}_\nu$. By (6.12) this point determines a σ–conjugacy class $\bar{b}$ of an element $b \in G(K_0)$ such that $c(b) = p$.

Lemma 6.41 *The conjugacy class $\bar{b}$ is basic and does not depend on the choice of the point $(A_0, \bar{\lambda}_0, \bar{\eta}_0)$.*

Proof: The decomposition $F_p = \prod_{i=1}^{t} F_{\mathbf{p}_i}$ induces an orthogonal decomposition $V \otimes \mathbf{Q}_p = \oplus V_{\mathbf{p}_i}$ and moreover an injection with torus cokernel,

$$G_{\mathbf{Q}_p} \longrightarrow \Pi\, G'_{\mathbf{p}_i}. \tag{6.12}$$

Here $G'_{\mathbf{p}_i}$ is the algebraic group over $\mathbf{Q}_p$ given by

$$G'_{\mathbf{p}_i}(\mathbf{Q}_p) = \{g \in \mathrm{End}_{D_{\mathbf{p}_i}} V \otimes_F F_{\mathbf{p}_i}; \psi(gv, gw) = c_i(g)\psi(v, w),\ \ c_i(g) \in \mathbf{Q}_p^\times\}.$$

We will show in fact a stronger assertion than (6.41), which will be explained now. Let $G_1 \subset G$ be the subgroup given by the condition $c(g) = 1$. The group $G_1(K_0)$ acts by σ–conjugacy on the set of elements $g \in G(K_0)$ such that $c(g) = p$,

$$g \longmapsto h^{-1} g \sigma(h),\ \ h \in G_1(K_0).$$

Let $B_1(G) = \{g \in G(K_0); c(g) = p\}/G_1(K_0)$ be the orbit space. Let us fix a principal polarization $\lambda_0 \in \bar{\lambda}_0$ (6.6). The class in $B_1(G)$ of the element

$b \in G(K_0)$ defined by (6.12) depends only on the pair (A_0, λ_0) and not on the choice of the isomorphism (6.2).

Since $B_1(G_{\mathbf{Q}_p}) \subset \prod B_1(G'_{\mathbf{p}_i})$ it suffices to show that the components $\bar{b}_i \in B_i(G'_{\mathbf{q}_i})$ of b do not depend on the choice of (A_0, λ_0) and that the images of $\bar{b}_i$ in $B(G'_{\mathbf{p}_i})$ are basic.

In the case where $\mathbf{p}_i$ splits in K there is a further decomposition into totally isotropic subspaces

$$V_{\mathbf{p}_i} = V_{\mathbf{q}_i} \oplus V_{\bar{\mathbf{q}}_i}.$$

For the remaining primes $\mathbf{p}_i$ we have by definition $V_{\mathbf{p}_i} = V_{\mathbf{q}_i} = V_{\bar{\mathbf{q}}_i}$. Similar decompositions are obtained for the isocrystal $(N, \mathbf{F})$ associated to (A_0, λ_0). By (6.25) b_i is basic if all isocrystals $(N_{\mathbf{q}_i}, \mathbf{F})$ and hence by duality also $(N_{\bar{\mathbf{q}}_i}, \mathbf{F})$ are isoclinic. For the second assertion of the lemma we need moreover to verify that for $i = 1, \ldots, s$ the isomorphism classes of the isocrystal $(N_{\mathbf{q}_i}, \mathbf{F})$ and for $i = s+1, \ldots, t$ the isomorphism classes of the polarized isocrystals $(N_{\mathbf{p}_i}, \mathbf{F})$ do not depend on (A_0, λ_0).

The cocharacter μ defines a decomposition for each $i = 1, \ldots, t$ into weight spaces

$$V_{\mathbf{p}_i} \otimes \mathbf{C}_p = V_{\mathbf{p}_i,0} \oplus V_{\mathbf{p}_i,1}. \tag{6.13}$$

Of course for the Shimura variety and the associated moduli problem only the conjugacy class of μ respectively of (6.13) matters. More directly we obtain the decomposition (6.13) as follows.

For each $\alpha : K \longrightarrow \mathbf{C}_p$ we choose a $D \otimes_{K,\alpha} \mathbf{C}_p$–submodule V_α of $V \otimes_{K,\alpha} \mathbf{C}_p$ of dimension $\dim_{\mathbf{C}_p} V_\alpha = r_\alpha \cdot d$. Then $V_{\mathbf{p}_i,0} = \oplus V_\alpha$, where the direct sum is over all α which induce on F the valuation given by $\mathbf{p}_i$.

Let $X = \prod_{i=1}^{s} (X_{\mathbf{q}_i} \times X_{\bar{\mathbf{q}}_i}) \times \prod_{i=s+1}^{t} X_{\mathbf{q}_i}$ be the p–divisible group of A_0. The condition (i) of 6.9 implies

$$\det_{\bar{\kappa}_\nu}(\ell; \operatorname{Lie} X_{\mathbf{q}_i}) = \det_{\mathbf{C}_p}(\ell; V_{\mathbf{q}_i,0}).$$

For $i = 1, \ldots, r$ this implies that $X_{\mathbf{q}_i}$ is a special formal $O_{D_{\mathbf{q}_i}}$–module for the $O_{K_{\mathbf{q}_i}}$–algebra structure on $\bar{\kappa}_\nu$ given by $\alpha_i : O_{K_{\mathbf{q}_i}} \longrightarrow O_{E_\nu}$ (3.59). For $i = r+1, \ldots, s$ this implies that $X_{\mathbf{q}_i}$ is étale. Hence for $i = 1, \ldots, s$ the component of $\bar{b}$ in $G_{\mathbf{p}_i}$ is a basic conjugacy class and the isomorphism class of the polarized isocrystal $(N_{\mathbf{p}_i}, \mathbf{F})$ is independent of the choice $(A, \lambda_0, \bar{\eta}_0)$.

It remains to treat the primes $\mathbf{p}_i$ for $i > s$. We fix an isomorphism $O_{D_{\mathbf{q}_i}} \simeq M_d(O_{K_{\mathbf{q}_i}})$ and a perfect hermitian pairing inducing the involution $*$,

$$H : O_{K_{\mathbf{q}_i}}^d \times O_{K_{\mathbf{q}_i}}^d \longrightarrow O_{K_{\mathbf{q}_i}},$$

which is linear in the first variable and antilinear in the second.
Let M' be the Cartier module of $X_{\mathbf{q}_i}$. There is a Cartier module M with an action of $O_{K_{\mathbf{q}_i}}$ and an isomorphism of Cartier modules

$$M' \simeq O_{K_{\mathbf{q}_i}}^d \otimes_{O_{K_{\mathbf{q}_i}}} M$$

which respects the $M_d(O_{K_{\mathbf{q}_i}})$-module structure on both sides. Let Ψ' be a perfect polarization form on M' belonging to $\bar{\lambda}$. Then there is a perfect polarization form Ψ on M, defined by

$$\Psi'(x \otimes m,\ y \otimes n) = \Psi(H(x, y)m, n).$$

We have to show that M is isoclinic and that (M, Ψ) is unique up to isogeny.
Let $W = W(\bar{\kappa}_\nu)$ be the Witt ring and $\varphi : O_{K_{\mathbf{q}_i}^t} \to W$ an embedding. We set $L_\varphi = K_{\mathbf{q}_i} \otimes_{K_{\mathbf{q}_i}^t, \varphi} W_{\mathbf{Q}}$ and let $O_{L_\varphi} = O_{K_{\mathbf{q}_i}} \otimes_{O_{K_{\mathbf{q}_i}^t}, \varphi} W$ be its ring of integers.
We have the decomposition

$$O_{K_{\mathbf{q}_i}} \otimes_{\mathbf{Z}_p} W = \prod_{\varphi : O_{K_{\mathbf{q}_i}^t} \to W} O_{L_\varphi}.$$

Hence we get for the Cartier module

$$M = \bigoplus_\varphi M_\varphi.$$

Because of the equation

$$\Psi(km, n) = \Psi(m, \bar{k}n), \quad k \in O_{K_{\mathbf{q}_i}},$$

M_φ and $M_{\varphi'}$ are orthogonal unless $\varphi' = \bar{\varphi}$.
We denote by σ the Frobenius automorphism of W. The Verschiebung induces maps $\mathbf{V} : M_{\sigma\varphi} \to M_\varphi$. By the condition on the determinants (6.9 (i)) the cokernel of the last map is a $\bar{\kappa}_\nu$–vector space of dimension 0 if

$\varphi \in \Phi^t_{\mathbf{q}_i}$ and of dimension de for $\varphi \in \overline{\Phi^t_{\mathbf{q}_i}}$, where $e = [K_{\mathbf{q}_i} : K^t_{\mathbf{q}_i}]$ is the index of ramification of $K_{\mathbf{q}_i}$. Since M_φ is a free O_{L_φ}–module of rank d, we obtain

$$\begin{aligned} \mathbf{V}M_{\sigma\varphi} &= M_\varphi \quad \text{for } \varphi \in \Phi^t_{\mathbf{q}_i} \\ \mathbf{V}M_{\sigma\varphi} &= pM_\varphi \quad \text{for } \varphi \in \overline{\Phi^t_{\mathbf{q}_i}} \end{aligned} \tag{6.14}$$

Let $f_i = [F^t_{\mathbf{p}_i} : \mathbf{Q}_p] = \#\Phi^t_{\mathbf{q}_i}$. We fix an embedding $\varphi_0 : K^t_{\mathbf{q}_i} \to W_{\mathbf{Q}}$ and we set $M_j = M_{\sigma^{-j}\varphi_0}$.
It follows by (6.14) that the map $U = p^{-f_i}\mathbf{V}^{2f_i} : M_0 \to M_0$ is an isomorphism. This shows that M is isoclinic of slope 1/2. We will verify that the polarized crystal (M, ψ) with its $O_{K_{\mathbf{q}_i}}$–action is uniquely determined up to isomorphism by the conditions (6.14).
There is an integer a, such that $p^a\mathbf{V}^f : M_0 \to M_f$ is an isomorphism. Consider the perfect pairing

$$\Omega : M_0 \times M_0 \longrightarrow W$$

defined by the equation

$$\Omega(m, m') = \psi(m, p^a\mathbf{V}^f m'), \quad m, m' \in M_0.$$

Let $\tau = \sigma^{f_i}$ the Frobenius automorphism of L_0 relative to $F_{\mathbf{p}_i}$. Then Ω satisfies the following relations

$$\begin{aligned} \Omega(\ell m, m') &= \Omega(m, \ell^\tau m'), \quad \ell \in O_{L_0} \\ \Omega(Um, Um') &= \Omega(m, m')^{\tau^{-2}} \\ \Omega(m, m') &= -\Omega(m', m) \text{ for } m, m' \in M_0. \end{aligned} \tag{6.15}$$

The crystal (M, ψ) is uniquely determined by (M_0, U, Ω). Let $\Gamma_0 = M_0^U$ be the $O_{K_{\mathbf{q}_i}}$–module of invariants. Then Ω induces a perfect pairing

$$\Psi_{\Gamma_0} : \Gamma_0 \times \Gamma_0 \longrightarrow O_{K^t_{\mathbf{q}_i}},$$

which is $O_{K^t_{\mathbf{q}_i}}$–linear in the first and τ–linear in the second variable. Again $(\Gamma_0, \Psi_{\Gamma_0})$ determines (M, ψ) uniquely.
Clearly Ψ_{Γ_0} is antisymmetric and satisfies

$$\Psi_{\Gamma_0}(\ell m, m') = \Psi_{\Gamma_0}(m, \ell^\tau m'), \quad m, m' \in \Gamma_0, \quad \ell \in O_{K_{\mathbf{q}_i}}.$$

By lemma (6.39) the pair $(\Gamma_0, \Psi_{\Gamma_0})$ is uniquely determined up to isomorphism. The lemma is proved. □

Let $i > s$. Let us set $\Gamma = O^d_{K_{\mathbf{q}_i}} \otimes_{O_{K_{\mathbf{q}_i}}} \Gamma_0$ with its natural $O_{D_{\mathbf{p}_i}} \cong M_d(O_{K_{\mathbf{q}_i}})$–module structure. On Γ we define the perfect antisymmetric $\mathbf{Z}_p$–bilinear form

$$\psi_\Gamma(x \otimes m,\ y \otimes m') = Tr_{K^t_{\mathbf{q}_i}/\mathbf{Q}_p} \psi_{\Gamma_0}(H(x,y)m, m').$$

Then the symplectic module (Γ, ψ_Γ) is isomorphic to $(\Lambda_{\mathbf{p}_i}, \psi)$.

6.42 Lemma (6.41) shows that the assumptions of theorem (6.30) are fullfilled in our situation, and that Z consists of the whole special fibre of $\mathcal{A}_{C^p}$. Let us describe now more explicitly the group $I(\mathbf{Q})$ and the formal scheme $\breve{\mathcal{M}}$ over $Spf\ O_{\breve{E}_\nu}$ with its Weil descent datum.
We consider the orthogonal decomposition with respect to $\psi \otimes \mathbf{Q}_p$,

$$V \otimes \mathbf{Q}_p = \bigoplus_{i=1}^{t} V \otimes_F F_{\mathbf{p}_i}.$$

Correspondingly (cf. (6.12))

$$G(\mathbf{Q}_p) \subset \prod_{i=1}^{t} G'_{\mathbf{p}_i}(\mathbf{Q}_p).$$

The data b, μ decompose naturally as a product $b = \prod_i b_i, \mu = \prod \mu_i$. The decomposition

$$V_{\mathbf{p}_i} \otimes \bar{\mathbf{Q}}_p = V_0 \oplus V_1$$

defined by μ_i is up to conjugation determined as follows. Consider the decomposition

$$V_{\mathbf{p}_i} \otimes \bar{\mathbf{Q}}_p = \bigoplus_{\alpha : F_{\mathbf{p}_i} \to \bar{\mathbf{Q}}_p} V_{\mathbf{p}_i} \otimes_{F_{\mathbf{p}_i}, \alpha} \bar{\mathbf{Q}}_p = \bigoplus_\alpha V_\alpha.$$

Then V_0 has a decomposition

$$V_0 = \bigoplus_\alpha V_{0,\alpha},$$

where $V_{0,\alpha} \subset V_0$ is a $D \otimes_{F,\alpha} \bar{\mathbf{Q}}_p$–submodule of dimension $r_\alpha \cdot d$. The local Shimura field associated to

$$(K_{\mathbf{p}_i}, D_{\mathbf{p}_i}, O_{D_{\mathbf{p}_i}}, V_{\mathbf{p}_i}, b_i, \mu_i, \Lambda_{\mathbf{p}_i}) \tag{6.16}$$

is $E_{\nu,\mathbf{q}_i}$. Let us denote this field by E_i. Let $\breve{\mathcal{M}}_i$ be the associated formal scheme over $Spf\ O_{\breve{E}_i}$. It is equipped with a Weil descent datum relative to $\breve{E}_i/E_i$. On each $\breve{\mathcal{M}}_i$ we have defined a function $\breve{c}_i : \breve{\mathcal{M}}_i \rightarrow \mathbf{Z}$, comp. proof of (3.53).
One sees easily that $\breve{\mathcal{M}}$ is the formal subscheme of

$$(\breve{\mathcal{M}}_1 \times_{Spf\ O_{\breve{E}_1}} Spf\ O_{\breve{E}_\nu}) \times_{Spf\ O_{\breve{E}_\nu}} \cdots \times_{Spf\ O_{\breve{E}_\nu}} (\breve{\mathcal{M}}_t \times_{Spf\ O_{\breve{E}_t}} Spf\ O_{\breve{E}_\nu}),$$

where the functions $\breve{c}_i$ agree.

6.43 We consider the formal schemes $\breve{\mathcal{M}}_i$ more closely. Let us begin with the cases $i = 1, \ldots, r$. The group $X^*_{\mathbf{Q}_p}(G'_{\mathbf{p}_i})$ has three generators $\mathbf{n}, \mathbf{n}^*, c$, satisfying the relation $\mathbf{n} \cdot \mathbf{n}^* = c^{m_i}$, where $m_i = d[F_{\mathbf{p}_i} : \mathbf{Q}_p]$. We have defined a morphism

$$\varkappa : \breve{\mathcal{M}}_i \longrightarrow \mathrm{Hom}\,(X^*_{\mathbf{Q}_p}(G'_{\mathbf{p}_i}), \mathbf{Z})$$

by giving the $\mathbf{Z}$–valued maps

$$\breve{\mathbf{n}} = <\varkappa, \mathbf{n}>, \ \breve{\mathbf{n}}^* = <\varkappa, \mathbf{n}^*>, \ \breve{c} = <\varkappa, c> .$$

Let us denote by $\delta : G'_{\mathbf{p}_i}(\mathbf{Q}_p) \rightarrow \mathrm{Hom}\,(X^*_{\mathbf{Q}_p}(G'_{\mathbf{p}_i}), \mathbf{Z})$ the map defined by

$$< \delta(g), \chi >= \mathrm{ord}_p\, \chi(g), \ \chi \in X^*_{\mathbf{Q}_p}(G'_{\mathbf{p}_i}), \ g \in G'_{\mathbf{p}_i}(\mathbf{Q}_p).$$

We define an action of $G'_{\mathbf{p}_i}(\mathbf{Q}_p)$ on $\breve{\mathcal{M}}_i$ from the right, which commutes with the action of $J'_{\mathbf{p}_i}(\mathbf{Q}_p)$ from the left. Let $g \in G'_{\mathbf{p}_i}(\mathbf{Q}_p)$ and (X, ϱ) be a point of $\breve{\mathcal{M}}_i$ with values in some $O_{\breve{E}_i}$–scheme S. There is an element $b \in D_{\mathbf{p}_i}$, such that $bb^* \in \mathbf{Q}_p$ and $bg^{-1}\Lambda_{\mathbf{p}_i} = \Lambda_{\mathbf{p}_i}$. We define the action by

$$(X, \varrho)g = (X^b, \iota_X(b^{-1})\varrho).$$

This action and δ make $\varkappa$ into a $G'_{\mathbf{p}_i}(\mathbf{Q}_p)$–equivariant map.
Let (X, ϱ) be a point of $\breve{\mathcal{M}}_i$. Because of the decomposition $O_{D_{\mathbf{p}_i}} = O_{D_{\mathbf{q}_i}} \times O_{D_{\bar{\mathbf{q}}_i}}$ we have the induced decomposition $X = X_1 \times X_2$. The condition

on the determinants (3.21, (iv)) implies that X_1 is a special formal $O_{D_{\mathbf{q}_i}}$–module. Moreover X_2 has to be isomorphic to the dual of X_1. More precisely the involution on $O_{D_{\mathbf{p}_i}}$ induces an isomorphism $O_{D_{\mathbf{q}_i}} \simeq O^{opp}_{D_{\mathbf{q}_i}}$. With this identification X becomes isomorphic to the $O_{D_{\mathbf{p}_i}}$–module $X_1 \times \hat{X}_1$ and the polarization on X becomes the tautological polarization on $X_1 \times \hat{X}_1$. Hence in the definition of the functor $\breve{\mathcal{M}}_i$ we may take for $\mathbf{X}$ a p–divisible group of the form $\mathbf{X} = \mathbf{Y} \times \hat{\mathbf{Y}}$, where $\mathbf{Y}$ is a special formal $O_{D_{\mathbf{q}_i}}$–module and the polarization is the obvious one.
Then the fibre of $\varkappa$ over zero consists of points (X, ϱ), where $\varrho = \varrho_1 \times \varrho_2 : \mathbf{Y}_{\bar{S}} \times \hat{\mathbf{Y}}_{\bar{S}} \rightarrow (X_1 \times X_2)_{\bar{S}}$ is such that ϱ_1 and ϱ_2 are quasi–isogenies of height 0. Such a point is isomorphic to $(X_1 \times \hat{X}_1, \varrho_1 \times \hat{\varrho}_1^{-1})$. Hence the map $(X, \varrho) \rightarrow (X_1, \varrho_1)$ defines an isomorphism of the fibre of $\varkappa$ over zero and Drinfeld's formal scheme $\hat{\Omega}^d_{F_{\mathbf{p}_i}} \times_{Spf\ O_{F_{\mathbf{p}_i}}} Spf\ O_{\breve{E}_i}$.
Let $\Pi \in O_{D_{\mathbf{q}_i}}$ be a prime element. Then $\Pi : X_1 \rightarrow X_1^{\Pi^{-1}}$ is a quasi–isogeny of height $f_i d$. It follows easily from (3.53) that the images of δ and $\varkappa$ coincide. Therefore we obtain an isomorphism

$$\breve{\mathcal{M}}_i \cong \left(\hat{\Omega}^d_{F_{\mathbf{p}_i}} \times_{Spf\ O_{F_{\mathbf{p}_i}}} Spf\ O_{\breve{E}_i}\right) \times G'_{\mathbf{p}_i}(\mathbf{Q}_p)/C_{\mathbf{p}_i}, \tag{6.17}$$

where $C_{\mathbf{p}_i} = ker\,\delta$ is the maximal compact subgroup of $G'_{\mathbf{p}_i}(\mathbf{Q}_p)$.
Let us compare the descent data on both sides. To formulate the result we write the group $G'_{\mathbf{p}_i}$ in a more suitable form. Let us fix an isomorphism $V \simeq D$ and hence a right action of D on V. As above the form ψ and the involution $*$ provide us with isomorphisms

$$V_{\mathbf{p}_i} \cong V_{\mathbf{q}_i} \oplus V^*_{\mathbf{q}_i}, \quad D_{\mathbf{p}_i} \simeq D_{\mathbf{q}_i} \times D^{opp}_{\mathbf{q}_i},$$

which take ψ respectively $*$ to the obvious alternating form respectively involution on the right hand sides.
From these isomorphisms we get an identification

$$G'_{\mathbf{p}_i}(\mathbf{Q}_p) = \{(b_1, b_2) \in D^{\times opp}_{\mathbf{q}_i} \times D^{\times}_{\mathbf{q}_i}; b_1 b_2 \in \mathbf{Q}_p\}.$$

Lemma 6.44 *The Weil descent datum on $\breve{\mathcal{M}}_i$ induces on the right hand side of (6.17) the canonical Weil descent datum on $(\hat{\Omega}^d_{F_{\mathbf{p}_i}} \times_{Spf\ O_{F_{\mathbf{p}_i}}} Spf\ O_{\breve{E}_i})$ times multiplication by $(\Pi, p^{f_i}\Pi^{-1})$ on the second factor, where $f_i = [F^t_{\mathbf{p}_i} : \mathbf{Q}_p]$ is the degree of inertia.*

Proof: Indeed, consider a point (X_1, ϱ_1) of $\hat{\Omega}^d_{F_{\mathbf{p}_i}} \times_{Spf\ O_{F_{\mathbf{p}_i}}} Spf\ O_{\breve{E}_i}$. Then the canonical descent datum is given by

$$(X_1, \varrho_1) \longmapsto (X_1^{\Pi^{-1}},\ \iota_{X_1}(\Pi)\varrho_1\ \mathrm{Frob}_{\mathbf{Y}}^{-1}),$$

where $\mathrm{Frob}_{\mathbf{Y}}$ is the Frobenius morphism relative to the residue class field of E_i.

The point (X_1, ϱ_1) is mapped by (6.17) to the point $(X_1 \times \hat{X}_1, \varrho_1 \times \hat{\varrho}_1^{-1})$ of $\breve{\mathcal{M}}_i$. The Weil descent datum on $\breve{\mathcal{M}}_i$ maps this point to

$$(X_1 \times \hat{X}_1, \varrho_1\,\mathrm{Frob}_{\mathbf{Y}}^{-1} \times \hat{\varrho}_1^{-1}\,\mathrm{Frob}_{\hat{\mathbf{Y}}}^{-1}) = \\ (X_1 \times \hat{X}_1, \varrho_1\,\mathrm{Frob}_{\mathbf{Y}}^{-1} \times (\varrho_1\,\mathrm{Frob}_{\mathbf{Y}}^{-1})^{\wedge -1} p^{-f_i}).$$

The comparison follows easily if one takes into account that $\widehat{X_1^{\Pi^{-1}}} = \hat{X}_1^{\Pi}$. □

Let J_i be the algebraic group associated to the data (6.16) by the definition (3.22). It acts on $(\hat{\Omega}^d_{F_{\mathbf{p}_i}} \times_{Spf\ O_{F_{\mathbf{p}_i}}} Spf\ O_{\breve{E}_i}) \times G'_{\mathbf{p}_i}(\mathbf{Q}_p)/C_{\mathbf{p}_i}$ by the isomorphism (6.17). Let us make this action more explicit.

Let us choose an isomorphism $\mathrm{Aut}^0_{D_{\mathbf{q}_i}} \mathbf{Y} \simeq GL_d(F_{\mathbf{p}_i})$ in such a way that the isomorphism of theorem (3.72) becomes equivariant. We find an inclusion

$$J_i(\mathbf{Q}_p) \subset \mathrm{Aut}^0_{D_{\mathbf{q}_i}} \mathbf{Y} \times \mathrm{Aut}^0_{D^{opp}_{\mathbf{q}_i}} \hat{\mathbf{Y}} \cong GL_d(F_{\mathbf{p}_i}) \times GL_d(F_{\mathbf{p}_i})^{opp},$$

which identifies $J_i(\mathbf{Q}_p)$ with

$$\{(a,b) \in GL_d(F_{\mathbf{p}_i}) \times GL_d(F_{\mathbf{p}_i})^{opp};\ ab \in \mathbf{Q}_p\}.$$

By definition the action of (a, b) on a point $(X_1 \times X_2,\ \varrho_1 \times \varrho_2)$ of $\breve{\mathcal{M}}_i$ gives the point $(X_1 \times X_2,\ \varrho_1 a^{-1} \times \varrho_2 \hat{b}^{-1})$.

We may rewrite this point as follows. Let $ab = c \in \mathbf{Q}_p$ and let $o(a)$ be the integer $\mathrm{ord}_{F_{\mathbf{p}_i}} \det a$. Then $g = (\Pi^{o(a)}, c\,\Pi^{-o(a)}) \in G'_{\mathbf{p}_i}(\mathbf{Q}_p)$ acts from the right on $\breve{\mathcal{M}}_i$ and we have

$$(X_1 \times X_2,\ \varrho_1 a^{-1} \times \varrho_2 \hat{b}^{-1}) \\ = (X_1^{\Pi^{-o(a)}} \times X_2^{c\,\Pi^{o(a)}},\ \iota_{X_1}(\Pi^{o(a)})\varrho_1 a^{-1},\ \iota_{X_2}(c^{-1}\Pi^{-o(a)})\varrho_2 \hat{b}^{-1}) \cdot g.$$

It follows from theorem (3.72) that the natural action of $GL_d(F_{\mathbf{p}_i})$ on $\hat{\Omega}^d_{F_{\mathbf{p}_i}}$ defined by (3.68) is given in terms of the modular interpretation by

$$\varrho_1 \longmapsto \iota_{X_1}(\Pi^{o(a)})\varrho_1 a^{-1}.$$

We obtain the following

Lemma 6.45 *An element* $(a,b) \in J_i(\mathbf{Q}_p)$ *acts on*

$$(\hat{\Omega}^d_{F_{\mathbf{p}_i}} \times_{Spf\, O_{F_{\mathbf{p}_i}}} Spf\, O_{\breve{E}_i}) \times G'_{\mathbf{p}_i}(\mathbf{Q}_p)/C_{\mathbf{p}_i}$$

by the natural action of a *on the first factor and multiplication by* $g = g(a,b) = (\Pi^{o(a)}, c\,\Pi^{-o(a)})$ *on the second factor.*

We note that the image $\delta(g)$ coincides with $\omega_{J_i}((a,b))$ (see (3.52)). Hence we have described $\breve{\mathcal{M}}_i$ with its Weil descent datum and the action of $J_i(\mathbf{Q}_p)$ in terms of Drinfeld's Ω, for $i = 1, \ldots, r$.

6.46 Next we consider $\breve{\mathcal{M}}_i$ for $i = r+1, \ldots, s$. Consider a point (X, ϱ) of $\breve{\mathcal{M}}_i(\bar{\mathbf{F}}_p)$. We have decompositions $D_{\mathbf{p}_i} = D_{\mathbf{q}_i} \times D^{opp}_{\mathbf{q}_i}$, $X = X_1 \times X_2$. The p–divisible group X_1 is étale by the condition on the determinants, and X_2 is isomorphic to $\hat{X}_1$. Let $\mathbf{Y}$ be the étale p–divisible group with $O_{D_{\mathbf{q}_i}}$–action whose Tate module $T_p(\mathbf{Y}) = \Lambda_{\mathbf{q}_i}$. We may take the $\mathbf{X}$ in the definition of $\breve{\mathcal{M}}_i$ to be $\mathbf{Y} \times \hat{\mathbf{Y}}$. This identifies $J_i(\mathbf{Q}_p)$ with a subgroup of $\mathrm{Aut}_{D_{\mathbf{q}_i}} V_{\mathbf{q}_i} \times \mathrm{Aut}_{D^{opp}_{\mathbf{q}_i}} V^*_{\mathbf{q}_i}$, which is equal to $G'_{\mathbf{p}_i}(\mathbf{Q}_p)$. The stabilizer of $\Lambda_{\mathbf{p}_i} \oplus \Lambda^*_{\mathbf{p}_i}$ is a maximal compact subgroup $C_{\mathbf{p}_i} \subset J_i(\mathbf{Q}_p)$. It follows that $\breve{\mathcal{M}}_i$ is the constant étale scheme $G'_{\mathbf{p}_i}(\mathbf{Q}_p)/C_{\mathbf{p}_i}$ and that $J_i(\mathbf{Q}_p)$ acts naturally from the left by the identification $J_i(\mathbf{Q}_p) \simeq G'_{\mathbf{p}_i}(\mathbf{Q}_p)$ given above. One easily checks

Lemma 6.47 *The Weil descent datum on* $\breve{\mathcal{M}}_i \cong G'_{\mathbf{p}_i}(\mathbf{Q}_p)/C_{\mathbf{p}_i}$ *relative to* $E_i = \mathbf{Q}_p$ *is given by multiplication with* $(1,p) \in G'_{\mathbf{p}_i}(\mathbf{Q}_p) \subset \mathrm{Aut}_{D_{\mathbf{q}_i}} V_{\mathbf{q}_i} \times \mathrm{Aut}_{D^{opp}_{\mathbf{q}_i}} V^*_{\mathbf{q}_i}$.

6.48 Finally we consider the primes $\mathbf{p}_i$ for $i = s+1, \ldots, t$. We keep the notation of the proof of lemma (6.41). In particular we make the identification

$$\begin{array}{ccc} \mathbf{Z}/2f_i\mathbf{Z} & \cong & \mathrm{Hom}\,(K^t_{\mathbf{q}_i}, \bar{\mathbf{Q}}_p) \\ m & \longmapsto & \varphi_0\sigma^{-m}. \end{array}$$

Let u be the smallest natural number, such that $\Phi_{\mathbf{q}_i} + u = \Phi_{\mathbf{q}_i}$. Then u divides $2f_i$. All the intervalls $[mu, (m+1)u]$ contain the same number of elements of $\Phi_{\mathbf{q}_i}$ respectively of $\bar{\Phi}_{\mathbf{q}_i}$. Hence u is an even number. The local Shimura field $E_i = E_{\nu,\mathbf{q}_i}$ is the unramified extension of degree u of $\mathbf{Q}_p$.

Let (X, ϱ) be a point of $\breve{\mathcal{M}}_i(\bar{\mathbf{F}}_p)$. In the proof of lemma (6.41) we have associated to X in a functorially a symplectic $(O_{D_{\mathbf{p}_i}}, *)$–module $(\Gamma(X), \Psi_{\Gamma(X)})$. In particular the polarized $(O_{D_{\mathbf{p}_i}}, *)$–module X is uniquely determined up to isomorphism.
Let us fix an isomorphism $(\Gamma(\mathbf{X}), \Psi_{\Gamma,(\mathbf{x})}) \simeq (\Lambda_{\mathbf{p}_i}, \psi)$. It defines an isomorphism $J_i(\mathbf{Q}_p) \simeq G'_{\mathbf{p}_i}(\mathbf{Q}_p)$. By the deformation theory of Grothendieck–Messing one easily checks that $\breve{\mathcal{M}}_i$ is étale. We get an $J_i(\mathbf{Q}_p)$–equivariant isomorphism

$$\begin{array}{ccc} G'_{\mathbf{p}_i}(\mathbf{Q}_p)/C_{\mathbf{p}_i} & \simeq & \breve{\mathcal{M}}_i \\ g & \longmapsto & (\mathbf{X},\ \mathrm{id}_{\mathbf{X}} \cdot g^{-1}) \end{array} \tag{6.18}$$

By (6.14) the Weil–descent datum on $\breve{\mathcal{M}}_i$ relative to $\breve{E}_i/E_i$ induces on the left hand side of (6.18) multiplication by $p^{u/2}$.

Proposition 6.49 *$J(\mathbf{Q}_p)$ is the inverse image of the diagonal by the map*

$$\prod c_i : \prod_{i=1}^{t} J_i \longrightarrow \prod_{i=1}^{t} \mathbf{G}_{m,\mathbf{Q}_p}.$$

Similarly $G(\mathbf{Q}_p)$ is the inverse image of the diagonal by

$$\prod c_i : \prod_{i=1}^{t} G'_{\mathbf{p}_i} \longrightarrow \prod_{i=1}^{t} \mathbf{G}_{m,\mathbf{Q}_p}.$$

The actions of $J_i(\mathbf{Q}_p)$ on $G'_{\mathbf{p}_i}(\mathbf{Q}_p)/C_{\mathbf{p}_i}$, which we described induce an action of $J(\mathbf{Q}_p)$ on $G(\mathbf{Q}_p)/C_p$. We have an $J(\mathbf{Q}_p)$-equivariant isomorphism of formal schemes

$$\breve{\mathcal{M}} \simeq \prod_{i=1}^{r} (\hat{\Omega}^d_{F_{\mathbf{p}_i}} \times_{Spf\ O_{F_{\mathbf{p}_i}}} Spf\ O_{\breve{E}}) \times G(\mathbf{Q}_p)/C_p.$$

The action of $J(\mathbf{Q}_p)$ on the first r factors on the right hand side is via the projections $J(\mathbf{Q}_p) \to J_i(\mathbf{Q}_p) \to GL_n(F_{\mathbf{p}_i})$ and on the last factor as described above.
The Weil descent datum on $\breve{\mathcal{M}}$ relative to $\breve{E}/E$ induces on the right hand side the natural descent datum on the first r factors multiplied with the action of the element $g \in G(\mathbf{Q}_p)$ on the second factor, where g is given by

$$g = \prod_{i=1}^{r}(\Pi_i^{f/f_i},\ p^f\Pi_i^{-f/f_i}) \times \prod_{i=r+1}^{s}(1,p^f) \times \prod_{i=s+1}^{t} p^{f/2},$$

where the right hand side is viewed in $\Pi\, G'_{\mathbf{p}_i}(\mathbf{Q}_p)$. *Here* Π_i *is a prime element in* $D_{\mathbf{q}_i}$ *for* $i = 1, \ldots, r$, *and* f_i *respectively* f *are the index of inertia of* E_i *respectively* E_ν.

Proof: This follows by what we have said about $\breve{\mathcal{M}}_i$.

Theorem 6.50 *The Shimura variety of* (G, h) *and level* C *has under the assumptions made on* $\nu : \bar{\mathbf{Q}} \to \bar{\mathbf{Q}}_p$ *and* C *a model* Sh_C *over* O_{E_ν}. *The action of the Hecke-algebra* $\mathcal{H}(G(\mathbf{A}_f^p)//C)$ *extends to* Sh_C. *There is an* $\mathcal{H}(G(\mathbf{A}_f^p)//C)$*-equivariant isomorphism of formal schemes :*

$$I(\mathbf{Q})\backslash\prod_{i=1}^{r}(\hat{\Omega}^d_{F_{\mathbf{p}_i}} \times_{Spf\ O_{F_{\mathbf{p}_i}}} Spf\ O_{\breve{E}_\nu}) \times G(\mathbf{A}_f)/C \xrightarrow{\sim} Sh^\wedge_C \times_{Spf\ O_{E_\nu}} Spf\ O^n_{\breve{E}_\nu} \tag{6.19}$$

Here $Sh^\wedge_C$ *is the p-adic completion of* Sh_C. *The group* I *is an inner form of* G, *such that* $I(\mathbf{Q}_p)$ *is the group* $J(\mathbf{Q}_p)$ *defined above and* $I(\mathbf{A}_f^p) \simeq G(\mathbf{A}_f^p)$. *This defines the action of* $I(\mathbf{Q})$ *used in forming the quotient above.*
The natural descent datum on the right hand side of (6.19) induces on the left hand side the natural descent datum on the first r *factors multiplied with the action of* $g \in G(\mathbf{Q}_p)$ *on* $G(\mathbf{A}_f)/C$ *defined by (6.49).*

Proof: By the general theorem 6.30 the left hand side of (6.19) with the descent datum given is $\mathcal{A}_{C^p/Z'} \times_{Spf\ O_{E_\nu}} Spf\ O_{\breve{E}_\nu}$. Since the uniformization morphism is compatible with the Hecke operators it follows that the scheme theoretic closure of Sh_C in $\mathcal{A}_{C^p}$ has the special fibre Z'. In fact, to see this it is enough to verify that $G(\mathbf{A}_f^p)$ acts transitively on the connected components of $\varprojlim Sh_C$. Consider the map $\gamma : G \to F^\times \times K^\times$ whose first component c is the multiplier and whose second component $\mathbf{n}$ is given by (3.52). The kernel is the derived group G^{der} and the image is a torus T. Since G^{der} is simply connected we have by Deligne [De], that $\pi_0(Sh_C) = T^+(\mathbf{Q}) \setminus T(\mathbf{A}_f)/\gamma(C)$. Hence we only need to verify that $T(\mathbf{A}_f^p)$ acts transitively on $\pi_0(Sh_C)$. This follows since $T(\mathbf{Q})$ is dense in

$T(\mathbf{R}) \times T(\mathbf{Q}_p)$. Indeed, for $K^\times$ and $F^\times$ in place of T this is well known. For T itself we obtain it from the exact sequence

$$\begin{array}{ccccccc} 1 \longrightarrow & F^\times \longrightarrow & F^\times \times K^\times & \longrightarrow & T \longrightarrow 1 \\ & & (f,k) & \longmapsto & (fk\bar{k}, fk^2). \end{array}$$

□

For the convenience of the reader we formulate separately a special case of the previous theorem.

Corollary 6.51 *Let $(D, K, *, F, V, \psi)$ and the associated algebraic group G over $\mathbf{Q}$ be as in (6.37). We assume that the D–module V is of rank 1. We also assume that there is precisely one prime ideal $\mathbf{p}$ above p in F and that $\mathbf{p} = \mathbf{q} \cdot \bar{\mathbf{q}}, \mathbf{q} \neq \bar{\mathbf{q}}$, splits in K. We assume that*

$$\begin{aligned} \mathrm{inv}_{\mathbf{q}} D &= 1/d \\ \mathrm{inv}_{\bar{\mathbf{q}}} D &= -1/d. \end{aligned}$$

We fix an embedding $\nu : \bar{\mathbf{Q}} \to \bar{\mathbf{Q}}_p$. Let Φ be the CM–type of K such that under the identification (6.11) all elements of Φ induce the place $\mathbf{q}$ of K above p. We fix an element $\alpha \in \Phi$. We make the following assumption on the signature (cf. (6.37))

$$\begin{aligned} r_\alpha &= 1 \\ r_\varepsilon &= 0, \quad \varepsilon \in \Phi \setminus \{\alpha\}. \end{aligned}$$

Let Sh be the associated Shimura variety which is defined over the Shimura field E, cf. (6.37). Let $C_p \subset G(\mathbf{Q}_p)$ be the unique maximal compact subgroup and let $C^p \subset G(\mathbf{A}_f^p)$ be a sufficiently small open compact subgroup. Then there is a model Sh_C of the Shimura variety of level C over O_{E_ν}. There is a $\mathcal{H}(G(\mathbf{A}_f^p)//C)$–equivariant isomorphism of p–adic formal schemes

$$I(\mathbf{Q}) \setminus (\hat{\Omega}^d_{F_\mathbf{p}} \times_{Spf\ O_{F_\mathbf{p}}} Spf\ O_{\breve{E}_\nu}) \times G(\mathbf{A}_f)/C \simeq Sh^\wedge_C \times_{Spf\ O_{E_\nu}} Spf\ O_{\breve{E}_\nu} \qquad (6.20)$$

Here $I(\mathbf{Q})$ is the group of $\mathbf{Q}$–rational points of an inner form of G such that $I(\mathbf{Q}_p) \simeq \{(a,b) \in GL_d(F_\mathbf{p}) \times GL_d(F_\mathbf{p})^{opp};\ \ ab \in \mathbf{Q}_p\}$ and with $I(\mathbf{A}_f^p) \simeq G(\mathbf{A}_f^p)$. We used α to identify $F_\mathbf{p}$ with E_ν. The natural descent datum on *the right hand side induces on the left hand side the natural descent datum on the first factor multiplied with the action of*

$$g = (\Pi, p^f \Pi^{-1}) \in G(\mathbf{Q}_p) \subset D_{\mathbf{q}}^\times \times D_{\mathbf{q}}^{\times\, opp},$$

on $G(\mathbf{A}_f)/C$. *Here* Π *is a uniformizing element of* $D_{\mathbf{q}}$ *and* f *is the index of inertia of* $F_{\mathbf{p}}$.

We note that $C_p \subset G(\mathbf{Q}_p)$ is a normal subgroup under the assumptions of (6.51). Hence $G(\mathbf{Q}_p)$ acts by right translation on the left hand side of (6.20). The reader checks that this is compatible with the action of $G(\mathbf{Q}_p)$ on the general fibre of Sh_C.

Bibliography

[BBM] Berthelot, P., Breen, L., Messing, W.: Théorie de Dieudonné cristalline, vol. 2., LNM **930**, Springer 1982.

[BC] Boutot, J.-F., Carayol, H.: Uniformisation p-adique des courbes de Shimura, Asterisque **196 - 197**, 45 - 149 (1991).

[BT] Bruhat, F., Tits, J.: Groupes algébriques sur un corps local, chap. III, J. Fac. Sci. Univ. Tokyo Sect IA, Math. **34**, 671 - 698 (1987).

[Ber] Berthelot, P.: Cohomologie rigide et cohomologie rigide à support propre, in preparation.

[BL] Bosch, S., Lütgebohmert, W.: Formal and rigid geometry, I. Flattening techniques, Math. Ann. **295**, 291 - 317 (1993).

[CN1] Chai, C.L., Norman, P.: Singularities of the $\Gamma_0(p)$-level Structure, J. Alg. Geom. **1**, 251 - 278 (1992).

[CN2] Chai, C.L., Norman, P.: Bad reduction of the Siegel moduli scheme of genus two with $\Gamma_0(p)$-level structure, Amer. J. of Math. **112**, 1003 - 1071 (1990).

[Ca] Carayol, H.: Non-abelian Lubin-Tate theory, in: Clozel, L., Milne, J.S. (ed), Automorphic forms, Shimura varieties and L-functions vol. II, Persp. in Math. **11**, 15 - 40, Acad. Press Boston (1990).

[Ch] Cherednik, I.V.: Uniformization of algebraic curves by discrete subgroups of $PGL_2(k_w)\ldots$, Math. USSR Sbornik, **29**, 55 - 78 (1976).

[DP] Deligne, P., Pappas, G.: Singularités des espaces de modules de Hilbert, en les caractéristiques divisant le discriminant, Comp. math. **90**, 59 - 79 (1994).

[DI] Deligne, P., Illusie, L.: Cristaux ordinaires et coordonées canoniques, in: Surfaces Algébriques, 80 - 137, SLN **868**, Springer 1981.

[De] Deligne, P.: Travaux de Shimura, Sem. Bourbaki 389. LNM **244**, Springer 1971.

[Dr1] Drinfeld, V.G.: Elliptic modules, Math. USSR Sbornik **23**, 561 - 592 (1974).

[Dr2] Drinfeld, V.G.: Coverings of p–adic symmetric regions, Funct. Anal. and Appl. **10**, 29 - 40 (1976).

[Fa1] Faltings, G.: Mumford–Stabilität in der algebraischen Geometrie, Proceedings ICM 1994, Zürich, to appear.

[Fo1] Fontaine, J.-M.: Groupes p–divisibles sur les corps locaux, Astérisque **47-48** (1977).

[Fo2] Fontaine, J.-M.: Modules galoisiens, modules filtrés at anneaux de Barsotti–Tate, Astérisque **65**, 3 - 80 (1979).

[Fo3] Fontaine, J.-M.: Sur certains types de représentations p–adiques du groupe de Galois d'un corps local; construction d'un anneau de Barsotti - Tate, Ann. of Math. **115**, 529 - 577 (1982).

[Gr1] Grothendieck, A.: Groupes de Barsotti-Tate et cristaux de Dieudonné, Sém. Math. Sup. **45**, Presses de l'Univ. de Montreal, 1970.

[Gr2] Grothendieck, A.: Groupes de Barsotti-Tate et cristaux, Actes du Congr. Int. Math., Nice (Paris), **1**, Gauthier-Villar, 431 - 436 (1971).

[HG1] Hopkins, M., Gross, B.: The rigid analytic period mapping, Lubin-Tate space, and stable homotopy theory, Bull. Amer. Math. Soc. **30**, 76 - 86 (1994).

[HG2] Hopkins, M., Gross, B.: Equivariant vector bundles on the Lubin-Tate moduli space, Contemp. Math. **158**, 23 - 88 (1994).

[HN] Harder, G., Narashimhan, M.S.: On the cohomology groups of moduli spaces of vector bundles on curves, Math. Ann. **212**, 215 - 248 (1975).

[Hu] Huber, R.: A generalization of formal schemes and rigid–analytic varieties, Math. Zeitschr. **217**, 513 - 551 (1994).

[IM] Iwahori, N., Matsumoto, H.: On some Bruhat decomposition and the structure of the Hecke ring of p–adic Chevalley groups, Pub. Math. IHES **25**, 5 - 48 (1965).

[I] Iwahori, N.: Generalized Tits system on p–adic semisimple groups, Proc. Symp. Pure Math **9**, 71 - 83 (1966).

[dJ1] Jong de, A.J.: The moduli spaces of principally polarized abelian varieties with $\Gamma_0(p)$–level structure, J. Alg. Geo. **2**, 667 - 688 (1993).

[dJ2] Jong de, A.J.: Crystalline Dieudonné module theory via formal and rigid geometry. Preprint Utrecht 1993.

[dJ3] Jong de, A.J.: Etale fundamental groups of non–archimedean analytic spaces. Preprint Utrecht 1994.

[Ka1] Katz, N.M.: Travaux de Dwork, Sem. Bourbaki 409, 431 - 436, LNM **417**, Springer 1973.

[Ka2] Katz, N.M.: Slope filtration of F–crystals, Astérisque **63**, 113 - 164 (1979).

[Ka3] Katz, N.M.: Serre - Tate local moduli, in: Surfaces Algèbriques, 138 - 202, SLN **868**, Springer 1981.

[Ko1] Kottwitz, R.E.: Isocrystals with additional structure, Comp. Math. **56**, 201 - 220 (1985).

[Ko2] Kottwitz, R.E.: Shimura varieties and λ–adic representations, in: Clozel, L., Milne, J. (ed): Automorphic forms, Shimura varieties and L–functions, vol. I, Persp. in Math. **10**, 161 - 209, Acad. Press 1990.

[Ko3] Kottwitz, R.E.: Points on some Shimura varieties over finite fields, J. AMS, **5**, 373 - 444 (1992).

[Kn] Knutson, D.: Algebraic spaces, LNM **203**, Springer 1971.

[LT] Lubin, J., Tate, J.: Formal moduli for one-parameter formal Lie groups, Bull. Soc. Math. France **94**, 49 - 60 (1966).

[L] Laffaille, G.: Construction de groupes p–divisibles: le cas de dimension 1; Astérisque **65**, 103 - 124 (1979).

[La] Langlands, R.P.: Sur la mauvaise réduction d'une variété de Shimura, Astérisque **65**, 125 - 154 (1979).

[Me] Messing, W.: The crystals associated to Barsotti–Tate groups ..., LNM **264**, Springer 1972.

[Mi] Milne, J.: Canonical models of (mixed) Shimura varieties and automorphic vector bundles, in: Clozel, L., Milne, J.S. (ed), Automorphic forms, Shimura varieties and L–functions vol. I, Persp. in Math. **10**, 283 - 414, Acad. Press Boston (1990).

[M1] Mumford, D.: Abelian varieties, Oxford 1970.

[M2] Mumford, D.: An analytic construction of degenerating curves over complete local rings, Comp. math. **24**, 129 - 174 (1972).

[Mu] Mustafin, G.A.: Nonarchimedean uniformization, Math.USSR Sbornik **34**, 187 - 214 (1978).

[OT] Oort, F., Tate, J.: Group schemes of prime order, Ann. Sci. ENS **3**, 1 - 21 (1970).

[PV] van der Put, M., Voskuil, H.: Symmetric spaces associated to split algebraic groups over a local field, J. reine angew. Math. **433**, 69 - 100 (1992).

[RR] Rapoport, M., Richartz, M.: On the classification and specialization of F–isocrystals with additional structure, preprint Wuppertal 1994.

[RZ] Rapoport, M., Zink, Th.: A finiteness theorem in the Bruhat-Tits building: an application of Landvogt's embedding theorem, Preprint Wuppertal 1995.

[R1] Rapoport, M.: On the bad reduction of Shimura varieties, in Clozel, L., Milne, J.S. (ed), Automorphic forms, Shimura varieties and L–functions vol. II, Persp. in Math. **11**, 253 - 321, Acad. Press Boston (1990).

[R2] Rapoport, M.: Non–archimedean period domains, Proceedings ICM 1994 Zürich, to appear.

[Ra1] Raynaud, M.: Géométrie analytique rigide d'après Tate, Kiehl,...; Bull. Soc. Math. France, Mém. **39/40**, 319 - 327 (1974).

[Ra2] Raynaud, M.: Schémas en groupes de type $(p, \ldots, p)$; Bull. Soc. Math. France **102**, 241 - 280 (1974).

[Ro] Roubaud, J.: Morphismes rigides étales, thèse 3eme cycle, Orsay 1970.

[Rou] Rousseau, G.: Exercices métriques immobiliers, Preprint Wuppertal 1995.

[Se] Serre, J.-P.: Local class field theory, in: J. Cassels, A. Fröhlich (ed), Algebraic number theory, 129 - 162, Academic Press, London 1967.

[SS] Schneider, P., Stuhler, U.: The cohomology of p–adic symmetric spaces, Invent. math. **105**, 47 - 122 (1991).

[St] Stamm, H.: On the reduction of the Hilbert–Blumenthal moduli scheme with $\Gamma_0(p)$–level structure, Preprint Wuppertal 1993.

[T] Tits, J.: Reductive groups over local fields, Proc. Symp. Pure Math. **33** (vol 1), 29 - 70 (1977).

[To] Totaro, B.: Tensor products of weakly admissible filtered isocrystals, Preprint Chicago 1994.

[V] Varshavsky, Y.: P-adic uniformization of unitary Shimura varieties, Preprint Jerusalem 1995.

[W] Wintenberger, J.-P.: Torseur entre cohomologie étale p-adique ..., Duke Math. J. **62**, 511 - 526 (1991).

[Yu] Yu, J.-K.: A-divisible modules, period maps, and quasi-canonical liftings, Thesis Harvard University (1994).

[Z1] Zink, Th.: Über die schlechte Reduktion einiger Shimuramannigfaltigkeiten, Comp. Math. **45**, 15 - 107 (1981).

[Z2] Zink, Th.: Cartiertheorie kommutativer formaler Gruppen, Teubner Texte zur Mathematik **68**, Leipzig 1984.

[Z3] Zink, Th.: Isogenien formaler Gruppen über einem lokal noetherschen Schema, Math. Nachr. **99**, 273 - 283 (1980).

[Z4] Zink, Th.: Cartiertheorie über perfekten Ringen, I, II; Preprints Akademie der Wiss. Berlin, 1986.

Index

GPSR Authorized Representative: Easy Access System Europe - Mustamäe tee 50, 10621 Tallinn, Estonia, gpsr.requests@easproject.com

www.ingramcontent.com/pod-product-compliance
Ingram Content Group UK Ltd.
Pitfield, Milton Keynes, MK11 3LW, UK
UKHW031853030425
457052UK00004B/102

* 9 7 8 0 6 9 1 0 2 7 8 1 4 *